DATA ACCESS AND STORAGE MANAGEMENT FOR EMBEDDED PROGRAMMABLE PROCESSORS

T0180529

Data Access and Storage Management for Embedded Programmable Processors

by

Francky Catthoor
IMEC, Leuven, Belgium

Koen Danckaert
IMEC, Leuven, Belgium

Chidamber Kulkarni
IMEC, Leuven, Belgium

Erik Brockmeyer
IMEC, Leuven, Belgium

Per Gunnar Kjeldsberg
Norwegian Univ. of Sc. and Tech. (NTNU), Trondheim, Norway

Tanja Van Achteren
Katholieke Universiteit Leuven, Leuven, Belgium

and

Thierry Omnes
IMEC, Leuven, Belgium

KLUWER ACADEMIC PUBLISHERS
BOSTON / DORDRECHT / LONDON

A C.I.P. Catalogue record for this book is available from the Library of Congress.

ISBN 978-1-4419-4952-3

Published by Kluwer Academic Publishers,
P.O. Box 17, 3300 AA Dordrecht, The Netherlands.

Sold and distributed in North, Central and South America
by Kluwer Academic Publishers,
101 Philip Drive, Norwell, MA 02061, U.S.A.

In all other countries, sold and distributed
by Kluwer Academic Publishers,
P.O. Box 322, 3300 AH Dordrecht, The Netherlands.

Printed on acid-free paper

Preface

The main intention of this book is to give an impression of the state-of-the-art in data transfer/access and storage exploration[1] related issues for complex data-dominated embedded real-time processing applications. The material is based on research at IMEC in this area in the period 1996-2001. It can be viewed as a follow-up of the earlier "Custom Memory Management Methodology" book that was published in 1998 based on the custom memory style related work at IMEC in the period 1988-1997. In order to deal with the stringent timing requirements, the cost-sensitivity and the data dominated characteristics of our target domain, we have again adopted a target architecture style and a systematic methodology to make the exploration and optimization of such systems feasible. But this time our target style is oriented mostly to (partly) predefined memory organisations as occurring e.g. in instruction-set processors, cores or "platforms".

Our approach is also very heavily application-driven which is illustrated by several realistic demonstrators, partly used as red-thread examples in the book. Moreover, the book addresses only the steps above the traditional compiler tasks, even prior to the parallelizing ones. The latter are mainly focussed on scalar or scalar stream operations and other data where the internal structure of the complex data types is not exploited, in contrast to the approaches discussed in this book. The proposed methodologies are nearly fully independent of the level of programmability in the data-path and controller so they are valuable for the realisation of both instruction-set and custom processors, and even reconfigurable styles. Some of the steps do depend on the memory architecture instance and its restrictions though.

Our target domain consists of real-time processing systems which deal with large amounts of data. This happens both in real-time multi-dimensional signal processing (RMSP) applications like video and image processing, which handle indexed array signals (usually in the context of loops), in wired and wireless terminals which handle less regular arrays of data, and in sophisticated communication network protocols, which handle large sets of records organized in tables and pointers. All these classes of applications contain many important applications like video coding, medical image archival, advanced audio and speech coding, multi-media terminals, artificial vision, Orthogonal Frequency Domain Multiplex (OFDM), turbo coding, Spatial Division Multiplex Access (SDMA), Asynchronous Transfer Mode (ATM) networks, Internet Protocol (IP) networks, and other Local/Wide Area Network (LAN/WAN) protocol modules. For these applications, we believe (and we will demonstrate by real-life experiments) that the organisation of the global communication and data storage, together with the related algorithmic transformations, form the dominating factors (both for system power and memory footprint) in the system-level design decisions, with special emphasis on source code transformations. Therefore, we have concentrated ourselves mainly on the effect of system-level decisions on the access to large (background) memories and caches, which require separate access cycles, and on the transfer of data over long "distances" (i.e. which have to pass between source and destination over long-term main memory storage). This is complementary to and not in competition with the existing compiler technology. Indeed, our approach should be

---

[1] a more limited term used in literature is memory management but the scope of this material is much broader because it also includes data management and source code transformations

considered as a precompilation stage, which precedes the more conventional compiler steps that are essential to handle aspects that are not decided in "our" stage yet. So we are dealing with "orthogonal" exploration choices where the concerns do not overlap but that are still partly dependent though due to the weak phase coupling. From the precompilation stage, constraints are propagated to the compiler stage. The effectiveness of this will be demonstrated in the book.

The cost functions which we have incorporated for the storage and communication resources are both memory footprint and power oriented. Due to the real-time nature of the targeted applications, the throughput is normally a constraint. So performance is in itself not an optimisation criterion for us, it is mostly used to restrict the feasible exploration space. The potential slack in performance is used to optimize the real costs like power consumption, and (on- or off-chip) memory foot-print.

The material in this book is partly based on work in the context of several European and national research projects. The ESPRIT program of Directorate XIII of the European Commission, though the ESD/LPD Project 25518 DAB-LP has sponsored the SCBD step and the DAB application work. The Flemish IWT has sponsored some of the demonstrator work under SUPERVISIE project, and part of the loop transformation and caching related work under the MEDEA SMT and SMT2 projects (System level methods and Tools), including partial support of the industrial partners Alcatel Telecom, Philips, CoWare and Frontier Design. The Flemish FWO foundation and the Belgian inter-university attraction pole have sponsored the data reuse related work under FWO-G.0036.99 resp. IUAP4/24. The FWO has also sponsored part of loop transformation work in a Ph.D. fellowship. The Norwegian Research Council and the European Marie Curie Fellowship funding have sponsored part of the size estimation work through research project 131359 CoDeVer resp. MCFH-1999-00493.

A major goal of the system synthesis and compilation work within these projects has been to contribute systematic design methodologies and appropriate tool support techniques which address the design/compilation trajectory from *real system behaviour* down to the detailed platform architecture level of the system. In order to provide complete support for this design trajectory, many problems must be tackled. In order to be effective, we believe that the design/compilation methodology and the supporting techniques have to be (partly) domain-specific, i.e. targeted. This book illustrates this claim for a particular target application domain which is of great importance to the current industrial activities in the telecommunications and multi-media sectors: data-dominated real-time signal and data processing systems. For this domain, the book describes an appropriate systematic methodology partly supported by efficient and realistic compilation techniques. The latter have been embedded in prototype tools to prove their feasibility. We do not claim to cover the complete system compilation path, but we do believe we have significantly contributed to the solution of one of the most crucial problems in this domain, namely the ones related to data transfer and storage exploration (DTSE).

We therefore expect this book to be of interest in academia, both for the overall description of the methodology and for the detailed descriptions of the compilation techniques and algorithms. The priority has been placed on issues that in our experience are crucial to arrive at industrially relevant results.

All projects which have driven this research, have also been application-driven from the start, and the book is intended to reflect this fact. The real-life applications are described, and the impact of their characteristics on the methodologies and techniques is assessed. We therefore believe that the book will be of interest as well to senior design engineers and compiler/system CAD managers in industry, who wish either to anticipate the evolution of commercially available design tools over the next few years, or to make use of the concepts in their own research and development.

It has been a pleasure for us to work in this research domain and to co-operate with our project partners and our colleagues in the system-level synthesis and embedded compiler community. Much of this work has been performed in tight co-operation with many university groups, mainly across EUrope. This is reflected partly in the author list of this book, but especially in the common publications. We want to especially acknowledge the valuable contributions and the excellent co-operation with: the ETRO group at the V.U.Brussels, KTH-Electrum (Kista), the ACCA and DTAI groups at K.U.Leuven (Leuven), INSA (Lyon), Patras Univ., Norwegian Univ. of Sc. and Tech. (Trondheim), Democritus University of Thrace (Xanthi).

In addition to learning many new things about system synthesis/compilation and related issues, we have also developed close connections with excellent people. Moreover, the pan-European aspect of the projects has allowed us to come in closer contact with research groups with a different background and "research culture", which has led to very enriching cross-fertilization.

We would like to use this opportunity to thank the many people who have helped us in realizing these results and who have provided contributions in the direct focus of this book, both in IMEC and at other locations. In particular, we wish to mention: Einar Aas, Javed Absar, Yiannis Andreopoulos, Ivo Bolsens, Jan Bormans, Henk Corporaal, Chantal Couvreur, Geert Deconinck, Eddy De Greef, Kristof Denolf, Hugo De Man, Jean-Philippe Diguet, Peeter Ellervee, Michel Eyckmans, Antoine Fraboulet, Frank Franssen, Thierry Franzetti, Cedric Ghez, Costas Goutis, Ahmed Hemani, Stefaan Himpe, Jos Huisken, Martin Janssen, Stefan Janssens, Rudy Lauwereins, Paul Lippens, Anne Mignotte, Miguel Miranda, Kostas Masselos, Lode Nachtergaele, Martin Palkovic, Rainer Schaffer, Peter Slock, Dimitrios Soudris, Arnout Vandecappelle, Tom Van der Aa, Tycho van Meeuwen, Michael van Swaaij, Diederik Verkest, Sven Wuytack.

We finally hope that the reader will find the book useful and enjoyable, and that the results presented will contribute to the continued progress of the field of system-level compilation and synthesis for both instruction-set and custom processors.

> Francky Catthoor, Koen Danckaert, Chidamber Kulkarni, Erik Brockmeyer,
> Per Gunnar Kjeldsberg, Tanja Van Achteren, Thierry Omnes
> October 2001.

Glossary

1-D	One-Dimensional
ACU	Address Calculation Unit
ACROPOLIS	A Caching and data Reuse Optimizer exploiting Parallelism choices, including access Ordering and data Locality Improvement for multimedia Systems
ADT	Abstract Data Type
ADOPT	ADdress OPTimisation
ASIC	Application Specific Integrated Circuit
ASIP	Application Specific Instruction-set Processor
ATOMIUM	A Tool-box for Optimizing Memory and I/O communication Using geometric Modelling.
ATM	Asynchronous Transfer Mode
BG	Basic Group
BOAT(D)	Binary Occupied Address/Time (Domain)
BTPC	Binary Tree Predictive Coding
CAD	Computer Aided Design
CDFG	Control/Data-Flow Graph
CDO	Conflict Directed Ordering
CG	Conflict Graph
DAB	Digital Audio Broadcasting decoder
DFG	Data-Flow Graph
DPCM	Differential Pulse Code Modulation
DRAM	Dynamic Random-Access Memory
DRD	Data Reuse Dependency
DSP	Digital Signal Processing
DTSE	Data Transfer and Storage Exploration
ECG	Extended Conflict Graph
FDS	Force Directed Scheduling
FU	Functional Unit
GCDO	Global Conflict Directed Ordering
GCDO (k)	Multi-level Global Conflict Directed Ordering
GCP	Global Constraint Propagation
ILP (1)	Integer Linear Program
ILP (2)	Instruction-Level Parallelism
I/O	Input/Output
IP (1)	Internet Protocol
IP (2)	Intellectual Property
LAN	Local Area Network
LB	Lower Bound
LBL	Linear Bounded Lattice
LCDO	Localized Conflict Directed Ordering

LCDO (k)   Multi-level Localized Conflict Directed Ordering
LPC   Linear Predictive Coding
M-D   Multi-Dimensional
MAA   Memory Allocation and signal-to-memory Assignment
ME   Motion Estimation
MHLA   Memory Hierarchy Layer Assignment
MILP   Mixed Integer Linear Program
MIMD   Multiple Instruction, Multiple Data
MMU   Memory Management Unit
MPEG   Moving Picture Experts Group
OAT(D)   Occupied Address/Time (Domain)
OFDM   Orthogonal Frequency Division Multiplex
PDG   Polyhedral Dependence Graph
QSDPCM   Quad-tree Structured Differential Pulse Code Modulation
RAM   Random-Access Memory
RMSP   Real-time Multi-dimensional Signal Processing
ROM   Read-Only Memory
RSP   Real-time Signal Processing
SBO   Storage Bandwidth Optimisation
SCBD   Storage Cycle Budget Distribution
SDMA   Spatial Division Multiplex Access
SDRAM   Synchronous Dynamic Random-Access Memory
SIMD   Single Instruction, Multiple Data
SPP   Segment Protocol Processor
SRAM   Static Random-Access Memory
UB   Upper Bound
USVD   Updating Singular Value Decomposition
VLIW   Very Large Instruction word
VoCoder   Voice Coder

1 DATA TRANSFER AND STORAGE EXPLORATION IN PROGRAMMABLE ARCHITECTURES

This introductory chapter will contain the problem context that we focus on, including the target application domain and architectural style. Then the global approach for reducing the cost of data access and storage is described. Next, a summary is provided of all the steps in the Data Transfer and Storage Exploration (DTSE) script. Finally the content of the rest of the book is briefly outlined.

1.1 PROBLEM CONTEXT AND MOTIVATION

Embedded systems have always been very cost-sensitive. Up till recently, the main focus has been on area-efficient designs meeting constraints due to performance and design time. During the last couple of years, power dissipation has become an additional major design measure for many systems next to these traditional measures [441, 52]. More and more applications become portable because this is felt as an added value for the product (e.g. wireless phones, notebook computers, multimedia terminals). The less power they consume, the longer their batteries last and the lighter their batteries can be made. Higher power consumption also means more costly packaging and cooling equipment, and lower reliability. The latter has become a major problem for many high-performance (non-portable and even non-embedded) applications. As a consequence, power efficient design has become a crucial issue for a broad class of applications.

Many of these embedded applications turn out to be data-dominated, especially in the multimedia and network protocol domains. Experiments have shown that for these applications, a very large part of the power consumption is due to data transfer and storage. Also the area cost (on-chip and on the board) is for a large part dominated by the memories. So storage technology "takes the center stage" [311] in more and more systems because of the eternal push for more complex applications with especially larger and more complicated data types. This is true both for custom hardware [365, 385] and for programmable processors [503]. Hence, we believe that the data storage and transfer should be optimized as a first step in the design methodology for this type of applications. This has been formalized the Data Transfer and Storage Exploration (DTSE) methodology that is under development at IMEC since 1988 [519, 87, 93].

Experience shows that the initial algorithmic specification heavily influences the outcome of the underlying processor compilation tools (Fig. 1.1). Indeed, many different alternatives are available in terms of algorithm variants and behaviour-preserving source code transformations. The final source code after this system design exploration stage then has to pass through more conventional (and

1

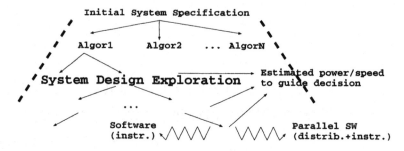

Figure 1.1.   System exploration environment for data-dominated applications on instruction-set (SW) processors: envisioned situation

more supported) compilation paths. The specification is typically also very complex, e.g. 100K lines of MPEG4 code in C. Therefore, transforming this specification is one of the most time-consuming tasks during the early system-level exploration of cost measures. This task is both tedious and error-prone if done fully manually. For most real-time multi-dimensional signal processing (RMSP) applications, the manipulation of the data (storage and transfer) is the dominant cost factor in the global (board-level) system, and this in terms of system bus load/power consumption, memory size and even pure processor speed [93, 296, 65]. Indeed, the performance mismatch between processing units and memory systems exists since the beginning of the computer history, and has been growing larger ever since. Fig. 1.2 shows the evolution of CPU and DRAM speeds during the last decades [424]. It is not surprising that the memory gap is currently being referred to as the "memory wall" [547].

For many years, an intense focus in the compiler and architecture community has been put on ameliorating the impact of *memory latency* on performance. This work has led to extremely effective techniques for reducing and tolerating memory latency, primarily through caching, data prefetching and software pipelining. Since the late 90's, it is however the *effectively exploitable memory bandwidth* which is becoming the main bottleneck. Due to caching the potential memory bandwidth has increased significantly but without appropriate source code transformations most applications can only exploit a fraction of the full potential and that is now a major limitation. In [374, 155], it is shown for a variety of applications that the required (exploited) bandwidth between the main memory and the cache hierarchy is between five and ten times higher than the available bandwidth on state-of-the-art systems. Also within the cache hierarchy a bandwidth mismatch exists, although less severe than for the main memory. The latency issue is even more problematic for applications that have real-time constraints on the system delay.

The main focus of the DTSE project at IMEC until 1998 has been energy consumption and memory size reduction, but since then an equal emphasis has been placed on the system bus load (Fig. 1.3) and removing the bottlenecks to increase the raw processor speed [131, 353, 356]. This is crucial because even when the bottleneck is still within the CPU for a single application, the huge bandwidth/latency requirements will cause performance problems when this application has to share the system bus with other applications or background processes running on the same board. Optimizing the data transfer and storage of a data-dominated application will thus also enhance its overall system performance on bandwidth/latency-constrained platforms.

While the performance of silicon systems has increased enormously during the last decade(s), the performance of a single processing unit still lags behind the ever increasing complexity of the applications. Therefore, parallelism exploitation has been traditionally considered as a major means to increase the performance even further. The correspondig mapping problem has mainly focussed on the parallelizing compiler aspects. However, parallelizing data-dominated applications also heavily complicates the data transfer and storage problem, since the data cannot only flow between a processor and its memory hierarchy, but also between the different processors and memories in the multiprocessor system [188]. Therefore, we have also spent considerable effort on incorporating the influence of a multi-processor target and the parallelization issues into the DTSE methodology.

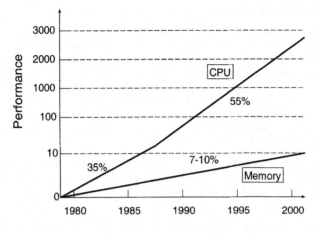

Figure 1.2.   The memory gap.

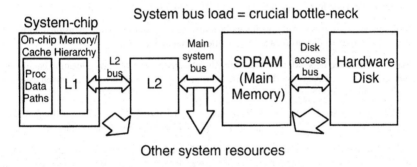

Figure 1.3.   System bus load issue at the board level

Up to now, no real design support was available in this embedded cost-sensitive real-time context for data-dominated RMSP designs, neither in terms of systematic design methodologies nor in terms of automatable tool support. Therefore, currently, due to design time restrictions, the system designer has to select – on an ad-hoc basis and almost without any accurate feedback – a single promising path in the huge decision tree from abstract specification to more refined code specification. To remove this design time bottle-neck, we have developed formalized methodologies to allow a systematic and thorough exploration of the available transformation space within a global scope. The steps in this methodology are becoming supported with compiler techniques, for the set of system-level code transformations that have the most crucial effect on the system exploration decisions. These source-to-source compiler optimisations are applied to C code specifications[1] in our prototype ACROPOLIS environment, under construction at IMEC. These tools augment the tools in the mature ATOMIUM environment [90, 96, 93, 84, 61] that has been under development since 1994 for customizable memory architectures (e.g. in ASICs). Several of the steps and tools in this extended DTSE/ACROPOLIS project have become mature enough to try out realistic application experiments[2], and the focus in the rest of this book will be mainly on these.

---

[1] Other languages like C++, Fortran or Java could also be supported when the front-end parsing is modified, though significant engineering effort is required then.
[2] Which does not mean that they are ready for use by industrial designers

1.2 TARGET APPLICATION DOMAIN

The above mentioned ambitious objectives can be achieved by focusing ourselves on a specific target application domain and target architectural styles. As mentioned, our main target domain consists of embedded signal and data processing systems which deal with large amounts of data (data-dominated). This target domain can be subdivided into three classes: multimedia applications (including video processing, medical imaging, multi-media terminals, artificial vision, ...), front-end telecommunication applications (e.g. OFDM, turbo coding, SDMA, ...) and network component applications (e.g. ATM or IP network protocols, and other LAN/WAN protocol modules). Our main focus in this text lies on multimedia applications. The main characteristics of this class of applications are the following:

- Loop nests. Multimedia applications typically contain many (deep) loop nests for processing the multidimensional data. Due to the rectangular shape of images and video streams, the loop bounds are often constant, or at least manifest (i.e. only dependent on the other iterators). The loops are typically quite irregularly nested though.

- Mostly manifest affine conditions and indices. Most of the conditions and indexing are manifest. They are a function of the loop iterators like (i :: a.k+b.l+c .. d.k+e.l+f) whre (k, l) are iterators of enclosing loops. However, more and more multimedia applications (e.g. MPEG-4) also contain data dependent conditions and indexing, and WHILE loops.

- Multidimensional arrays. Multimedia applications mainly work on multidimensional arrays of data. Also large one-dimensional arrays are often encountered.

- Statically allocated data. Most data is statically allocated. Its exact storage location can be optimized at compile time. Also this is partly starting to change however, but the mixed case is a topic of future research.

- Temporary data. Most of the data is temporary, only alive during one or at most a few iterations of the algorithm. When the data is not alive anymore, it can be overwritten by other data.

In addition to the above, our applications are typically characterized by a combination of: high volume and low power/size requirements so cost-driven design: real-time processing so timing and ordering constraints have to be included; the presence of several sampling rates; many pages of complex code; data-dominated modules with large impact on cost issues.

Front-end telecommunication applications have generally the same characteristics as multimedia applications, with the exception that the loop bounds are partly varying. Network component applications, on the other hand, contain almost no deep loop nests and have mainly data dependent conditions and indices. Moreover, most of the data is non-temporary and organized into dynamically (run-time) allocated tables and lists. Most of the DTSE methodology described here is applicable to all three classes, but some additional stages in the system design flow are needed to deal with the dynamic data (see [565]).

1.3 TARGET ARCHITECTURE STYLE

Many programmable processor architectures have been proposed for video and image processing. A good survey oriented to MPEG video processors is available in [450]. More general-purpose processor architectures issues are summarized in [423].

Several multi-media oriented processors have been introduced in the past decade [219], including the well-known TI C6x (see Fig. 1.4), and Philips-TriMedia (see Fig. 1.5). The latter are both VLIWs with many different units and a (partly) software controllable memory hierarchy. Several other Super-scalar/VLIW processors have been announced with an extended instruction-set for multi-media applications: Intel (MMX), SGI/MIPS (MDMX), HP (MAX), DEC (MVI), Sun (VVIS), AMD (MMX), IBM (Java), ... Also many dedicated domain-specific ASIP processors have been proposed (see recent ISSCC and CICC conference proceedings).

Fig. 1.7 shows a relatively general abstraction of the current state-of-the-art multimedia type processors [502, 507]. The following key features are typically observed in these architectures :

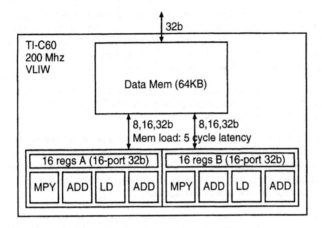

Figure 1.4.   TI C6x processor architecture template

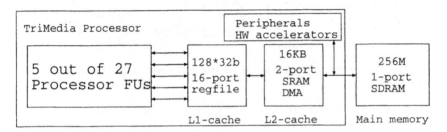

Figure 1.5.   Philips-TriMedia/TM1 processor architecture

1. Multiple functional units (FU) are present which constitute a superscalar [427] or a very long instruction word (VLIW) [507, 502] processor or a multi-threaded multiprocessor type of core [493]. The programming happens mostly from the program memory (directed by a compiler or assembler). In the VLIW case, as in multi-media processors like the Philips TriMedia or the TI C6x series, nearly everything is compile time controlled in the FUs, including which ones are activated.

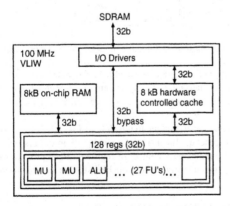

Figure 1.6.   Locking and bypassing in the Philips-TriMedia/TM1

2. One or more layers of cache memories are present, at least one of which resides on chip. The transfer and storage organization is left almost always to hardware up to now (hardware cache controller) [427]. Still, a clear trend towards more software control of data cache management is observed recently [502, 507]. This includes cache locking and cache bypass selection (see Fig. 1.6). We target both these types of control over data cache management.

3. A relatively centralized bus organization[3] is usually selected with one or two data busses. This is easier to control or program but forms a potential data transfer bottle-neck, especially for data-dominated multimedia applications [91].

4. The main memory resides off-chip, typically accessed over a single (wide) bus. The data organization inside this memory is usually determined at link time and the data transfer is controlled by the hardware (Memory Management Unit). Compile-time optimizations on this external transfer are hence left out in current practice. However, recent research has shown that this bottleneck can be heavily reduced at compile-time by code rewriting [131, 66].

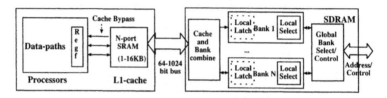

Figure 1.7.   Target architecture model for programmable superscalar/VLIW processor with the memory hierarchy.

Another abstract view including the program memory hierarchy is shown in Fig. 1.8.

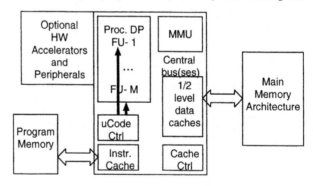

Figure 1.8.   Current architecture model: programmable processors

   In general, the trend towards a more distributed memory organisation to accommodate the massive amount of data transfers needed, is clearly present in commercial processors. The power cost for such (partly) multi-media application oriented processors is heavily dominated by storage and transfers [204, 503] (see Fig.1.9).

   For multi-processors, the generic target architecture of Fig. 1.10 will be used, which is parallel, programmable and heterogeneous. It consists of a number of processors, each of which has a data-path with some functional units and a number of registers. Each processor can also have a local memory hierarchy, consisting of (software controllable) SRAMs and (hardware controlled) caches.[4]

---

[3]As opposed to a very customized point-to-point communication network
[4]Hybrid cases are also possible, e.g. caches which can be partly locked to a certain address range, and which will then mainly behave as an SRAM.

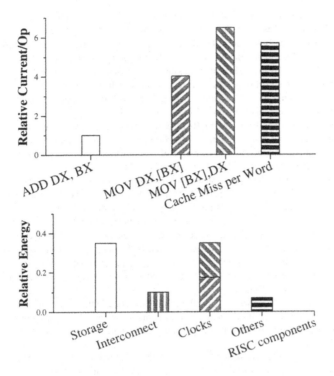

Figure 1.9.   Demonstration of dominance of transfer and storage over control and arithmetic operations in programmable processors: top result by Tiwari et al. [503] where both on-chip read and write
are included in the diagonally hashed bars; bottom result by Gonzales et al. [204]. Large memory
storage and global interconnect contributions are rising for deep submicron technologies due to the
dominance of wiring over transistors. The clock contribution is being significantly reduced by novel
circuit and architectural approaches (expressed by different shades in clock bar).

Furthermore, a global shared memory is present, which can itself be hierarchically organized. In
hardware, the shared memory can be implemented as a NUMA (non-uniform memory access) architecture, but we do not use message passing at the software level (because of the associated power
and size overhead in embedded systems). The number of processors will usually be rather limited
(between 1 and 8).

In section 1.7, it will be motivated that the DTSE script contains both (memory) platform-
independent and -dependent stages. The important parameters in the platform independent steps
of the DTSE script are the size of each of the memories, and the *bandwidth/latency* between the
layers of the hierarchy.[5] The latency does not have to be taken into account at this level (for data-
dominated multimedia and communication systems). Indeed, by using software pipelining and data
prefetching techniques, this latency can in most cases be masked during the detailed instruction
scheduling at the end of the compilation process [155]. E.g., when the latency is predictable (i.e.
when we are not dealing with a cache), a number of delay slots are usually available after the memory load instruction. These delay slots can be used for instructions which do not depend on the
loaded data.

Evidently, a cache will not always be able to maintain its maximal bandwidth (i.e. the hit rate
will be less than 100%), but at the platform-independent stage we cannot take into account the cache

---

[5]For the global memory, these do not have to be fixed yet at the start of the DTSE loop transformation step.

strategy. However, the later steps of the DTSE script remove most of the capacity [296] and conflict [285] misses.[6] Since in most cases the hit rate is close to 100% after applying those optimizations, we will assume (in this early step of the DTSE script) a priori that the DTSE optimized code behaves optimally on the cache, and that the hit rate is thus 100%. Doing the opposite (ignoring the cache, i.e. assuming a hit rate of 0%) would lead to very unrealistic worst-case results. Note that on a super-scalar processor, only the load/store functional unit has to be stalled when a cache miss occurs. On a VLIW, all functional units have to be stalled, but prefetch instructions can take care of loading data into the cache before it is actually used (and thus avoiding stalls).

A similar discussion applies to the SDRAM usage. Bank and page misses can lead to large penalties if the code is not matched to the SDRAM organization. In this case the SCBD and MAA steps in our script (see subsection 1.7) take care of that however, so in the platform-independent stage we will assume that also the SDRAM is optimally used.

The bandwidth between the register file and the data-path is always determined by the number of functional units which can operate in parallel. No restriction exists on the communication between the functional units and the registers. For us it is only this bandwidth itself which is important as we do not take into account the type of operations or the instructions during the DTSE optimizations. At this level no power can be gained either, since all data on which operations are executed, have to be transferred between a register and the data-path. So our DTSE approach is in principle independent of the content of the data-path modules. The latter can be programmable (instruction-set based), custom (ASIC or parametrisable modules) or even reconfigurable (fine- or coarse-grain). That is why we call it *processor-independent*.

The available bandwidth between the local memory or cache and the register file is usually deter-mined by the number of load/store functional units in the processor. Here power optimizations are already possible, by minimizing the use of load/store instructions, which means maximizing the use of the register file.

Note that we do not directly take into account the program memory or instruction cache costs in the DTSE methodology. Ways to reduce also that part of the power and memory footprint (size) have been proposed in literature; see [91, 97] for a summary of these techniques and some extensions. Still, experiments (see e.g. section 8.2 and [300]) indicate that indirectly several DTSE steps actually reduce the I-cache miss rate too. Moreover, they only increase the total program size with a small percentage if applied correctly.

1.4   STORAGE COST MODELS

The storage power model used in our work comprises a power function which is dependent upon frequency of memory accesses, size of the memory, number of read/write ports and the number of bits that can be accessed in every access. The power is directly proportional to the number of memory accesses. Let P be the total power, N be the total number of accesses to a memory (can be SRAM or DRAM or any other type of memory) and S be the size of the memory. The total power in the memory is then given by:

$$P = N \times F(S) \tag{1.1}$$

Here F is a function[7] with the different coefficients representing a vendor specific energy model per access. F is completely technology dependent.

Also for area, similar vendor specific models have been used but these are usually in the form of tables parametrized with the appropriate levels of freedom in the memory generator. Apparently no accurate analytical area formulas can be derived in practice for realistic modern memories that have a very complex internal structure.

Due to confidentiality agreements with these vendors we cannot provide information on the exact models. Some more information on models that we have used in the past is available in [93].

---

[6]The best results are achieved when some software control over the cache is possible [295].

[7]Between sublinear polynomial and logarithmic behaviour

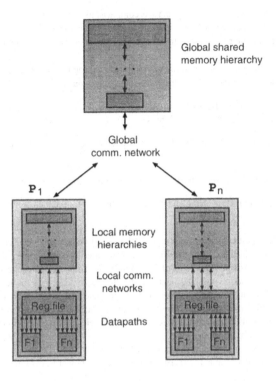

Global shared
memory hierarchy

Global
comm. network

$\mathbf{P}_1$

$\mathbf{P}_n$

Local memory
hierarchies

Local comm.
networks

Datapaths

Figure 1.10.   Parallel target architecture.

For an architecture with no cache at all, that is only a CPU and (off chip) main memory, power can be approximated as below:

$$P_{total} = P_{main} = N_{main} \times F(S_{main})$$

If a level-one cache is present, then we observe that the total number of accesses is distributed into the number of cache accesses (N_{cache}) and number of main memory accesses (N_{main}). The total power is now given by:

$$P_{total} = P_{main} + P_{cache} \tag{1.2}$$
$$P_{main} = (N_{mc} + N_{main}) \times F(S_{main})$$
$$P_{cache} = (N_{mc} + N_{cache}) \times F(S_{cache})$$

The terms S_{main} and S_{cache} represent the main memory and cache size. The term N_{mc} represents the amount of data traffic (in accesses) between the two layers of memory. This term is a useful indicator of the power overhead due to various characteristics like block size, associativity, replacement policy etc. for an architecture with memory hierarchy.

Let the reduction (ratio) in power due to an on-chip access as compared to an off-chip access for the same word length be a factor α. Here α is technology dependent and varies in the range of 3 to 5 for one access [117] [237]. Then we have:

$$P_{main} = \alpha \times P_{cache} \quad where \ \alpha > 1. \tag{1.3}$$

It is obvious from equation 1.3 that better usage of the cache reduces the power consumption for a fixed number of CPU accesses. In the literature several more detailed cache power models have been reported recently (for example [255]) but we do not need these detailed models during our exploration because we only require relative comparisons. Also, these models require low-level information from the cache usage which we do not have available at higher abstraction levels during our overall system exploration. Nevertheless they are useful to make a detailed analysis for the final set of alternatives.

Some academic groups have tried to built parametrized cache models. They assume then that the cache power can be sub-divided in to the following components:

1. Memory cell power, which usually dominates for most cache structures and includes the power to drive the bit lines of the cache columns;

2. Row decoding power - which decodes the address supplied by the CPU into tag, index and offset parts;

3. Row driving power - once the address is decoded, the corresponding cache line must be selected;

4. Sense amplifier activation at the end of each column bit line, which is also very costly in many modern SRAMs.

The sum of these four components is equal to the energy consumed by one cache access operation. Interconnect capacitance, cell size, and transistor size are constants while parameters such as the cache block size, words per block, word size, number of address bits, cache size, and cache-CPU bus width are independent variables. These independent variables determine the number of index and tag bits, and the number of rows and columns in the cache. Thus, increasing the associativity increases the number of tag bits and the number of parallel tag comparators. In practice it is often even worse: usually all banks are accessed speculatively. Hence this results in increased row driving power (and also the total power).

A detailed discussion on cache power modeling is available in [255]. Another power model that is often used in the literature is [277]. Issues related to low power RAM's are discussed in [243].

1.5   OBJECTIVES AND GLOBAL APPROACH

It can be shown that the evolution in memory circuit technology or programmable processor architectures, even though major improvements have been obtained in the last decades, will on its own not solve the system design problem in terms of band-width and storage bottlenecks [91, 97]. The same is true for the evolution in hardware support to tackle the data memory bottleneck in programmable processor architectures [79]. So also major design technology improvements and compiler modifications will have to be introduced in future system design.

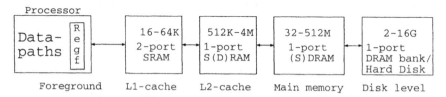

Figure 1.11.   Hierarchical memory pipeline

The access bottle-neck is especially severe for the main memories in the hierarchical pipeline (see Fig. 1.11). This is due to the large sizes of these main memories and partly also because up to now they are always situated off-chip which causes a highly capacitive on-off chip bus. Due to the use of embedded (S)DRAM (eDRAM) technology, the latter problem can be solved for several new applications. However, even for eDRAM based main memory solutions, the central shared memory would still imply heavy restrictions on the latency and exploitable bandwidth. In addition, the large energy consumption issues would not be sufficiently solved either. Indeed, because of the heavy push towards lower power solutions in order to keep the package costs low, and recently also for mobile applications or due to to reliability issues, the power consumption of such external DRAMs has been reduced significantly [568, 244]. It is expected that not much more can be gained because the "bag of tricks" is now containing only the more complex solutions with a smaller return-on-investment.

Hence, modern stand-alone DRAM chips, which are often of this SDRAM type, potentially offer low power solutions but this comes at a price. Internally they contain banks and a small cache with a (very) wide width connected to the external high-speed bus (see Fig. 1.12) [269, 271]. So the low power operation per bit is only feasible when they operate in burst mode with entire (parts of) memory column(s) transferred over the external bus. This is not directly compatible with the actual use of the data in the processor data-paths[8] so without a buffer to the processors, most of the data words that are exchanged would be useless (and discarded). Obviously, the effective energy consumption per useful bit becomes very high in that case and also the effective bandwidth is low.

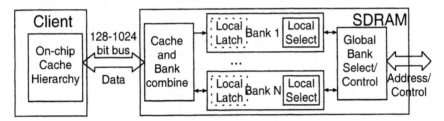

Figure 1.12.   External data access bottle-neck illustration with SDRAM

Therefore, a hierarchical and typically much more power-hungry intermediate memory organization is needed to match the central DRAM to the data ordering and effective bandwidth/latency requirements of the processor data-paths. The reduction of the power consumption in fast random-access memories is not as advanced yet as in DRAMs but also that one is saturating, because many circuit and technology level tricks have already been applied also in SRAMs. As a result, fast SRAMs keep on consuming on the order of Watt's for high-speed operation around 500 MHz. So the memory related system power bottle-neck remains a very critical issue for data-dominated applications.

From the process technology point of view this is not so surprising, especially for (deep) submicron technologies. The relative power cost of interconnections is increasing rapidly compared to the transistor (active circuit) related components. Clearly, local data-path and controllers themselves contribute little to this overall interconnect compared to the major data/instruction busses and the internal connections in the large memories. Hence, if all other parameters remain constant, the energy consumption (and also the delay or area) in the storage and transfer organization will become even more dominant in the future, especially for deep submicron technologies. The remaining basic limitation lies in transporting the data and the control (like addresses, instruction bits and internal array signals) over large on-chip distances, and in storing them.

From the above it is clear that on the mid-term (and even the short term already for heavily data-dominated applications), the access and power bottle-necks should be broken also by other means not related to process technology. It will be shown that this is feasible with quite spectacular effects by a more optimal design of or mapping on the memory organisation and system-level code transformations applied to the initial application specification. The price paid there will be an increased system design complexity, which can be offset with appropriate design methodology support tools. It requires another view on platform architecture design and mapping methodologies. The architectures also have to become more compiler-friendly (see e.g. [373]).

At IMEC, these issues have been in the forefront of several of our internal projects (see above). In contrast to traditional parallelizing or instruction-level compilers our techniques work on a truly global scope in the algorithm and also in the search space available for the DTSE steps. This is feasible, by ignoring the issues which are tackled in traditional compilers and by only focusing

---

[8]Except in some regular media processing kernels (if the source code is explicitly written to exploit the packing and bus modes) and some numerical vector processing algorithms

on the global data transfer and storage. In that sense, our approach should be really seen as a source-to-source precompiler for the traditional approach. Hence, we believe that the organisation of the global communication and data storage forms the dominating factor (both for area and power) in the system-level design decisions. This can be especially improved by focusing on the related algorithmic code transformations. Therefore, we have concentrated ourselves mainly on the effect of system-level decisions on the access to (on- of off-chip) background memories (which requires separate access cycles as opposed to the foreground registers), and on the transfer of data over long "distances" (e.g. over long-term main memory storage). In order to assist the system designer in this, we are developing a formalized system-level DTSE methodology. This is complementary to and not in competition with the existing compiler technology. Indeed, our approach should be considered as a precompilation stage, which precedes the more conventional compiler steps that are essential to handle aspects that are not decided in "our" stage yet. So we are dealing with "orthogonal" exploration choices where the concerns do not overlap but that are still partly dependent though due to the weak phase coupling. From the precompilation stage, constraints are propagated to the compiler stage. The effectiveness of this will be substantiated in the application demonstrators described in this book.

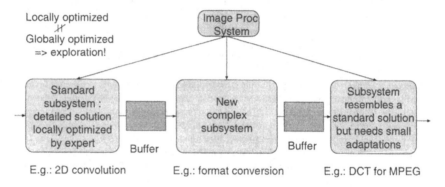

Figure 1.13.   Data transfer and storage exploration for heterogeneous data-dominated multi-process systems aims at removing the large buffers introduced between the subsystems due to production-consumption mismatches for large data streams.

We have demonstrated that for our target application domain, it is best to optimize the storage/transfer related issues **before** the (task and data) parallelizing and VLIW/superscalar compilation steps involving control and arithmetic operations [128, 129]. In conventional approaches (see Fig. 1.13), the subsystems of a typical image processing system are treated separately and each of these will be compiled in an optimized way onto the parallel processor. This strategy leads to a good parallel solution but unfortunately, it will typically give rise to a significant buffer overhead for the mismatch between the data produced and consumed in the different subsystems. To avoid this, we propose to address the system-level storage organization (see subsection 1.7) for the multi-dimensional (M-D) arrays in the background memories, as a first step in the overall methodology to map these applications, even before the parallelization and load balancing [128, 132], or the software/hardware partitioning [129, 130]. The amount of parallelism which can be achieved is hardly affected by this. Even within the constraints resulting from the memory and communication organisation decisions, it is then still possible to obtain a feasible solution for the (parallel) processor organisation, and even a near-optimal one if the appropriate transformations are applied at that stage also [128, 129]. This new system-level design methodology is illustrated in Fig. 1.14.

1.6   BRIEF STATE OF THE ART

A much more detailed literature survey will follow in chapter 2 but a very brief summary is already provided here.

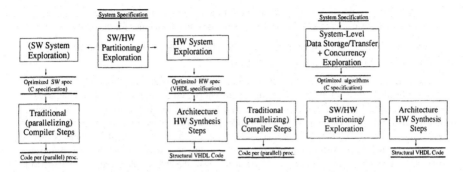

Figure 1.14.   Traditional SW/HW co-design methodology (left) and proposed system-level design methodology (right) where SW/HW partitioning is postponed in the system design trajectory and where a common DTSE stage is introduced prior to it.

In traditional (parallel) compiler research [9, 381], memory related issues have partly been addressed in the past, but they have concentrated almost solely on performance in individual loop nests. Very few approaches directly address energy and system bus load reduction, or more aggressive speed improvements obtainable by fully global algorithm transformations. In this parallelizing compiler community, work has mainly focused on loop transformations to improve the locality in individual (regular) loop nests, major efforts have taken place e.g. at Cornell [320], at Illinois [43, 402], at Univ. of Versailles [174], and in the well-known SUIF project at Stanford [18]. These typically do not work globally across the entire system, which is required to obtain the largest energy impact for multi-media algorithms, as we will show further on. Partitioning or blocking strategies for loops to optimize the use of caches have been studied in several flavours and contexts, e.g. at HP [169] and at Toronto [293, 341]. More recently, also multi-level caches have been investigated [250]. The main focus of these memory related transformations has been on performance improvement in individual loop nests though and not on power or on global algorithms. As shown in Fig. 1.15, the speed depends on the critical path in the global condition tree, which can be viewed as a concatenation of local critical sections (bold line segments in Fig. 1.15) in separate loop nests. For the power however, other weights are placed on the data-flow arrows and the global data-flow of complex data type accesses across conditional and concurrent paths has to be considered during the exploration for cost-effective code transformations. As a result, mostly different optimisation and exploration techniques have to be developed.

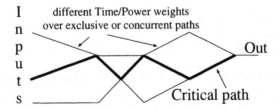

Figure 1.15.   Illustration of difference between speed and power oriented optimisation. The speed exploration focuses on the critical path reduction, which operates on the worst-case path (line segments represent individual loop nests) in the global condition tree (with mutually exclusive branches). (Average) power optimisation should act on the global data flow over all concurrent and exclusive path.

Compiling for embedded processors differs from that for general purpose processors in the following ways:

1. We now have a known (set of) application(s) running on the processor when the product is shipped. Hence we are permitted longer analysis and compilation times for the application.

2. Cost issues like power consumption, memory bandwidth, latency and system bus load are very important for cost efficient global system design.

These features present many (new) opportunities and pose interesting problems which are unique to the embedded processor environment. In the embedded system-level synthesis domain, several groups have started looking at these challenges and opportunities. Especially the work at Irvine, which started relatively early on custom (embedded) memory organisations [279], should be mentioned here. Since 1996 they have started working on a compiler for embedded contexts that is "memory-aware" [412]. Also several other academic groups have started working in this promising area since 1996. We refer to [404] for a good overview of the state-of-the-art and the comparison with our own work. In contrast to the work in that domain our techniques work on a global scope in the search space available for the DTSE steps. But this is only feasible by only focusing on the global data transfer and storage issues and not on all problems at once (including the normal compilation issues). So our approach is complementary to the more conventional approaches and it should be really seen as a source-to-source precompiler for the parallelizing or instruction-level compilers. These source-to-source precompiler transformations should also proceed most other recent work in the embedded compiler or system-level synthesis domains. Experiments further on will demonstrate that this decoupling is very well feasible and it leads to a nearly platform-independent precompilation stage in front of the conventional compilers [131]. Our approach also has a strong enabling effect on the use of subword parallel instructions in modern multi-media processors [355, 350], so also for this purpose it is a crucial preprocessing step.

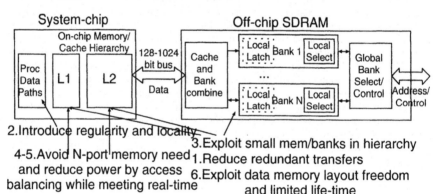

Figure 1.16.   Steps used in ACROPOLIS methodology for reducing the data transfer and storage cost in the background memory hierarchy.

1.7   THE DATA TRANSFER AND STORAGE EXPLORATION APPROACH AT IMEC

In order to achieve a significant drop of the actual energy consumed for the execution of a given application, we have to look again into the real power formula: $F_{real} \cdot V_{dd} \cdot V_{swing} \cdot C_{load}$. Note that the real access rate F_{real} should be provided and not the maximum frequency F_{cl} at which the RAM can be accessed. The maximal rate is only needed to determine whether enough bandwidth is available for the investigated array signal access. Whenever the RAM is not accessed (and is idle), it should be in power-down mode. Modern memories support this low-power mode (see e.g. [471]).

We can now reduce both F_{real} and the effective C_{load} by the way we use the storage and interconnect devices. The goal should be to arrive at a memory and transfer organisation with the following characteristics (see Fig. 1.16):

1. Reduce the redundancy in data transfers.

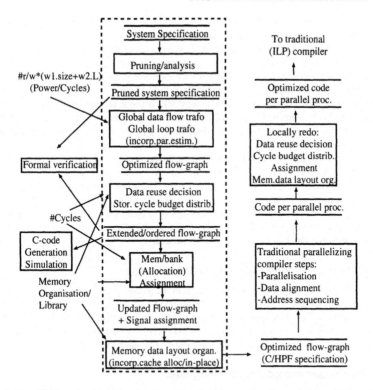

Figure 1.17.   DTSE methodology for data transfer and storage exploration as front-end to parallelizing compilers and/or to (single-processor oriented) VLIW or super-scalar compilers. This methodology is partly supported with prototype tools in the ACROPOLIS environment.

2. Introduce more locality in the accesses so that more data can be retained in registers local to the data-paths.

3. Use a hierarchical memory organisation where the smaller memories (with reduced C_{load}) are accessed the most and the larger ones are accessed the least.

4. Avoid the use of N-port memories if 1-port alternatives with a reasonable cost can be used instead, because more area but also more interconnect (and hence energy-delay) is required for these multi-port components.

These principles can be enabled by applying an appropriate system design methodology with "orthogonal" appropriately ordered steps (Fig. 1.16). A script for supporting our objective has been developed at IMEC since 1996 for this purpose [87, 96, 93, 561].
The current starting point of this DTSE methodology for partly predefined platform architectures [96, 128], is a system specification with accesses on multi-dimensional (M-D) array signals which can be statically ordered. The output is a transformed source code specification, potentially combined with a (partial) netlist of memories which is the input for the final platform architecture design/linking stage when partly customisable memory realizations are envisioned. The transformed source code is input for the software compilation stage in the case of instruction-set processors. The address code (for instruction-set processors) or address calculation unit (ACU) generators (for custom targets) are produced by a separate methodology (ADOPT) [371, 372]. This DTSE methodology is partly supported with prototype tools developed in our ACROPOLIS research project, Reuse is also made of the tools in our more mature ATOMIUM research project [87, 93, 561].

Since 1996, we have obtained several major contributions on the envisioned precompiler methodology for data-dominated real-time signal processing applications.

The research results on techniques and prototype tools [9] which have been obtained within the ACROPOLIS project [90, 96, 98, 93, 85, 86] in the period January 1996 - Spring 2001 are briefly discussed now (see also Fig. 1.17). More details are available in the cited references and for several steps also in the subsequent chapters (as indicated). A more detailed tutorial on the different substeps and what their effect is in a global context will become available in the autumn of 2002 as a companion book to this one (also available from Kluwer). That book will focus on self-study course material for designers or researchers who want to apply the methodology and techniques on real-world application codes.

The platform dependent DTSE stage start from the SCBD step onwards. Most of the material that is presented in this book for this stage is useful in a source-to-source precompilation context with a predefined memory organisation. But then also some pragma's need to be added to the application source code to express some of the decisions such as memory (bank) assignment, or cache bypassing or SDRAM modes that are exploited. Some substeps only apply for an (embedded) customisable memory organisation which is becoming available on several platforms by partly powering down overdimensioned memory blocks that are not fully needed.

1.7.1   Reducing the required data bit-widths for storage

This step belongs to the data-type refinement stage in the global system design trajectory [94, 83]. So it is not part of the DTSE stage as such. The objective of reducing the required data bit-widths for storage can be achieved by applying optimizing algorithmic transformations. These can change the actual algorithm choice, e.g. in digital filters or in the choice DFT versus FFT. Much can be done also in the step from floating-point data types to fixed-point types (see e.g. [96]).

1.7.2   Pruning and related preprocessing steps

This step precedes the actual DTSE optimizations; it is intended to isolate the data-dominant code which is relevant for DTSE, and to present this code in a way which is optimally suited for transformations [93]. All freedom is exposed explicitly, and the complexity of the exploration is reduced by hiding constructs that are not relevant. This step also contains extensive array data-flow analysis techniques to arrive at useful information for the subsequent steps. This includes splitting up the initially defined data types (arrays and sets) into basic groups (BGs).

Full reuse is made here of the ATOMIUM related methodology work [93]. Automation is a topic of current research, so up to now this step is applied manually on the source code with a systematic method.

1.7.3   Global data flow trafo

We have classified the set of system-level data-flow transformations that have the most crucial effect on the system exploration decisions [89, 95], and we have developed a systematic methodology for applying them. Up to now, full reuse is made here of the previous ATOMIUM related work. Two main categories exist. The first one directly optimizes the important DTSE costs factors especially by removing redundant accesses. The second category serves as enabling transformation for the subsequent steps because it removes the data-flow bottlenecks wherever required. An important example of this are advanced look-ahead transformations. Automated design tool support is a topic of future research.

---

[9]All of the prototype tools operate on models which allow run-time complexities which are dependent in a limited way on system parameters like the size of the loop iterators, as opposed to the scalar-based methods published in conventional high-level synthesis literature. Several of them are almost fully reusable from the ATOMIUM research project.

1.7.4 Global loop and control flow trafo

The loop and control flow transformations in this step of the script aim at improving the data access locality for M-D array signals and at removing the system-level buffers introduced due to mismatches in production and consumption ordering (regularity problems). They are applied globally across the full code, also across function scopes because of the selectively inlining applied in the preprocessing step. This is different from the inter-procedural analysis approaches that are commonly used now in recent parallelizing compiler research projects. But our task is simplified still, because we only have to incorporate the constructs related to the loops (and other manifest control flow) and the array type data, i.e. not the code that is "hidden" in a separate function layer after the preprocessing step. The effect of these system-level transformations is shown with two simple examples below:

Ex.1: effect on storage cycle budget distribution and allocation (assume 2N cycles):

```
for (i=1;i¡=N;i++)              for (i=1;i¡=N;i++)
    B[i]=f(A[i]);                  {
for (i=1;i¡=N;i++)                 B[i]=f(A[i]);
    C[i]=g(B[i]);                  C[i]=g(B[i]);
                                   }
```

$\Rightarrow$ 2 background memory ports $\Rightarrow$ 1 background port + 1 foreground register

In this first example, the intermediate storage of the $B[]$ signals in the original code is avoided by merging the two loop bodies in a single loop nest at the right. As a result, the background access bandwidth is reduced and only 1 background memory port is required, next to the register to store the intermediate scalar results.

Ex.2: effect on potential for in-place mapping (assume 1 memory and function g() initializes first element of each row A[i][] to 0):

```
for (j=1;j¡=M;j++)             for (i=1;i¡=N;i++)
for (i=1;i¡=N;i++)                {
    {                             FOR j:= 1 TO M DO
    A[i][j]=g(A[i][j-1]);             A[i][j]=g(A[i][j-1]) ;
    };                            OUT[i]= A[i][M];
for (i=1;i¡=N;i++)                };
    OUT[i]= A[i][M];
```

$\Rightarrow$ N locations (background) $\Rightarrow$ 1 location (foreground)

In this second example, a function of the $A[]$ values is iteratively computed in the first loop nest, which is expressed by the j loop iterating over the columns in any row of $A[][]$. However, only the end results (N values in total) have to be stored and retrieved from background memory for use in the second loop nest. This intermediate storage can be avoided by the somewhat more complex loop reorganisation at the right. Now only a single foreground register is sufficient, leading to a large reduction in storage size.

It has to be stressed that loop and control flow trafo occur also in many other steps of the DTSE script. However, in those cases they either focus on a totally different cost functions (e.g. at the SCBD step) or they serve as enabling substeps for the actual optimizing trafo that succeed them e.g. in the data-flow and data-reuse decision steps).

In the ATOMIUM context, an interactive loop transformation prototype engine (SYNGUIDE) has been developed that allows both interactive and automated (script based) steering of language-coupled source code transformations [456]. It includes a syntax-based check which captures most simple specification errors, and a user-friendly graphical interface. The transformations are applied by identifying a piece of code and by entering the appropriate parameters for a selected transformation. The main emphasis lies on loop manipulations including both affine (loop interchange, reversal and skewing) and non-affine (e.g. loop splitting and merging) cases.

In addition, research has been performed on loop transformation steering methodologies. For power, we have developed a script oriented to removing the global buffers which are typically present between subsystems and on creating more data locality [142, 149]. This can be applied manually.

In the ACROPOLIS context, an automatable compiler technique is currently being developed. The original work of [539, 183] has been used as a basis. That technique has been further extended with more refined cost estimators of the other memory related optimisation steps (especially data reuse and locality impacts) in the placement phase of our two-phase approach [124, 123, 127]. Methods to automate this step in a precompiler context will be discussed in detail in chapter 3.

Also pure performance improvement is being stressed (in addition to energy reduction) with very promising results [389, 355, 352]. Note that the system performance is not only expressed in the raw cycle count, but also by the as important system bus load at the board level [125, 126]. The latter is highly related to the energy of the data transfers so this external bus load is heavily reduced simultaneously with our power gains. We have investigated systematic techniques needed to steer the high-level loop and data-reuse transformations in the platform independent steps. Significant progress towards a formalization of the combined optimization (and partial trade-off) for the power/system bus load and processor speed has been made, as demonstrated in [352, 353]. We have shown also that the program code size and instruction cache miss effects of our DTSE transformations are very acceptable [354, 349].

1.7.5   Array-oriented memory size estimation

This has been an active topic of recent research in the system-level synthesis community [523, 572]. Most of the approaches only give an idea on the size of the memories involved for a given procedural loop ordering. An upper bound estimate without any ordering to be imposed is proposed in [34, 39]. Lower and upper bounds for a partially defined loop organisation and order that are needed to really steer the many possible loop transformations for a given application, have been proposed in [274, 275]. They include both size estimation technique for individual dependencies [273] and one for inter-array dependencies with overlapping life-times [276]. In addition, guidelines for loop transformation steering have been developed [272]. See chapter 4.

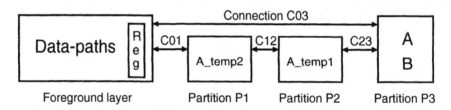

Figure 1.18.   Memory hierarchy illustration: the foreground memory next to the data-paths consists of registers and register files; the intermediate buffers (partition $P1$ and $P2$) typically consist of fast embedded synchronous SRAMs with many ports to the lower level where potential "array copies" from the higher levels are stored; the top layer (partition $P3$ in this example) consists of a slower mass storage unit with low access bandwidth. Virtual connections (bypasses) can also be made between the memories and the common data-path box.

1.7.6   Data reuse decision in a hierarchical memory context

In this step, we have to decide on the exploitation of the given memory hierarchy, including potential bypasses wherever they are useful [561, 563, 564]. Important considerations here are the distribution of the data (copies) over the hierarchy levels as these determine the access frequency and (where freedom still exists) the size of each of the resulting memories. Obviously, the most frequently accessed memories should be the smallest ones reside the closest to the data-paths and which consume the least power per access. This can be fully optimized only if a (potentially predefined) memory hierarchy is present. We have proposed a formalized methodology to steer this, which is driven by estimates on bandwidth and high-level in-place cost [154, 563]. Based on this, the background transfers are partitioned over several hierarchical memory levels, including the exploitation of bypasses, to reduce the power and/or access (cycle) cost.

This memory hierarchy exploitation objective is broken up over two steps in the DTSE methodology: a platform-independent one where the signal copies are distributed over a virtual memory hierarchy and a memory hierarchy layer assignment substep (see chapter 5) where the actual physical memory hierarchy is decided (see section 1.7.7).

An illustration of this data reuse decision step is shown in Fig. 1.18, which shows two arrays, A and B, stored in background memory. Array signal A is accessed via two intermediate array copies (A_temp1 and A_temp2) so its organisation involves 3 levels (including the initial array), while array B is accessed directly from its storage location, i.e. located at level 1. Remark that A and B are stored in the same memory. This illustrates that a certain memory can store arrays at a different level in the hierarchy. Therefore we group the memories in partitions ($P1$, $P2$, and $P3$) and do not use the term "levels" which is reserved for the level of the array signal copy. Our data reuse and memory hierarchy decision step [154] selects how many partitions are useful in a given application, in which partition each of the arrays (most of which are only intermediate data in data-dominated applications) and its temporary copies will be stored, and the connections between the different memory partitions (if e.g. bypassing is feasible). An important task at this step is to perform code transformations which introduce extra transfers between the different memory partitions and which are mainly reducing the power cost. In particular, these involve adding loop nests and introducing temporary values – to be assigned to a "lower level" – whenever an array in a "higher level" is read more than once. This clearly leads to a trade-off between the power (and partly also memory size) lost by adding these transfers, and the power gained by having less frequent access to the larger more remote memories at the higher level(s). This step also includes some early in-place decision on the foreground storage that is implied by the data reuse decisions.

An effective exploration technique and a prototype tool to automate this has been proposed in [512]. This systematic exploration step is not present in conventional (parallelizing) compilers. See chapter 5.

Here ends the platform-independent stage of the DTSE script.

1.7.7 *Storage cycle budget distribution*

In this step, we mainly determine the bandwidth/latency requirements and the balancing of the available cycle budget over the different memory accesses in the algorithmic specification [562, 564, 566].

1.7.7.1 Memory hierarchy layer assignment. After the platform independent transformation steps, the algorithm has to be examined for memory transfers that are best moved to foreground memory. These memory transfers have to be flagged as being foreground memory transfers (level 0), so that other tasks that concentrate on background memory transfers don't have to take them into account.

In addition, based on band-width and high-level in-place storage cost, the background transfers should be partitioned over several hierarchical memory layers[10], to reduce the power and/or area cost. Every original signal is assigned to a memory partition. In order to get the signals to the data-path however, usually intermediate copies have to be created which are stored in other memory partitions. The decision to create these intermediate layers in the hierarchy and the signal copies should be based on high-level cost estimates for access band-width and storage requirements (including in-place effects) [274]. Also the memory class of the different memory layers should be decided (e.g. ROM, SRAM or DRAM and other RAM "flavors").

1.7.7.2 Loop transformations for SCBD. In order to meet the real-time constraints, several loop transformations still need to be applied. These mainly involve loop merging of loops without dependencies (where locality of access was not improved), software pipelining (loop folding) and (if all else fails) partial loop unrolling. This step will not be further formalized in this book, though preliminary results are available [473].

---

[10]Within a range of 1 to a maximal memory depth *MaxDepth*, to be provided by the designer

1.7.7.3   BG structuring.   In this SCBD context, also a methodology to optimize the way the initial data types (arrays or sets) are grouped/partitioned and mapped into the memory has been proposed [166, 164, 165]. We call this basic group (BG) structuring, which is an essential substep as shown e.g. in the DAB application of chapter 8. In this book it will not be dealt with in any detail however.

1.7.7.4   Storage bandwidth optimization.   Storage Bandwidth Optimization (SBO) performs a partial ordering of the flow graph at the BG level [560, 562]. It tries to minimize the required memory bandwidth for a given cycle budget. This step produces a conflict graph (see chapter 6 that expresses which BGs are accessed simultaneously and therefore have to be assigned to different memories or different ports of a multi port memory.

Techniques have been proposed to steer this step for typical RMSP code with promising results [483, 566]. Originally this was only for code without nested loops [483, 562]. But since 1999 our prototype tool now also achieves a balanced distribution of the globally available cycle budget over the loop nests such that the maximally required band-width is reduced (full storage *cycle budget distribution* step) [566, 66, 64]. This ensures that fewer multi-port memories are required to meet the timing constraints and also fewer memories overall. Very fast techniques have been developed to make this step feasible in an interactive way with different variants [394, 398] including more adaptive [395] and space-time extensions [396, 397] of the underlying access ordering technique. See chapter 6.

For a fully predefined memory organisation, this SCBD step is as important to achieve the maximal exploitation of the available memory/bank ports and this with a minimized amount of memory bank activation (especially within the external memories, which are typically based on the SDRAM concept). So also here our prototype tools are extremely useful. See chapter 6.

1.7.8   Memory/bank allocation and signal assignment

The goal in the memory/bank allocation and signal-to-memory/bank assignment step (MAA) is to allocate memory units and ports (including their types) from a memory library and to assign the data to the best suited memory units, given the cycle budget and other timing constraints [40, 483]. Clearly, the actual memory allocation is usually fully predefined on the processor chip. However, the external memory organisation is at least partly free. Indeed, even for fully predefined memory allocations, freedom remains present in the assignment of data to the different memory alternatives compatible with the real-time access constraints defined in the previous SCBD step, and this especially in the banks that are powered up in e.g. the on-chip cache or the external SDRAM (or other fast DRAM styles). The latter assignment has major impact on power consumption. Also for a heavily distributed (see e.g. several of the recent TI C5x processors) or reconfigurable (see e.g. the on-chip memories in the Xilinx Virtex series) on-chip SRAM memory organisations, assignment freedom exists

The memory interconnect problem that arises for a (heavily) distributed memory organisation, has been solved also [526]. In these substeps, mostly ATOMIUM results are reused but new developments have been added too. Especially the assignment to SDRAM and on-chip memory banks is a major new issue. This step is not really incorporated in this book (see the book on custom memory management [93] and [524] for more details).

The combination of the SCBD and MAA tools allows us to derive real Pareto curves of the background memory related cost (e.g. power) versus the cycle budget [63]. This systematic exploration step is not present in conventional compilers. See chapter 6 for more details.

1.7.9   Memory data layout optimization

In the memory allocation and signal-to-memory assignment step, 'abstract' signals were assigned to physical memories or to banks within predefined memories. However, the signals are still represented by multi-dimensional arrays, while the memory itself knows only addresses. In other words, the physical address for every signal element still has to be determined. This transformation is the data layout decision.

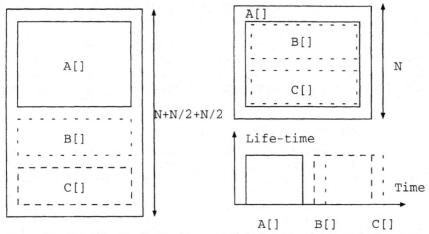

Figure 1.19.   Illustration of in-place mapping task. In the worst-case, all arrays require separate storage locations (left-hand side solution). However, if the life-times of arrays $B[]$ and $C[]$ are not overlapping with the life-time of array $A[]$, the space reserved in the memory for these groups can be shared (right-hand side, top).

This involves several substeps and focuses both on the cache(s) and the main memory. One of the main issues involves *in-place mapping* of arrays, which is illustrated in Fig. 1.19. In the ATOMIUM context initial methods have been investigated for deciding on in-place storage of M-D arrays [520, 521, 532]. Recently, further experiments have been performed to allow even more reduced storage requirements and to identify extensions for predefined memory organisations as in programmable DSP's [141, 143, 142, 149]. This has lead to a solid theoretical foundation for the in-place mapping task and the development of promising heuristics to solve this very complex problem in a reasonable CPU time in a 2-stage approach for M-D array signals [144, 146, 148]. These techniques have been implemented in a prototype tool for evaluating in-place storage of M-D arrays. Very promising results have been obtained with this approach for realistic applications [147, 145, 146, 148].

Since 1997, in the ACROPOLIS context, we have obtained new insights in the problem of reducing the storage size for statically analyzable programs that are being mapped onto several classes of parallel architectures [144, 296]. These insights and the accompanying mathematical descriptions have allowed us to employ more aggressive data transformations than previously possible with existing work, largely based however on our in-place mapping optimisations for custom memory organisations [149]. These transformations can result in a considerable reduction of the storage size by reusing memory locations several times. This is especially important when mapping data-intensive algorithms (e.g. in the multimedia domain) to programmable (embedded) processors. Typically, these processors rely on a limited size cache to obtain sufficient memory bandwidth for the parallel arithmetic units. Our in-place mapping approach enables a better exploitation of the caches and it allows to remove nearly all capacity misses, as illustrated in our experiments. The results are very promising both for software controlled caches [296] as in the Philips TriMedia processor, and for more traditional hardware controlled ones like the Pentium [295, 298]. At the same time, this technique reduces the overall size of the required (external) memories (and hence the capacitive load for off-chip accesses). After the in-place data mapping step we now decide on the signals which will be locked in the data cache, in case of a software controlled cache. See chapter 7.

Also the offsets and organisation of data in the main memory can be affected at this substage in a more aggressive way then the usual padding, even when extended for multi-media applications [405]. We have developed more advanced main memory layout organisation techniques (which go beyond padding) and which allow to remove also most of the present conflict misses due to the limited cache associativity [285]. The overall result is a clear reduction in the cache miss penalty, as demonstrated in [131]. Significant reductions of the bandwidth between the L1 and L2 cache (up to 250%) and between the L2 cache and the main memory (an order of magnitude) have been achieved for realistic imaging and voice coding applications [285, 288]. In combination with address optimisation also a significant speed up of the code can be achieved [289, 287, 288]. See chapter 7.

Finally, the data layout determines the spatial locality of the memory accesses, which is important for exploiting e.g. page-mode and burst-mode memory access sequences in (S)DRAMs. We will not discuss the (S)DRAM issues further in this book, because that research is not mature enough yet to include here.

1.7.10   Other related methodologies and stages

1. **Further optimisation of DTSE script for parallel processor targets:** As indicated in this introductory chapter, all the DTSE concepts remain in principle valid also for a parallel processor context [351, 347, 348], whether it is a data-/task-parallel hybrid combination [132] or a heterogeneous parallel processor [130]. It is however possible to break up the DTSE problems in several separate stages that are dealt with sequentially in the complete system design trajectory. In that case they are separated by concurrency or parallelisation management stages and they are, linked through constraint propagation. This is supported by unified system design flow work that is partly described for data-dominated multi-media and telecom applications in [81, 83]. In order to optimize our DTSE approach for these separate stages, where even more exploration freedom is achievable but where also specific constraints have to be incorporated, extensions to our methods and techniques are needed. Therefore we have started to work on extensions both at the task-level DTSE [342, 343] and data-level DTSE [461, 460] levels of abstraction.

    For complex concurrent applications involving multiple threads of control and a large amount of concurrent processes, we are splitting up the DTSE stage in a two-layer approach, where the problem is partly solved globally first at the more abstract task level, and then on a processor-scope level in a second more detailed stage. Experiments on e.g. an MPEG-4 demonstrator application have shown very promising results [94, 70]. This also allows to couple the DTSE approach to a system-level abstraction stage dealing with a set of dynamic concurrent tasks [500]. That flow includes a data management stage at the concurrent task level [342, 82] where both an extended DTSE flow and dynamic memory oriented transformations [343] should be applied.

    For the (regular) data-level parallelism stage it has been shown also that the methodology has a strong enabling effect when integrated in more conventional array processor mapping methods [461, 460]. This includes the data management related to the (un)packing of the subwords for effectively exploiting the sub-word parallel instructions in modern multimedia processors [355]. So also for this purpose data-level DTSE is a crucial preprocessing step.

2. **Dynamic memory management for dynamically allocated data:** Many modern multi-media applications and most network protocols are based on complex structured data types that are (at least partly) dynamically allocated at run-time. The DTSE approach can then not be directly applied. However, in [565] we show how a preprocessing stage in our global system design trajectory can solve this issue. At the highest level, the application is specified in terms of *abstract data types* (ADTs). The *ADT refinement step* refines these ADTs into a combination of concrete data structures, such as linked lists or pointer arrays with one or more access keys. The *virtual memory management step*, defines a number of *virtual memory segments* and their corresponding custom memory managers. Each virtual memory segment reserves an amount of memory to store all instances of one or more concrete data types. To this end, the virtual memory segment is divided into a number of slots, each capable of storing a single instance of the data type. The virtual memory management step determines, via analysis or simulation of a number of scenarios, the amount of slots that is required to store all instances of that data type. For dynamically allocated data types, the virtual memory management step also determines a custom memory manager for the corresponding virtual memory segment, implementing the allocation and deallocation functions. The ADT and virtual memory management refinement are combined in the dynamic memory management stage of the MATISSE project at IMEC [565, 569].

3. **Address optimization:**   Also this stage is not part of the DTSE methodology itself, but is vital to deal with the addressing and control flow overhead that is introduced by the DTSE steps. The methodology to deal with this is incorporated into another system design stage developed at IMEC, namely the ADOPT (ADdress OPTimization) project [372]. That methodology has been extended to programmable processor contexts [216], including modulo addressing reduction [201].

   The DTSE steps described above will typically significantly increase the addressing complexity of the applications, e.g. by introducing more complex loop bounds, index expressions, conditions, ... When this computational overhead is neglected, the optimized system will, although being more power efficient, suffer a severe performance degradation.

   Most of the additional complexity introduced by DTSE, can however be removed again by source code optimizations such as constant propagation, code hoisting, modulo addressing reduction, strength reduction and others [216, 201]. Moreover, the fact that it originates from DTSE optimizations makes it easily recognizable for the ADOPT tool set. For custom processors, address generation hardware can be synthesized [371]. The final result is that for most applications not only the power consumption is reduced, but also the system performance is increased.

4. **Formal verification techniques for system-level transformations:**   In addition to these results on exploration and optimisation methodologies, we have also continued our earlier work on system-level validation by formal verification of global data-flow and loop transformations [457, 458]. We have developed a complete methodology to tackle this crucial validation task for a very general context including procedural (e.g. WHILE) loops, general affine indices (including the so-called "lattices" of the form $[a.i + b]$), and data-dependent array indices. Such a formal verification stage avoids very CPU-time and design time costly resimulation. We have extended our initial prototype loop transformation validation tool with very effective heuristics to deal with complex real-life applications [120]. We have demonstrated the feasibility of our approach on global transformations applied to several pages of complex code for a H.263 video conferencing decoder [119] and also for caching oriented optimisations in a voice coder application [121].

5. **Platform architecture issues:**   The importance of our approach in a design reuse context for embedded data-dominated applications has been demonstrated in [537, 536, 533]. Also a flexible architecture approach that combines the cost-efficiency of custom architecture with the flexibility of a programmable processor has been proposed [534, 535]. It has the flexibility of a programmable memory and processor architecture organisation but inherits much of the cost efficiency of custom architectures by introducing a novel synchronisation scheme and underlying protocol.

   Most of the steps and tools in the platform-dependent stage of the DTSE approach can also be used as a way to perform "what-if-experiments" in a memory platform definition context. Instead of mapping one specific application instance, the techniques should then be applied to a representative set of applications selected from the target domain. The data on the resulting memory organisations can then be used to steer the definition of the memory architecture parameters like cache memory and line size, associativity, number of memory hierarchy layers, number and size of SDRAM banks, number and size of on-chip SRAMs.

Even though many results have been obtained, as motivated above, the global problem of DTSE is a huge one, and many issues still remain (at least partly) unresolved [91, 92, 97] (see also chapter 9). So a significant future research activity is clearly motivated on a world-wide scale on this important domain.

1.8   OVERVIEW OF BOOK

Chapter 2 gives an general overview of the related work in the domain covered by this book. Specific directly related work is discussed in the corresponding chapters further on.

   Then the most crucial of the platform-independent steps of the script are discussed in more detail. Chapter 3 presents a systematic methodology for global loop transformations. The size estimation techniques that are needed to give feedback are addressed in chapter 4. Chapter 5 presents a systematic methodology for data reuse exploration and decisions.

This is followed by the most crucial platform-dependent steps in the overall script. Chapter 6 explains what storage cycle budget distribution is, with emphasis on the (hierarchical) storage-bandwidth optimization step and the exploitation of Pareto curves for system-wide trade-off exploration. In addition, our automated storage-bandwidth optimization techniques are proposed.

Chapter 7 reviews the data layout organisation related optimizations, with emphasis on the caching related issues, and illustrates its effect and importance on examples. In addition, our automated conflict miss removal techniques are proposed.

Chapter 8 substantiates the claims made in this book on several real-life application demonstrators for platforms with (partly) predefined memory organisations.

Chapter 9 concludes the book and points out the most interesting areas for future research.

A companion book that will be published in 2002, will provide a tutorial self-study course text. It will introduce all the steps in the DTSE methodology in a more intuitive way. Moreover, it will illustrate these in source codes that are documented for a single representative real-life test-vehicle: a cavity detector kernel used in the context of a medical imaging application.

2 RELATED COMPILER WORK ON DATA TRANSFER AND STORAGE MANAGEMENT

In this chapter, an extensive summary is provided of the main related compiler work in the domain of this book. It is organized in a hierarchical way where the most important topics receive a separate discussion, with pointers to the available literature. Wherever needed, a further subdivision in subsections or paragraphs is made.

It should be stressed that the specific related work for the DTSE steps that are treated in detail in this book, is discussed only in the respective chapters where these steps are dealt with. So here only the general related work context is provided.

2.1 PARALLELISM AND MEMORY OPTIMIZING LOOP AND DATA FLOW TRANSFORMATIONS

It has been recognized quite early in compiler theory (for an overview see [43]) and high-level synthesis [519] that in front of the memory organisation related tasks, it is necessary to perform transformations which optimize mainly the loop control flow. Otherwise, the memory organisation will be heavily suboptimal.

2.1.1 Interactive loop transformations

Up to now most work has focused on the loop control flow. Work to support this crucial transformation task has been especially targeted to interactive systems of which only few support sufficiently general loop transformations. Very early work on this has started already at the end of the 70's [332] but that was only a classification of the possible loop transformations.

In the *parallel compiler domain*, interactive environments like Tiny [556, 557], Omega at U.Maryland [265], SUIF at Stanford [18, 23, 220], the Paradigm compiler at Univ. of Illinois [45] (and earlier work [430]) and the ParaScope Editor [360] at Univ. of Rice have been presented. Also non-singular transformations have been investigated in this context [265, 320]. These provide very powerful environments for interactive loop transformations as long as no other transformations are required.

In the *regular array synthesis domain*, an environment on top of the functional language *Alpha* at IRISA [152] has been presented.

In the *high-level synthesis and formal languages community*, several models and environments have been proposed to allow incremental transformation design (see e.g. SAW at CMU [499] and Trades at T.U.Twente [368]). These are however not oriented to loop or data-flow transformations on loop nests with indexed signals, and they assume that loops are unrolled or remain unmodified. An early exception to this is SynGuide [456], a flexible user-friendly prototype tool which allows to interactively perform loop transformations. It has been developed at IMEC to illustrate the use of such transformations on the applicative language Silage and its commercial version DFL (of EDC/Mentor). Also more recent work has followed in this area.

2.1.2 Automated loop transformation steering

In addition, research has been performed on (partly) automating the steering of these loop transformations. Many transformations and methods to steer them have been proposed which *increase the parallelism*, in several contexts. This has happened in the array synthesis community (e.g. at Saarbrucken [498] (mainly intended for interconnect reduction in practice), at Versailles [172, 174] and E.N.S.Lyon [136] and at the Univ. of SW Louisiana [474]) in the parallelizing compiler community (e.g. at Cornell [320], at Illinois [402], at Stanford [552, 18], at Santa Clara [477], and more individual efforts like [517] and [111]) and finally also in the high-level synthesis community (at Univ. of Minnesota [416] and Univ. of Notre-Dame [418]). Some of this work has focussed even on run-time loop parallelisation (e.g. [317]) using the Inspector-Executor approach [455]. None of these approaches work globally across the entire system, which is required to obtain the largest impact for multi-media algorithms.

Efficient parallelism is however partly coupled to *locality of data access* and this has been incorporated in a number of approaches. Therefore, within the parallel compiler community work has been performed on improving data locality. Most effort has been oriented to dealing with *pseudo-scalars* or signal streams to be stored in local caches and register-files. Examples of this are the work at INRIA [162] in register-file use, and at Rice for vector registers [13].

Some approaches are dealing also with *array signals* in loop nests. Examples are the work on data and control-flow transformations for distributed shared-memory machines at the Univ. of Rochester [112], or heuristics to improve the cache hit ratio and execution time at the Univ. of Amherst [361]. At Cornell, access normalisation strategies for NUMA machines have been proposed [319]. Rice Univ. has recently also started looking at the actual memory bandwidth issues and the relation to loop fusion [155]. At E.N.S.Lyon the effect of several loop transformation on memory access has been studied too [185]. A quite broad transformation framework including interprocedural analysis has been proposed in [362]. It is focused on parallelisation on a shared memory multiprocessor. The memory related optimisations are still performed on a loop nest basis (so still "local") but the loops in that loop nest may span different procedures and a fusing preprocessing step tries to combine all compatible loop nests that do not have dependencies blocking their fusing.

Finally, also in the context of DSP data-flow applications, work has been performed on "loop" transformations to optimize the buffer requirements between interacting data-flow nodes. This has been in particular so for the SDF model [55] but it is based on data flow production/consumption sequence analysis and sequence merging which poses restrictions on the modeling. Also "buffer merging" has been applied in that context [383]. A more recent approach [261] uses a shared-memory organisation as target. The main focus of these memory related transformations has been on performance improvement in individual loop nests though and not on power or on global algorithms.

Some work is however also directly oriented to storage or transfer optimisation *between the processor(s) and their memories* to reduce the memory related cost (mainly in terms of area and power). Very early work in the compiler community has focussed on loop fusion and distribution of the arrays over multiple pages in the memories [1]. That was intended to numerical loop nests with limited indexing.

The recent work at Ecole des Mines de Paris [22] on data mapping is also relevant in the context of multi-media applications. It tries to derive a good memory allocation in addition to the transformation (affine scheduling and tiling) of loop nests on a distributed memory MIMD processor. It does not support optimisations inside the processor memory however, so the possibility to map signals in-place on top of one another in the memory

In the *high-level and embedded system-level synthesis community*, also some other recent work has started in this direction. An example is the research at U.C.Irvine on local loop transformations to reduce the memory access in procedural descriptions [279, 280]. In addition, also at the Univ.

of Notre Dame, work has addressed multi-dimensional loop scheduling for buffer reduction [420, 422]. The link with the required compile- and run-time data dependence analysis to avoid false dependencies leading to unnecessary memory accesses has been addressed in [445]. Within the Phideo context at Philips and T.U.Eindhoven, loop transformations on periodic streams have been applied to reduce an abstract storage and transfer cost [527, 529, 530, 531]. At Penn State [260, 256, 240] work on power oriented loop transformations has started. In the same context, at Louisiana State Univ., improving data locality by unimodular transformations is addressed in [443].

2.1.3 Data-flow transformations

Also data-flow transformations can effect the storage and transfer cost to memories quite heavily. This has been recognized in compiler work [8, 367] but automated steering methods have not been addressed there. The specific subject of handling associate scans or chains has been addressed in somewhat more depth in the context of systolic array synthesis [452] and recently also in parallel array mapping [47], but mainly with increased parallelism in mind. Look-ahead transformations have been studied in the context of breaking recursive loops in control [416], algebraic processing [335] and (scalar) signal processing [417], but not in the context of memory related issues. At IMEC, we have classified the set of system-level data-flow transformations that have the most crucial effect on the system exploration decisions and we have developed a formalized methodology for applying them in real-time image and video processing applications [89, 95, 352]. The effect of signal recomputation on memory transfer and data-path related power has been studied too but only in the context of flow-graphs with fixed power costs associated to the edges [196].

2.2 MIMD PROCESSOR MAPPING AND PARALLEL COMPILATION APPROACHES

In most work on compilation for parallel MIMD processors, the emphasis has been on parallelisation and load balancing, in domains not related to multi-media processing. The storage related issues are assumed to be mostly solved by hardware mechanisms based on cache coherence protocols (see other section). Several classes of parallel machines can be identified, depending on several classification schemes [224, 236].

Several manual methodologies have been proposed to map specific classes of multi-media applications to multi-processors in terms of communication patterns [122, 364, 487]. These are however not automatable in a somewhat more general context. Whether this can/should be automated and how far is still under much discussion in the potential user community [116].

In terms of design automation or parallel compilation, especially parallelism detection [58] and maximalisation [402] (see loop transformation section) have been tackled. For this purpose, extensive data-flow analysis is required which is feasible at compile-time or at run-time (see other section). Once introduced, this parallelism is exploited during some type of scheduling stage which is NP-complete in general [54]. A distinction is made between distributed memory machines where usually a data parallel approach is adopted and shared-memory machines where usually more coarse-grain tasks are executed in parallel. In principle though, both these levels can be combined. In addition, there is a lower, third level where parallelism can be exploited too, namely at the instruction level. Also here several approaches will be reviewed below.

Recently, also the mapping on (heterogeneous) clusters of workstations has attracted much attention. Here the focus is still mainly on the general programming environment [75, 138, 179] and ways to ease the communication between the nodes in such a network, like the PVM and MPI libraries. The importance of multi-media applications for the future of the parallel processing research community has been recognized also lately [188, 283], because the traditional numerical processing oriented application domains have not had the originally predicted commercial break-through (yet).

2.2.1 Task-level parallelism

In terms of task parallelism, several scheduling approaches have been proposed some of which are shared-memory oriented [15, 325], some are not assuming a specific memory model [167, 199, 313, 282], and some incorporate communication cost for distributed memory machines [230, 403].

In addition, a number of papers address task-level partitioning and load balancing issues [6, 105, 187, 229]. Interesting studies in image processing of the potential effect of this data communication/storage versus load balancing trade-off at the task level are available in [128, 215, 490]. At IMEC, work has started to extend the (instruction/operation-level oriented) DTSE approach to higher levels of abstraction so that it fits in a more global design flow for data-dominated multi-media and telecom applications [81, 83]. That flow includes a data management stage at the concurrent task level where both DTSE and dynamic memory oriented transformations [343] are applied.

2.2.2 Data-level parallelism

Data parallelism (e.g. operating on image pixels with similar operations in parallel) in a distributed memory context requires an effective data parallelism extraction followed by a data partitioning approach which minimizes the data communication between processors as much as possible. This is achieved especially in relatively recent methods. Both interactive approaches oriented to difficult or data-dependent cases [103, 19] and more automated techniques focussed on regular (affine) cases [233] within predefined processor arrays have been proposed. The automated techniques are oriented to shared memory MIMD processors [5], systolic arrays [176] or to uniform array processors [475, 106, 476, 156]. The minimal communication cost for a given level of parallelism forms an interesting variant on this and is discussed in [153]. Also clustering of processes to improve the access locality contributes to the solution of this problem [333]. All the automatable approaches make use of regular "tiling" approaches on the data which are usually distributed over a mesh of processors. Also in the context of scheduling this tiling issue has been addressed, already since Lamport's hyperplane method [305]. Since then several extensions and improvements have been proposed including better approaches for uniform loop nests [135], affine loop nests [172, 173], non-singular cases [567], trapezoid methods [516], interleaving for 2D loop nests [419] and a resource-constrained formulation [157]. Also in the array synthesis context, such tiling approaches have been investigated for a long time already (see e.g. [72]). Recently, they also incorporate I/O and memory related costs (albeit crude) and are based on a combination of the Locally Sequential Globally Parallel (LSGP) and Locally Parallel Globally Sequential (LPGS) approaches [160, 161]. Most methods do this locally within a loop nest. Only some recent work also performs the required analysis across loop boundaries [357, 217, 235]. Within the Paradigm compiler at Univ. of Illinois also run-time support for exploiting the regularity in irregular applications has been addressed [304].

None of these approaches considers however the detailed communication between the processors and their memories (including e.g. the real physical realisation like the internal SDRAM organisation or the internal cache operation). Neither do they incorporate the limited size or the desired cost minimisation of the required storage units during the parallelisation. In some cases however, there is also a data locality improvement step executed during compilation (see other section). An interesting study in image processing of the potential effect of this data communication/storage versus load balancing trade-off at the data level is available in [128]. It has been shown also that the interaction between such data transfer and storage related transformations and the exploitation of data parallelism on subword parallel data-paths in modern multi-media oriented processors is very strong and positive [355]. For the (regular) data-level parallelism stage it has been shown also that the methodology has a pronounced effect when integrated in more conventional array processor mapping methods [461, 460]. This includes the data management related to the (un)packing of the subwords for effectively exploiting the sub-word parallel instructions in modern multimedia processors.

The final production of the communication code necessary to transfer the data in the parallel processor is addressed rarely [446].

2.2.3 Instruction-level parallelism

A huge amount of literature has been published on scheduling and allocation techniques for processors at the instruction level. Here, only a few techniques oriented to signal processing algorithms are mentioned. One of the first approaches is the work of Zeman et al. [571] which was very constrained. Later on extensions and improvements have been proposed for scheduling based on cyclo-static models [468], list scheduling including loop folding [206], pipeline interleaving [312], pipelined multi-processors [268], and the combination of scheduling and transformations [545, 544]. An evaluation of several scheduling approaches has been described in [225].

In addition, instruction-level parallelism (ILP) has become a major issue in the RISC compiler community. Examples of this can be found in the ILP workshop embedded in Europar'96 [193]. Notable issues at the compilation side are advanced software pipelining [17] global code motion on the inter-BB (basic block) level [331], and the fact that advanced code manipulation is required to obtain the required speed-up potential [432].

The interaction between RISC architecture features and the (instruction-level) parallelism and compilation issues have been emphasized heavily, with a range of optimisations from pure static compile-time transformations to run-time transformations executed on the processor hardware [4].

2.3 MEMORY MANAGEMENT APPROACHES IN PROGRAMMABLE PROCESSOR CONTEXT

The memory interaction has been identified as a crucial bottleneck [424]. Still, the amount of effort at the compiler level to address this bottleneck show a focus on a number of areas while leaving big holes in other (equally important) research issues. A summary about the different issues is provided in [97, 384].

2.3.1 Memory organisation issues

Several papers have analysed memory organisation issues in processors, like the number of memory ports in multiple-instruction machines [377], or the processor and memory utilisation [170, 505]. Also the deficiencies of RISC processor memory organisations have been analysed (see e.g. [31, 373]). This is however only seldom resulting in a formalizable method to guide the memory organisation issues. Moreover, the few approaches are usually addressing the "foreground" memory organisation issues, i.e. how scalar data is organized in the local register files. An example of this is a theoretical strategy to optimally assign scalars to register-file slots [28]. Some approaches are quite ad hoc, e.g. dedicated memory storage patterns for numerical applications on supercomputers have been studied in [336].

Some approaches address the data organisation in processors for programs with loop nests. Examples include a quantitative approach based on life-time window calculations to determine register allocation and cache usage at IRISA [60], a study at McMaster Univ. for the optimisation of multiple streams in heavily pipelined data-path processors [358], and work at Rice on vector register allocation [13]. Also the more recent work on minimizing access time or data pattern access in SIMD processors [16] is relevant in this context. It is however limited to the SIMD context and only takes into account the minimisation of the cycle cost. Moreover, it does not support data-dependent access. The recent work at Ecole des Mines de Paris [20, 22] on data mapping is closer already to what is desired for multi-media applications. It tries to derive a good memory allocation in addition to the partitioning of operations over a distributed memory MIMD processor. It does not support optimisations inside the processor memory however, so the possibility to map signals in-place on top of one another in the memory space is not explored yet and every processor has a single memory.

In a high-level or system synthesis context also several approaches have been introduced. A multimedia stream caching oriented approach is proposed in [222]. Interleaved memory organisations for custom memories are addressed in [104]. Memory bank assignment is addressed in [415]. Exploiting quadword access for interleaved memory bank access is supported in [194].

Also in an ASIP software synthesis context, the approaches are practically always "scalar", or "scalar stream" oriented. Here the usually heavily distributed memory organisation is fixed and the target is to assign the variables to registers or memory banks [323] and to perform data routing to optimize code quality (e.g. [26, 307, 491]).

In a software/hardware co-design context, several approaches try to incorporate the memory modules. Up till now however, they are restricting themselves mostly to modeling the memory transfers and analyzing them by software profiling (e.g. [223]). Also the cache behaviour is sometimes incorporated [321] but no real optimisations directed to the memory utilisation itself is integrated.

Preliminary work on memory allocation in a GNU compiler script for minimizing memory interface traffic during hardware/software partitioning is described in [248].

In a systolic (parallel) processor context, several processor allocation schemes have been proposed. Recently, they also incorporate I/O and memory related costs (albeit crude), e.g. with a solution approach based on a combination of the LSGP and LPGS approaches (see earlier).

Since 1998, this area has been the focus of much more effort, due to the growing importance of embedded software systems and their sensitivity to cost efficiency. As a result, the pure performance oriented approaches that had been the main focus for many decades are now augmented also with cost-aware approaches. A nice overview of the state-of-the-art related to the low power cost aspect can be found in section 4.2 of [52].

2.3.2   Data locality and cache organisation related issues

Data (re)placement policies in caches have been studied for a long time, especially the theoretically optimal ones [49]. Even today, extensions are published on this (see e.g. [504, 496]). Most of this work however requires a-priori knowledge of the data sequence and hardware overhead.

Also data locality optimizing algorithm transformations have been studied relatively well already in the past, e.g. at Illinois/INRIA [192] and Stanford [553]. The focus lies on the detection of spatial and temporal data reuse exploitation in a given procedural code. Tiling (blocking) combined with some local (applied within a single loop nest) uniform loop transformations is then used to improve the locality. Partitioning or blocking strategies for loops to optimize the use of caches have been studied in several flavours and contexts, in particular at HP [169] and at Toronto [291, 293, 341, 340]. The explicit analysis of caches is usually done by simulation leading to traces (see e.g. [158]), but some analytical approaches to model the sources of cache misses have been recently presented also [202, 255]. This allows to more efficiently steer the transformations. Also software approaches to analyse the spatial cache alignment and its possible improvement have been proposed [558].

Recently, this has been extended and implemented in the SUIF compiler project at Stanford [23, 220] to deal with multiple levels of memory hierarchy but it still is based on conventional loop nest models. Multi-level caches have also been investigated in [250]. The work at IRISA, Rennes [509] which distributes data over cache blocks is relevant here too. It uses a temporal correlation based heuristic for this but relies mostly on the hardware cache management still and is oriented mostly to scalars (or indivisible array units). The effect of spatial locality is studied e.g. in [254]. Prefetching (both in hardware and software) is also employed to achieve a better cache performance (see e.g. [525, 334, 326]). Pointer-oriented cache conscious data layout transformations that are very local in nature have been proposed in [110]. Program and related data layout optimisations based on a branch-and-bound search for the possible cache offsets (including "padding") have been proposed in [46].

Recent work has also focused on the multi-media and communication application domain related memory organization issues in an embedded processor context. This is the case for especially U.C. Irvine. At U.C.Irvine, the focus has been initially on [409]: distributing arrays over multiple memories with clique partitioning and bin packing [406], optimizing the main memory data layout for improved caching based on padding and clustering signals in the main memory to partly reduce conflict misses [405, 407, 414], selecting an optimal data cache size based on conflict miss estimates [411] and distributing data between the cache and a scratch-pad memory based on life-time analysis to determine access conflicts in a loop context [408, 413]. All of this is starting from a given algorithm code and the array signals are treated as indivisible units (i.e. similar as scalars but with a weight corresponding to their size). More recently they have added reordering of the accesses to reduce the misses [211]. This is based on accurate memory access models [212]. The exploration of the effect of the cache size and associativity and the related code transformations has been studied in [481, 480, 296]. The CPU's bit-width is an additional parameter that can be tuned during architectural exploration of customizable processors including the link to the memory subsystem [478].

Bypassing the cache for specific fetches is supported by several multi-media oriented processors. Exploiting this feature for instruction caches can be steered in the compiler based on simple usage counts of basic blocks [252]. Also for data caches, schemes have been proposed [251, 447, 515].

Related write buffer oriented compiler issues have been addressed in [345]. Even the disk access from processors is a topic of recent studies. Similar transformations as used for cache optimisation are applied here [259].

In addition, there is a very large literature on code transformations which can potentially optimize this cache behaviour for a given cache organisation which is usually restricted to 1 level of hardware caching (see other section).

Architecture level low-power or high-performance caching issues have been addressed in recent work also. Modified data caching architectures have been explored, such as victim buffers [10]. Both for instructions and data these are actually successors of the original idea of adding a small fully associative cache [253] to the memory hierarchy. In the SIMD context, a customized hardware caching scheme that separates "direct" from "indirect" accesses has been proposed to avoid the problem of cache state divergence [14]. But that is giving only a small performance increase. So the conclusion is still that hardware solutions do not on their own solve the data access problem to main memory. Also major software oriented (compiler-based) improvements are urgently required.

2.3.3 Data storage organisation and memory reuse

In the context of systolic arrays and later also for parallel compilation, the reduction of M-D signals in applicative languages to less memory consuming procedural code has been tackled, especially at IRISA [548, 549, 439, 437]. This is a restricted form of in-place mapping. Also in the work of PRISM on storage management for parallel compilation [314, 113], the emphasis is again on the removal of overhead for static control (applicative) programs. Similarly, in the context of parallel architectures and compilation, also some related work on so-called update-in-place analysis has been reported for applicative languages (see e.g. [178] and its refs) or for Prolog (see [214]). That research only focuses on the removal of redundant copy nodes introduced by the applicative language rewriting. At Penn State, in-place optimisations both between loop nests and procedures have been proposed in [257]. They focus on rectangular data shapes.

2.4 SUMMARY

It can be concluded that the memory organisation related issues have not been sufficiently addressed in the majority of the (parallel) compiler domain literature up till now. Selected recent work has started to look at remedies for this, but that will be addressed in more detail at the appropriate place in the subsequent chapters.

3 GLOBAL LOOP TRANSFORMATION STEERING

As motivated, the reorganisation of the loop structure and the global control flow across the entire application is a crucial initial step in the DTSE flow. Experiments have shown that this is extremely difficult to decide manually due to the many conflicting goals and trade-offs that exist in modern real-life multi-media applications. So an interactive transformation environment would help but is not sufficient. Therefore, we have devoted a major research effort since 1989 to derive automatic steering techniques in the DTSE context where both "global" access locality and access regularity are crucial.

This chapter is organized as follows. Section 3.1 gives an overview of related work in the loop transformation area. Section 3.2 presents the problem definition and the basic model and assumptions. Section 3.3 first presents the multi-step approach that we will adopt, and also the model we will use for representing the search space of the global loop transformation step. The first main phase considers only the geometrical properties of the algorithms (and not the timing aspects). It is divided further in a regularity and a locality improving step. The second main phase then maps the geometrically optimized description to time. In section 3.4, we develop the cost functions to be used for steering loop transformations when the main goal is data transfer and storage optimization. In section 3.5, the outline of a constraint-based exploration method to traverse the search space of all interesting loop transformations is presented. In [127] it is also described how the proposed global loop transformation approach can be combined with parallelization/partitioning.

3.1 RELATED WORK

This section we present related work in the loop transformation area (mainly for locality and regularity oriented loop transformations). First we will discuss related work on loop transformations in general; next we will compare it with our own work.

3.1.1 Loop transformation research

Given the very large body of work performed w.r.t. loop transformations, this overview cannot be extensive. Rather, we will sketch the history of loop transformations and give references to the most important developments.

3.1.1.1 The hyperplane method. When transforming loop nests for parallelization, the goal is to have some loops in the transformed program which can be executed fully in parallel. In general, the transformed loop nest can have its parallel loops innermost or outermost, which corresponds respectively to synchronous and asynchronous parallelism. A synchronous target program looks like this:

```
for time = 0 to N do
    for p in E(time) do in parallel
        P(p)
    enddo
enddo
```

The external loop corresponds to an iteration in time and *E(time)* is the set of all the points computed at the step *time*. An asynchronous target program looks like this:

```
for proc = 0 to M do in parallel
    for p in G(proc) do
        P(p)
    enddo
enddo
```

The external loop corresponds to the processor array, and *G(proc)* is the set of all the points computed by *proc*. In the absence of synchrony, the data dependencies must be enforced in some other way. In a distributed memory implementation, this is imposed by channel communications; in a shared memory implementation, one must take other measures (e.g. with semaphores, ...).

If we aim at minimizing the execution time, the best parallel execution of the loop nest is based on the *free schedule*: each instance of each statement is executed as soon as possible. However, such an implementation would be totally inefficient as soon as control cost (the loop structure becomes very irregular), communication cost (we lose time when transmitting data single item per single item) and synchronization cost are taken into account.

Another approach has emerged from the work done on compiling Fortran loops. The idea is to keep the structural regularity of the original loop nest and to apply regular loop transformations to extract some parallelism. Pioneering work on such loop transformations has been performed in [284] and in Lamport's hyperplane method [305], where the *polytope model* is first used. The idea is to search for a linear scheduling vector π, such that $E(time) = \{p : \pi p = time\}$. In this way, the set of iteration points executed at time step *time* constitutes a hyperplane of the iteration domain. The n-level iteration domain itself is represented geometrically as an n-dimensional polytope.

Consider the following perfect loop nest (in a perfect loop nest, all statements are at the innermost level):

```
for i₁ = l₁ to u₁ do
    for i₂ = l₂(i₁) to u₂(i₁) do
        for i₃ = l₃(i₁,i₂) to u₃(i₁,i₂) do
            ...
                for iₙ = lₙ(i₁,...,iₙ₋₁) to uₙ(i₁,...,iₙ₋₁) do
                    begin
                    Program P:
                        { Statement S₁ }
                        { Statement S₂ }
                        ...
                        { Statement Sₖ }
                    end
```

Here $l_j()$ (resp. $u_j()$) is the maximum (resp. minimum) of a finite number of affine functions of $i_1,...,i_{j-1}$. A particular execution of statement i is denoted by $S_i(I)$, where I is a vector containing the values of the loop iterators $(i_1,...,i_n)$. The semantics of the loop nest are given by the sequential execution. This sequential execution introduces dependencies between iterations. A dependency exists between iteration $S_i(I)$ and iteration $S_j(J)$ if:

- $S_i(I)$ is executed before $S_j(J)$.

- $S_i(I)$ and $S_j(J)$ refer to a memory location M, and at least one of these references is a write. If the program is in single assignment format, only flow dependencies can be present, i.e. $S_i(I)$ is a write and $S_j(J)$ is a read.

- The memory location M is not written to between iteration I and iteration J. (This condition is meaningless if the program is in single assignment format.)

The dependency vector between iteration $S_i(I)$ and iteration $S_j(J)$ is $d_{(i,I),(j,J)} = J - I$. The loop nest is said to be *uniform* if the dependency vectors do not depend on I or J, and we can denote them simply by $d_{i,j}$. Dependencies within one loop iteration are called *loop-independent dependencies*, and dependencies between different iterations are called *loop-carried dependencies*.

Initially, only perfect loop nests with uniform dependencies were targeted in the hyperplane model. Later on, extensions for affine dependencies have been considered [137, 156].

Many particular cases of the hyperplane method have been proposed. For example, selective shrinking and true dependency shrinking [474] are in fact special cases of the hyperplane method, in which a particular scheduling vector is proposed. In [133], a linear programming method is proposed to determine the optimal scheduling vector. In [134], it is shown that this leads in general to scheduling vectors that are dependent on the problem size. Therefore, a limit problem is proposed which determines the optimal scheduling vector for parametric domains. In this way, the scheduling vector is slightly non-optimal, but independent of the problem size.

Rather than considering the k statements of program P as a whole block, we can try to schedule them separately. A first technique consists in introducing a different constant for each statement: node p of statement S_i is executed at time $\pi.p + c_i$. This technique is also called the index shift method, and is in fact similar to the loop folding transformation. A second technique consists in using an affine scheduling for each statement: with a different timing vector for each statement, node p of statement S_i is executed at time $\pi_i.p + c_i$. This is called affine-by-statement scheduling in [133]; see also [174]. Note that in this case, also loop independent dependencies have to be taken into account.

Although the polytope model was first used by [305] for the parallelization of loop nests, it has further been developed in the systolic array synthesis world [375, 370, 438]. The formulation is slightly different from the hyperplane method: loop transformations are performed by transforming the source polytope into a target polytope, and by then scanning that target polytope. For a parallel execution, the polytope is transformed (using an affine mapping) into a new polytope in a new coordinate system. In this coordinate system, certain dimensions correspond to space and others to time. Therefore the mapping is called a *space-time mapping*. Later on, the polytope model has also been used for parallelizing loop nests for massively parallel architectures [316, 172, 175]. Also our own PDG (Polyhedral Dependency Graph) model has its roots in early polytope related work in the array synthesis domain at IMEC [538].

3.1.1.2 Dependency abstractions and dependency analysis. In the above work, the dependencies are modeled exactly. To make the optimization process less complex, loop nests with more general (affine) dependencies have mostly been targeted by using *direction vectors* [12] to approximate the dependencies. In these methods, the transformations are not performed in a geometrical model, but directly on the code itself. Therefore, different kinds of loop transformations (permutation, reversal, ...) usually have to be considered (and evaluated) individually.

In [552, 554], a theory is presented in which loop permutation, reversal and skewing transformations (and combinations of them) are unified, and modeled as *unimodular* transformations. These have the same range of application as hyperplane transformations. They also present a new technique to model dependencies, namely what they call "dependency vectors", which incorporate both distance and direction information. Using this theory, they have developed algorithms for improving parallelism as well as locality of a loop nest. This work has led to the SUIF environment for parallel compilation [18].

When non-uniform dependencies are present, it can become quite difficult to determine these dependencies. A lot of work has been presented on solving this *dependency analysis* problem, e.g. in [551, 41, 435].

However, in all of this non-polytope related work ([42, 44] gives a good overview), dependencies are approximated and modeled in a format that retains only their existence and not the exact number of dependency instances which exist. This is in contrast with our work, where not only the existence of dependencies is important, but also how many instances are present. It is clear that the amount of data which has to be stored (and thus the number of transfers) is indeed related to the number of dependencies. Thus when the input algorithm is converted into our model, a dependency analysis has to be performed to determine all dependencies. Furthermore, the dependencies have to be captured in a simple mathematical construct. In the polytope model this is possible because the dependencies can be represented as a relation between polytopes, i.e. the *dependency relation* or *dependency function* [435]. This work has been performed in [542] for our model; in this chapter we will make abstraction of the dependency analysis problem. Apart from the dependency function, we will also use a certain approximation, namely the *dependency cone* [239, 21]. This cone, which can be represented by its extremal rays, is the smallest cone in which all dependencies fit.

3.1.1.3 Locality optimization.   It has been recognized early on that an efficient parallelization is partly coupled to data locality. Therefore, the loop transformations described above have also been applied to optimize data locality [192, 431, 59], as well as to combine parallelization and locality optimization [554]. However, these techniques also use the dependency approximations described above, and thus the main focus has always been on performance, and not on data transfer and storage minimization as such.

Another important method to improve data locality is to perform data transformations [112]. However this will not be discussed here since it has to be dealt with in the lower steps of the DTSE methodology [93, 289].

3.1.1.4 Tiling the iteration/data space.   Loop tiling is a transformation which does not fit in the hyperplane or unimodular transformation techniques. However, it is one of the most applied loop transformations [550, 77, 71, 306, 554, 78, 278], as it offers better possibilities for handling load balancing, localization, synchronization and communication.

Important issues are reducing the communication time (e.g. through message vectorization) [451, 45], decreasing the number of synchronizations (e.g. using structural information to eliminate some tests), and diminishing memory access costs through cache reuse or tuned allocation strategies (see next paragraph).

In [45], block data partitions and cyclic data partitions are discussed. Wolf [554] discusses the combination of unimodular transformations and tiling to achieve improved parallelism and locality.

Several extensions of Fortran [102, 103], e.g. HPF [232], are also tuned to tile the iteration/data space by providing directives for data distribution. The corresponding iteration space partitions are then derived based on the "owner computes" rule. In [391], also explicit directives for iteration space partitioning are provided. Moreover, automatically inserting the directives is considered, based on static dependency analysis.

Loop tiling is *not* targeted by the loop transformation step in our script because it is not platform independent, and not intended to improve the global locality across loop nests; see also Section 3.4.4.2.

3.1.1.5 Communication-minimal loop partitionings.   Several research papers aim at deriving optimal iteration space partitionings to minimize interprocessor communication. In [233, 475, 106], communication-free hyperplane partitionings are constructed. However, it is clear that these communication-free partitions only exist in a few restricted cases. In [5], optimal tiling parameters for minimal communication between processors are derived. These papers however consider the communication independently from the parallelization. For example, [5] states "this paper is not concerned with dependency vectors as it is assumed that the iterations can be executed in parallel." And [135] states "The problems of mapping and scheduling are not independent, and this is the main difficulty that any compiling technique has to face. (...) At the moment, we do not know how to solve the two problems simultaneously."

3.1.1.6 Fine-grain scheduling. Some methods have been presented to optimize data transfer and storage based on fine-grain scheduling techniques [328, 22]. However these approaches do not really perform source code level loop transformations, since they do not lead to transformed code afterwards. Therefore they are difficult to use on programmable processors.

3.1.2 Comparison with the state of the art

The main differences between the work presented in this chapter and the related work discussed above are the following:

- **DTSE context and global loop nest scope:** our transformation strategy is a part of the DTSE methodology, which means that the main goal is not optimizing raw speed, but optimizing data storage and transfers. Therefore, we have to focus on transformations across all loop nests instead of in single loop nests.

- **Exact modeling:** our method uses an exact representation for affine dependencies, which has only been used by a few other methods in the research community. Even these do not have the same assumptions and restrictions.

- **Separate phases:** those who use an exact modeling, also differ in that they perform the transformations in one step, whereas we will use a multi-phase approach consisting of polytope placement and ordering phases. We will show that this is needed in a DTSE context to reduce the complexity.

In the following subsections, we will elaborate on these issues.

3.1.2.1 DTSE context and global loop nest scope. Our transformation strategy is a part of the DTSE methodology, which means that the main goal is not optimizing speed, but optimizing data transfer and storage.[1] To optimize these aspects, it is essential that *global* loop transformations are performed, over artificial boundaries between subsystems (procedures). It is exactly at these boundaries that the largest memory buffers are usually needed (see chapter 1). To optimize these buffers, global (inter-procedural) transformations are necessary. Most existing transformation techniques (described above) can only be applied locally, to one procedure or even one loop nest [42, 421]. Some groups are performing research on global analysis and transformations (e.g. [18, 118, 362]), but our approach is not about the inter-procedural analysis itself (which is performed prior to the steps described here). Instead we focus on a strategy to traverse the search space of all global transformations.

Lately, the gap between CPU speed and memory bandwidth has grown so large that also for performance, data transfers are becoming the determining factor. Therefore, the need for global transformations to optimize memory bandwidth has also been recognized in the compiler community very recently [374, 155]. However, they only identify some transformations with limited applicability, and do not present a systematic methodology.

Another important consequence of the DTSE context is that the initial algorithm is in single assignment style (or single definition style when also data dependent constructs and while loops are present). Most existing transformation methods assume a dependency between two statements which access the same memory location when at least one of these accesses is a write [59, 133, 474, 517, 477]. In this way, output-, anti- and flow-dependencies are all considered as real dependencies. By converting the code to single assignment, only the flow dependencies are kept (i.e. when the first access is a write and the second one is a read). Indeed, only these dependencies correspond to real data flow dependencies, and only these should constrain the transformations to be executed on the code. Converting the code to single assignment will of course increase the required storage size at

---

[1] However, because the performance in current microprocessors is largely determined by the ability to optimally exploit the memory hierarchy and the limited system bus bandwidth, our methodology usually also leads to an improved speed.

first, but methods to compact the data again in memory (after the loop transformations) have been developed by us [522, 146] and others [549]. In fact, our in-place mapping step of [146] will usually achieve significantly lower storage requirements compared to the initial algorithm.

As for parallel target architectures, the previously applied compiler techniques tackle the parallelization and load balancing issues as the only key point so they perform these first in the overall methodology [551, 430, 106, 43, 391, 18]. They ignore the global data transfer and storage related cost when applied on data-dominated applications like multimedia systems. Only speed is optimized and not the power or memory size. The data communication between processors is usually taken into account in most recent methods [106, 5, 156, 153, 217] but they use an abstract model (i.e. a virtual processor grid, which has no relation with the final number of processors and memories). In this abstract model, the real (physical) data transfer costs cannot be taken into account.

In our target architecture model, we consider the real multiprocessor architecture on which the algorithm has to be mapped, and also the real memory hierarchies of the processors. This will however not be very apparent in this chapter, because these real parameters are only considered during the ordering phase and not yet during the placement.

3.1.2.2 Exact modeling. To adequately express and optimize the global data transfers in an algorithm, an exact and concise modeling of all dependencies is necessary. Our PDG model provides such a modeling, i.e. the dependencies are represented as a mathematical relation between polytopes.

Most techniques which are currently used in compilers do not use an exact modeling of the dependencies, but an approximation (see Section 3.1.1.2). An example is [362], which combines data locality optimization and advanced inter-procedural parallelization. However, it does not use an exact modeling, and as a result it cannot analyze the global data flow.

Some techniques have been developed using an exact modeling, but most of these only target uniform loop nests. An exception is the method described in [137], which can in theory lead to the same solutions as our PDG based method, but the approach is mainly theoretical: the ILPs (Integer Linear Programs) to solve get too complicated for real world programs. Moreover they are only designed to optimize parallelism, and they neglect the data reuse and locality optimization problem.

3.1.2.3 Separate phases. The main difference between our approach and the existing polytope based methods ([267, 316, 175]) is the fact that we split the loop reordering step in two totally separate phases. During the first phase (placement), we map the polytopes to an abstract space, with no fixed ordering of the dimensions. During the second phase (ordering), we define an ordering vector in that abstract space. The advantage is that each of these phases is less complex than the original problem, and that separate cost functions can be used in each of the phases. This is crucial for our target domain, because global optimizations on e.g. MPEG-4 code involve in the order of hundreds of polytopes, even after pruning.

3.1.2.4 Comparison with earlier work at IMEC. The material in this chapter extends the work on control flow optimization which has been performed at IMEC since 1989, mainly by Michael van Swaaij [539, 541, 542] and Frank Franssen [181, 182, 184]. Much of that work was also summarized in [93], oriented to custom memory architectures. The main difference with the material presented here is the target architecture, which is now parallel and especially programmable. In this book we will mainly deal with the aspects of a (largely) predefined memory organisation for a programmable processor (see section 3.3). Our recent research is also focused on methodologies that deal with parallelization and processor partitioning [128, 132, 299, 351, 356] but these aspects are not covered here.

Furthermore, we propose several extensions w.r.t. the cost functions that will be used to estimate the quality of the intermediate and final results and to drive the optimization process. The main extension is that our cost functions will now deal with the exploitation of data reuse in the memory hierarchy of the target architecture. This is the topic of section 3.4.

Finally, we will develop a new strategy w.r.t. the optimization process, because the strategies used in previous work do not work for real-life examples. This is the topic of section 3.5.

3.2   PROBLEM DEFINITION

Since our intent is to perform loop transformations on an algorithm, we will give the problem definition in terms of the input and output of that loop transformation engine.

3.2.1   Input

The input will be a multidimensional signal processing algorithm, together with the target architecture on which it has to be mapped, and some cost functions for the memory organization. We will not perform a data dependency analysis. Instead we require that the input is an algorithm on which a full data flow analysis has already been done, and where the result is expressed explicitly in the code by using single assignment semantics, e.g. in the form of a Silage or DFL-program, or a semantically equivalent C-description. Therefore, we will assume here that the input is in the mathematical form of a set of Conditional Recurrence Equations (CREs) (see e.g. [438]).

In the problem definition we will not limit the recurrence equations to be affine, but later on this restriction will have to be imposed for most of our actual modeling and optimization steps. (With piecewise affine functions as an exception [181, 36].)

While a set of recurrence equations is not always procedurally executable, we will for the moment impose this constraint on the input. The reason is that it is very difficult, for humans as well as for tools, to reason on and manipulate programs which are not procedurally executable.

A CRE is defined here as a 6-tuple:

$$< D_k, f_k, U_k, \mathcal{V}_k, I_k, \mathcal{J}_k > \tag{3.1}$$

where:

- k is the identification number of the CRE.

- $D_k = \{x \in \mathbb{Z}^{n_k} \mid C_k x \geq c_k\}$ is the *iteration domain*: the domain over which the CRE is defined. C_k is an $m \times n_k$ matrix and c_k an $m \times 1$ vector, with m the number of inequalities which define the domain boundaries.

- f_k is the operation performed by the CRE.

- U_k is the n_{U_k}-dimensional signal being defined by the CRE.

- $\mathcal{V}_k = \{V_k^i\}$ is the set of operand signals of the CRE.

- I_k is the index function for the signal being defined U_k. It is a mapping $\mathbb{Z}^{n_k} \to \mathbb{Z}^{n_{U_k}}$.

- $\mathcal{J}_k = \{J_k^i\}$ are the index functions for the operands $\mathcal{V}_k$. They are mappings $\mathbb{Z}^{n_k} \to \mathbb{Z}^{n_{V_k^i}}$.

The signal U_k is thus defined as:

$$U_k(I_k(x)) = f_k(\ldots, V_k^i(J_k^i(x)), \ldots)$$

When the CRE is affine, the index functions can be written as:

$$I_k(x) = I_k x + i_k$$
$$J_k^i(x) = J_k^i x + j_k^i$$

Note that, when a CRE has a non-dense iteration domain (i.e. when the corresponding loops have non-unit strides), it does not fit in the model above. However, such a CRE can always be converted to a CRE with a dense iteration domain. Consider the following CRE, which is defined over a *lattice* described by an $n_k \times n_k$ matrix L and an $n_k \times 1$ vector l:

$$\{x \in \mathbb{Z}^{n_k} \mid \exists \alpha \in \mathbb{Z}^{n_k} : C_k \alpha \geq c_k \wedge x = L\alpha + l\} :$$

$$U_k(I_k(x)) = f_k(\ldots, V_k^i(J_k^i(x)), \ldots)$$

This can be converted to the following CRE, which is defined over a dense iteration domain:

$$\{\alpha \in \mathbb{Z}^{n_k} \mid C_k \alpha \geq c_k\} \ : \ U_k(I_k(L\alpha + l)) \ = \ f_k(\ldots, V_k^i(J_k^i(L\alpha + l)), \ldots)$$

The above does not mean that we will exclude polytopes to be stretched out (by non-unimodular mappings) during the mapping to the common iteration space. In the final placement, non-dense iteration domains are allowed. The conversion to dense iteration domains is only meant to simplify the analysis of the initial CREs.

3.2.2   Output

Very generally, the required output is the following:

- A transformed program, which has been optimized for locality and regularity, considering the target architecture. This will contain explicit annotations on which loop constructs are fixed.

- An estimation of the most suitable processor partitioning, preferably as a set of constraints. The way these constraints are formulated depends on the underlying parallel compiler.

- Possibly, also an estimation for the data reuse and in-place mapping (and thus for the memory hierarchy organization) can be passed to the other steps of the DTSE script.

Another issue is the format in which this output has to be produced. A transformed set of procedurally executable Conditional Recurrence Equations is clearly not enough, since this does not include an ordering. However, it is sufficient to add an ordering vector to each of these CREs. The final code generation will not be discussed here.

3.3   OVERALL LOOP TRANSFORMATION STEERING STRATEGY

3.3.1   Loop transformations in the PDG model

A Polyhedral Dependency Graph model [539, 148] will be used for our optimization problem. Our PDG model is based on the well known polytope model [267, 316, 175]. In this model, a loop nest of n levels is represented as an n-dimensional polytope, where each integer point of the   polytope represents an instance of an operation of the loop nest. The size of the polytope in each dimension is determined by the boundaries of the loop nest. In this model, all dependencies between the operations can be specified *exactly*, in contrast with the more conventional parallelization techniques [551, 554, 43, 18] that make use of direction vectors or other approximations. Although the polytope model is exact, it has a compact mathematical description.

The PDG model is based on the polytope model, but is not restricted to one loop nest or polytope. Instead, each loop nest of the algorithm is represented by a polytope. Dependencies between the operations are captured exactly in relations between the polytopes. The PDG is a directed graph, of which the nodes represent the polytopes and the edges represent the relations. So each node of the PDG is associated with a CRE as defined in Section 3.2.1, and each edge $e = (p, q)$ of the PDG is associated with a dependency function $D() : \mathbb{Z}^{n_p} \to \mathbb{Z}^{n_q}$, where n_x is the dimension of polytope x.

The PDG model is a pure data flow model: it tells nothing about the order in which the operations will be executed. However, a control flow model based on these constructs allows for a generalization and formalization of high-level control flow transformations and their steering. In this model, a control flow is defined in two phases [539]:

- **Placement phase:** First all polytopes are mapped to one *common iteration space* (in a non-overlapping way). This placement phase is the primary focus of this chapter.

- **Ordering phase:** Next, an ordering vector is defined in this common iteration space.

In this way, global control flow transformations are possible, which are not prohibited by artificial boundaries in the algorithm specification, or by the initial ordering in this specification.

This basic control flow model can be modified or extended based on the context in which it is used. For example, in the DTSE script the ordering should not be fixed completely after the loop transformations, since the SCBD step also modifies the ordering (but on a smaller scale). Therefore, it is better that only a partial ordering is defined in the common iteration space, i.e. that the necessary constraints to achieve an efficient (in terms of DTSE) control flow are passed to the next step of the script.

In a parallel processor context, an additional preliminary processor mapping can be defined for all operations. Also, the ordering can be split up into a global and a local ordering. The global ordering gives the synchronization constraints between processors, and the local orderings define the ordering of the operations executed by each processor. Both the processor mapping and the ordering are preliminary since the final decisions are taken by the (parallel) compiler, which has to take into account the constraints imposed by DTSE.

Furthermore, estimations for data reuse, memory (hierarchy) assignment and in-place mapping can be used during the control flow transformations. These estimations can be passed to the next steps of the DTSE script too.

Below, the different issues of optimizing in a parallel processor context with a memory hierarchy are described in more detail, along with some examples.

3.3.1.1 Polytope placement and ordering.   During the placement phase, the polytopes are mapped to the common iteration space by affine transformations. Since no ordering information is available yet at this stage, the concept of time cannot be used: only geometrical information is available to steer the mapping process. For DTSE, the main objective will be to make the dependencies as short (for locality) and as regular (for regularity) as possible. This means that the placement phase is fully heuristic, and that the final decisions will be taken during the ordering phase. Another way to say this is that during the placement phase, the *potential* locality and regularity of the code are optimized, such that we then have a good starting point to optimize the real locality and regularity during the ordering phase.

Although the placement alone does not define a complete transformation yet, local loop transformations will correspond more or less to individual polytope mappings, and global transformations will be performed by choosing a relative placement of the different polytopes.

When all polytopes have been mapped to the common iteration space, a linear ordering is chosen in this space by defining an ordering vector π. This vector must be legal: it has to respect the dependencies. This means that if a dependency x $\to$ y is present, then πd_{xy} should be positive, where d_{xy} is the dependency vector between x and y (pointing in the direction of the data flow).

It is important to realize that for a programmable processor (our target platform), this ordering vector does not have the same stringent meaning as for e.g. a systolic array. The ordering vector only defines the relative ordering of all operations, not the absolute ordering. So if holes are present in or between the polytopes, these do not correspond to time points. Depending on the quality of the code generation (which is done by scanning the polyhedral space [316, 210, 175]), they will be skipped in the end by if-statements or by modified loop bounds [266].

The execution order of the operations can be visualized as a hyperplane, orthogonal to the ordering vector, which traverses the iteration space. This hyperplane cuts the dependencies which are alive. The source data corresponding to these dependencies have to be stored in memory at that moment of time.

Example: loop merging.   In Fig.3.1, an algorithm consisting of two loops (polytopes) is shown: the first loop computes an array $A[]$ and the second computes an array $B[]$ out of $A[]$. The iteration domains of the loops are presented geometrically, and the dependencies between the operations are drawn. Two (out of many) possibilities to map these polytopes to a common iteration space are shown: at the left, they are mapped linearly after each other, and at the right, they are mapped above each other. Note that for this last mapping, the dimension of the common iteration space needs to be higher than the dimension of the individual polytopes. It is clear that in the first solution, the dependency vectors are very long, while they are very short in the second solution.

If we now have to define a linear ordering vector for the first solution, only one possibility exists: traversing the space from left to right. At most N dependencies will be crossed at the same time, which means that a memory of N locations is required to execute the algorithm.

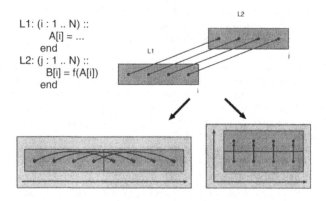

L1: (i : 1 .. N) ::
    A[i] = ...
  end
L2: (j : 1 .. N) ::
    B[i] = f(A[i])
  end

Figure 3.1.   Example of loop merging: placement phase.

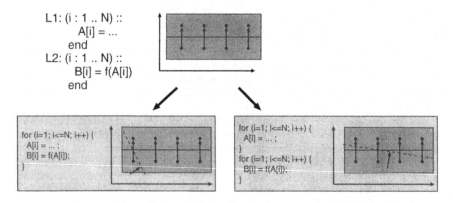

L1: (i : 1 .. N) ::
    A[i] = ...
  end
L2: (i : 1 .. N) ::
    B[i] = f(A[i])
  end

```
for (i=1; i<=N; i++) {
A[i] = ... ;
B[i] = f(A[i]);
}
```

```
for (i=1; i<=N; i++) {
A[i] = ... ;
}
for (i=1; i<=N; i++) {
B[i] = f(A[i]);
}
```

Figure 3.2.   Example of loop merging: ordering phase.

For the second solution, several possibilities exist to traverse the iteration space: see Fig.3.2 (which shows the ordering vector along with the corresponding procedural version of the algorithm). The ordering vector at the left represents in fact a loop merging of the initial algorithm. Only one register is required in this case. The solution at the right represents the procedural interpretation of the initial algorithm: a memory of N locations is required again in this case.

It is thus clear that in this example, the first placement will not lead to a good solution, while the second placement can potentially lead to a good solution. Therefore, we can say that the second placement has an optimal *potential* locality. Choosing a good ordering vector will then exploit this.

Example: loop folding.   Fig. 3.3 shows two ways in which a loop folding transformation can be performed: by choosing a different ordering vector, or by choosing a different placement.

Example: loop interchange.   Fig. 3.4 illustrates loop interchange. Also here two (out of many) possibilities for performing this transformation in the PDG model are presented, namely two combinations of placement and ordering.

3.3.1.2   Preliminary processor mapping.   Optionally, the preliminary processor mapping maps each operation of each polytope to a processor. For each polytope k, a function $P_k(x)$ has to be found, which defines the partitioning of the algorithm over the processors. This mapping should be

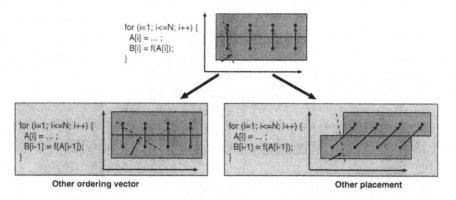

Figure 3.3.   Two ways to perform loop folding.

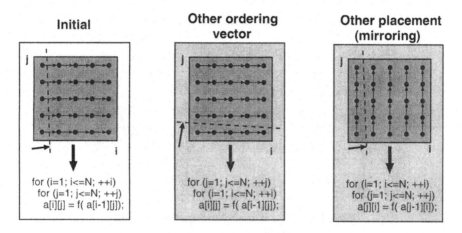

Figure 3.4.   Two ways to perform loop interchange.

reconsidered during the parallelization stage because much more detailed information is available then, but then within the imposed DTSE constraints.

Example.   Fig. 3.5 illustrates the processor mapping. The program consists again of two loops, the first one computing an array $A[]$ and the second one computing $B[]$ out of $A[]$. The iteration domains of the loops are shown again.

Two processor mappings are illustrated. At the left, the first loop is mapped to processor 1 (P_1) and the second loop to processor 2 (P_2). This results in a significant interprocessor communication: indeed all dependencies go from P_1 to P_2 now. At the right, half of each loop is mapped to P_1, and the other half of each loop to P_2. In this way, all dependencies are now inside the processors, and no interprocessor communication is needed anymore.

3.3.1.3   Global and local ordering.   When the placement and ordering results are combined with the processor mapping, it is clear that we have to split up the ordering into a local and a global ordering. Fig. 3.6 illustrates a trivial case of this, again for the same two-loop program as in Figs. 3.2 and 3.5. Here, the loops are merged and half of each loop is mapped to each processor. For each processor, the local ordering is indicated; no global ordering is necessary here.

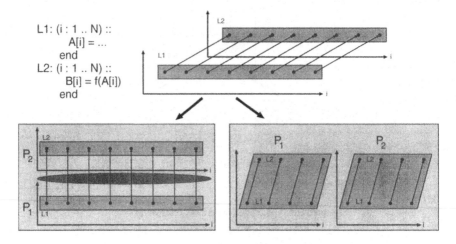

L1: (i : 1 .. N) ::
        A[i] = ...
    end
L2: (i : 1 .. N) ::
        B[i] = f(A[i])
    end

Figure 3.5.   Example of parallel processor mapping.

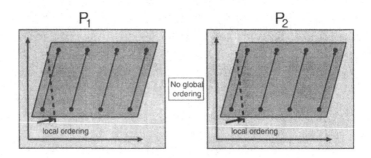

Figure 3.6.   Trivial example of local and global ordering.

Fig. 3.7(a) illustrates a more realistic case. Here, the loops are also merged but each loop is mapped to a different processor. The local ordering for each processor is indicated, as well as the global ordering (synchronization) constraints. Note however that this solution will prevent parallelism, because the global ordering constraints are too stringent. An additional loop folding transformation is needed to introduce parallelism. The modified global ordering constraints corresponding to a loop folding transformation, are shown in Fig. 3.7(b). Of course, storage for two signals of array $A[]$ will now be needed instead of one.

Decisions like this last one, involving a trade-off between parallelism and data transfer and storage, can better be postponed to the SCBD and parallelization stages. Therefore, we will not really focus on splitting up the ordering into a global and local ordering in this chapter.

3.3.1.4 Estimations for data reuse, memory (hierarchy) assignment and in-place mapping.
The most general problem is the following: at each moment of time, all signals which are alive have to be mapped to the memory hierarchy. Of course, this is far too complex, and moreover these issues are the object of later steps of the DTSE script. We only need estimations to drive the control flow optimization process.

Note that during the placement phase (which is the main focus of our work), the ordering of the statements is obviously not known yet, so only crude high-level estimates will be available there (see Sections 3.4.3 and 3.4.5). During the ordering phase, more accurate estimations can be used.

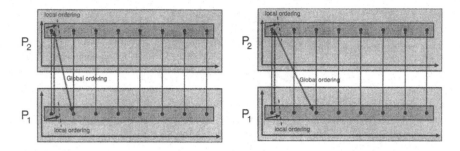

Figure 3.7.   Examples of local and global ordering.

3.3.1.5   Other estimations. Because the loop transformations are at the beginning of the DTSE script, they should in principle incorporate high-level estimates for *all* of the subsequent steps. Fortunately, it can be shown that some steps (like the storage cycle budget distribution (SCBD), memory allocation and assignment (MAA) and main memory data layout steps) have a negligible interaction with the choices made, because improving locality and regularity (nearly) only improves the input for these steps. Therefore, the estimations to be incorporated will be limited to the aspects mentioned above.

3.3.2   Example of optimizing an algorithm in the PDG model

In this section, a somewhat larger example will be discussed in order to illustrate how also global transformations can be performed by the placement and ordering phases.

The initial specification is the following:

```
A: (i: 1..N)::
      (j: 1..N-i+1)::
         a[i][j] = in[i][j] + a[i-1][j];

B: (p: 1..N)::
      b[p][1] = f( a[N-p+1][p], a[N-p][p] );

C: (k: 1..N)::
      (l: 1..k)::
         b[k][l+1] = g( b[k][l] );
```

Fig. 3.8 shows the common iteration space, as it could be when the polytopes corresponding to the loop nests are more or less placed arbitrarily (in the figures, N is assumed to be 5).

3.3.2.1   Placement phase. In a first step, we will remove the dependency cross, by mirroring polytope C around a horizontal axis. In the code, the corresponding transformation would be a loop reversal. This is however not completely true because, as explained above, we only have a valid transformation when placement as well as ordering have been chosen. The common iteration space now looks as follows: (Fig.3.9).[2]

In a second step, polytope B is skewed downwards, such that it closely borders polytope A. In this way, the length of the dependencies between A and B is strongly reduced. The common iteration space now looks as shown in Fig. 3.10.

Now we can see that all dependencies in A and B point upwards, while those in C point to the right. The dependencies between B and C have directions between these two extremes. However, by

---

[2]During the placement, all manipulations happen in polyhedral space, but to illustrate their effect, also equivalent loop descriptions are provided in the figures. These descriptions are however not maintained internally, and are neither really correct because no ordering has been defined yet.

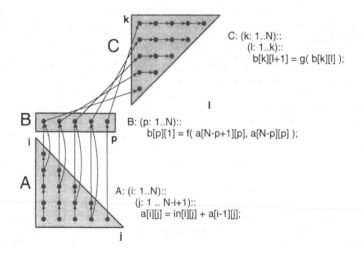

C: (k: 1..N)::
   (l: 1..k)::
      b[k][l+1] = g(b[k][l]);

B: (p: 1..N)::
      b[p][1] = f(a[N-p+1][p], a[N-p][p]);

A: (i: 1..N)::
   (j: 1 .. N-i+1)::
      a[i][j] = in[i][j] + a[i-1][j];

Figure 3.8.   Initial common iteration space.

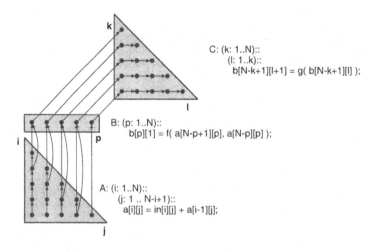

C: (k: 1..N)::
   (l: 1..k)::
      b[N-k+1][l+1] = g(b[N-k+1][l]);

B: (p: 1..N)::
      b[p][1] = f(a[N-p+1][p], a[N-p][p]);

A: (i: 1..N)::
   (j: 1 .. N-i+1)::
      a[i][j] = in[i][j] + a[i-1][j];

Figure 3.9.   Intermediate common iteration space (step 1).

rotating polytope C counterclockwise over 90 degrees , we can make all dependency vectors point in the same direction (upwards). This rotation of polytope C corresponds more or less to a loop interchange transformation in the code. See Fig. 3.11 for the common iteration space.

Next we are going to skew polytope C downwards, such that it also closely borders polytopes A and B. In this way, the dependency vectors between B and C become much shorter. This corresponds (again more or less) to a loop skewing transformation in the corresponding code. The final placement in the common iteration space is shown in Fig. 3.12.

3.3.2.2 Ordering phase.   The final placement looks optimal from a geometrical point of view: all dependencies are as short as possible, and they are very regular. We will now try two of the many possible orderings for the final placement. The first ordering, as well as the procedural code which corresponds to it, is depicted in Fig. 3.13.

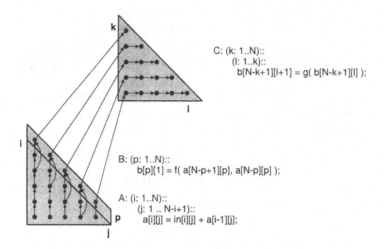

C: (k: 1..N)::
    (l: 1..k)::
        b[N-k+1][l+1] = g(b[N-k+1][l]);

B: (p: 1..N)::
    b[p][1] = f(a[N-p+1][p], a[N-p][p]);

A: (i: 1..N)::
    (j: 1 .. N-i+1)::
        a[i][j] = in[i][j] + a[i-1][j];

Figure 3.10.   Intermediate common iteration space (step 2).

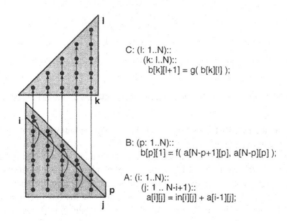

C: (l: 1..N)::
    (k: l..N)::
        b[k][l+1] = g(b[k][l]);

B: (p: 1..N)::
    b[p][1] = f(a[N-p+1][p], a[N-p][p]);

A: (i: 1..N)::
    (j: 1 .. N-i+1)::
        a[i][j] = in[i][j] + a[i-1][j];

Figure 3.11.   Intermediate common iteration space (step 3).

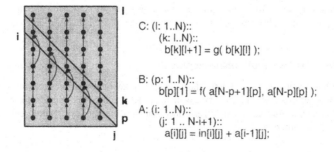

C: (l: 1..N)::
    (k: l..N)::
        b[k][l+1] = g(b[k][l]);

B: (p: 1..N)::
    b[p][1] = f(a[N-p+1][p], a[N-p][p]);

A: (i: 1..N)::
    (j: 1 .. N-i+1)::
        a[i][j] = in[i][j] + a[i-1][j];

Figure 3.12.   Final placement in the common iteration space.

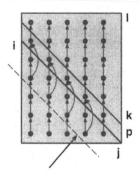

```
for (i=1; i<=N+2; ++i) {
  for (j=1; j<=N-i+1; ++j)
    a[i][j] = in[i][j] + a[i-1][j];
  if (...) b[N-i+2][1] = f( a[i-1][N-i+2],
                            a[i-2][N-i+2]);
  for (k=N-i+3; k<=N; ++k)
    b[k][i+k-N-1] = g( b[k][i+k-N-2] );
}
```

Figure 3.13.   Bad ordering choice.

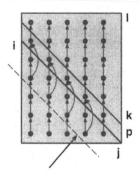

Figure 3.14.   Ordering with same storage requirements as the initial specification.

For this ordering, the required buffer space is $(N + 3)$. This is not optimal at all, but it is already much better than the initial specification which required $2N$ memory locations. By choosing a very bad ordering vector, we can however still obtain the same situation as in the initial specification. This ordering vector is shown in Fig. 3.14. It cuts $2N$ dependencies after having traversed the first "triangle" of the common iteration space. This shows that the placement phase can indeed only optimize the potential locality and regularity of the code, and that the real decisions are taken during the ordering phase. Also, this means that it will be difficult to obtain upper bounds for the storage requirements during the placement phase. Apparently, these upper bounds could even increase, as the enhanced regularity of the dependencies also allows to choose an ordering vector which cuts *more* of them, compared with the initial specification.[3]

Searching for an ordering vector which cuts the minimum number of dependencies, we find the solution indicated in Fig. 3.15. This solution requires only 2 registers of storage! In this solution, the optimized potential locality and regularity of the placement phase is fully exploited by the ordering vector.

3.3.3 Reasons for a multi-phase approach

In the previous section we have outlined the main idea of having two separate phases for the control flow transformation process, i.e. the placement and ordering phases. We have also shown that, because of this, all transformations can be performed in many different ways, by different combinations of placement and ordering.

---

[3]Therefore, it would be more relevant to obtain such an upper bound w.r.t. to a limited (estimated) ordering vector space. See Section 3.4.5.

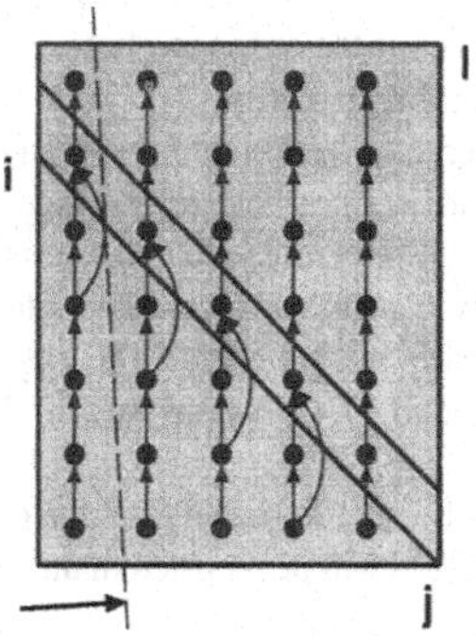

```
for (j=1; j<=N; ++j) {
  for (i=1; i<=N-j+1; ++i)
    a[i][j] = in[i][j] + a[i-1][j];
  b[j][1] = f( a[N-j+1][j], a[N-j][j] );
  for (l=1; l<=j; ++l)
    b[j][l+1] = g( b[j][l] );
}
```

Figure 3.15.   Best ordering choice.

Another way of working would be to place the polytopes in a common iteration space with a fixed ordering, i.e. in which the order and direction in which the dimensions will be traversed is fixed beforehand. Because of the fixed ordering, most transformations can then only be obtained in one way. This would be a much more direct way of working, since the concept of (relative) time is known during the placement in this case. In fact, this optimization strategy would be quite similar to the strategy used by [172, 267, 175]. These approaches have not led to an automated loop transformation methodology because of the high complexity of the solution process (even for relatively simple cases), which is in itself a reason to try another strategy. The originality of our approach lies exactly in the separation of the placement and ordering phases, and in the distinct cost functions used during these phases.

The advantage of decoupling placement and ordering is that it makes the problem less complex. In each phase, a cost function can be used which is more simple than a combined cost function. E.g. for the placement, purely geometrical cost functions (such as average dependency length and dependency cone) can be used which do not consider time or ordering.

On the other hand, the consequence of this is that the final decisions are all taken in the ordering phase. During placement, we can *only* use heuristic cost functions. The placement can be considered as a preliminary phase, which increases the *potential* locality and regularity of the code, and which removes a huge part of the ordering search space by eliminating non-promising choices.

To summarize, the advantages of our strategy are:

- The complexity is reduced. In each phase, a cost function can be used which is more simple than a combined cost function.

- It is a new approach compared to existing work. The latter has not led to an automated loop transformation methodology.

The disadvantages are:

- All transformations can be performed in many different ways, by different combinations of placement and ordering. Future work should investigate whether at least part of this redundancy can be removed.

- Only heuristic cost functions are possible during the placement, such that potentially a placement could be chosen which does not allow a good ordering (and final solution) anymore.

This last point will especially occur when after the placement, almost no freedom is available anymore for the ordering (or in the worst case when no ordering is possible at all). Thus to prevent this situation, we have to take care that enough freedom is still available after the placement. Fortunately, this is not conflicting with the goals of the placement itself (potential locality and regularity optimization). Indeed, optimizing the regularity means that the dependencies will point as much as possible in the same direction, resulting in maximal freedom for the ordering phase. Still, it means that we have to be careful when selecting a cost function for the regularity.

E.g. one dependency which points in a completely different direction than most others, will not have a strong influence on the potential locality and regularity (since it corresponds to at most one source signal for which storage is necessary). However, it will severely reduce the possible choices for the ordering vector. Therefore, a measure of regularity is needed which takes into account all dependencies: a statistical measure like the variance is clearly not sufficient. The extremes within the set of all dependencies would be a better choice. See Section 3.4.1, where we will define such a cost function.

The ordering phase will not be discussed further in this book. Preliminary results of algorithms that have been developed in co-operation with INSA Lyon [186] will be reported in the demonstrator section however.

3.4   POLYTOPE PLACEMENT: COST FUNCTIONS AND HEURISTICS

In the previous sections we have explained the PDG model and the basics of a multi-phase transformation technique in this model, consisting of a placement and an ordering (or space-time mapping) phase. From now on, we will concentrate on the placement phase. This section will focus on the cost functions to be used during the placement optimization process. It will not yet discuss the exploration of the search space itself, as this will be the topic of section 3.5.

In section 3.3, we have already pointed out that no information related to time is available yet during the placement phase, so only geometrical information can be used to steer the mapping process. We also pointed out that, to optimize the *potential* locality and regularity in the placement phase, the main objective will be to make the dependencies as short and as regular as possible. Thus the aspects which will be optimized, are basically the *dependency length* and the *dependency variance*:

- Indeed, dependency vectors which are geometrically shorter will generally also lead to shorter lifetimes of the associated signals, as has already been shown in the examples in the previous section. The locality cost function will be presented in subsection 3.4.2.

- The variance of the dependency vectors should be as small as possible, preferably with all dependency vectors being identical. Firstly, reducing the dependency variance will reduce the number of dependencies which cross each other (see the difference between Fig. 3.8 and 3.9 for an example), and this will in turn lead to better locality and shorter lifetimes of the involved signals. Secondly, reducing the dependency variance creates more freedom for choosing a linear ordering vector (in the second phase) which respects the dependencies. The associated regularity cost function will be discussed in subsection 3.4.1.

As for the dependency length, we can simply take the average to measure the locality, but it is also possible to construct a more complex cost function which takes potential *data reuse* into account (subsection 3.4.3 discusses this).

Apart from the cost functions for locality, regularity and data reuse, we will also discuss some other related aspects in this section, such as the interaction with the ordering phase (subsection 3.4.4), estimations for in-place mapping (subsection 3.4.5), and some ways to reduce the complexity and increase the freedom of the search space (subsection 3.4.6). In the end, all steps in the DTSE script that are affected by the placement are somehow incorporated. This allows to solve the chicken-egg problem that initially exists for this crucial early step of the script.

In a tool implementation, different weights will have to be assigned to the cost functions to arrive at an optimal solution. These weights have to be determined by exploration.

3.4.1   Cost functions for regularity

3.4.1.1   The dependency cone.   Optimizing the potential regularity means that the variance of the dependency vectors should be minimized, i.e. the dependencies should be made as regular as possible. We have already explained (subsection 3.3.3) that a good cost function should take into

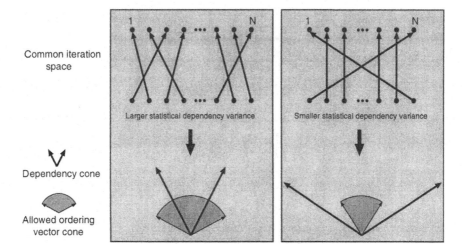

Figure 3.16.   Statistical dependency variance, compared to extreme dependency vectors expressed by the dependency cone.

account the extreme dependency directions. The variance in the statistical sense (square of standard deviation) is thus not a good choice.

Instead we will use the notion of *dependency cone*[4] to express the cost function. This is the cone in which all different directions of the dependency vectors fit. Since a cone can be represented by its extremal rays, the dependency cone is indeed an adequate representation of the extreme dependency vectors. Thus it is clear that the dependency cone contains enough information to know all valid ordering vectors [21].

The regularity cost function will be expressed in terms of the *sharpness* of the dependency cone. If a placement exists with a sharper cone than that of others, then this solution will have more freedom for the ordering vector, and it will in general lead to more regular and efficient solutions. This is illustrated in Fig. 3.16. The solution at the left has a larger statistical dependency variance than the one at the right, but its extreme directions are closer to each other. Therefore, it has a sharper dependency cone, and more freedom for the ordering vector (expressed by the allowed ordering vector cone, as defined in the next subsection).

3.4.1.2   The allowed ordering vector cone.   While the dependency cone adequately represents the extreme dependencies, its size (sharpness) is in fact not really significant. Imagine a very elliptic cone in three dimensions: such a cone only occupies a small portion of the angle-space in which vectors can point (i.e. it has a small solid angle). It can nevertheless significantly constrain the freedom for the ordering vector. This becomes more apparent when the extreme case is considered: a flat ellipse, with all dependencies lying in the same plane. This means that we have a two-dimensional cone in a three-dimensional space. The solid angle of such a cone is always zero steradians. This is even the case when two dependencies are pointing in opposite directions (which means that no ordering vector is possible at all).

Therefore the size of the dependency cone does not adequately represent the freedom for the ordering vector. Instead, another cone should be computed, the *allowed ordering vector cone*, which has already been illustrated in Fig. 3.16 for the two-dimensional case.

---

[4]In this context, the definition of a cone as given in [11] is used, i.e. a cone is defined as the intersection of a number of half-spaces bounded by hyperplanes going through a given point (the vertex of the cone). Some would call this a pyramid (without base). It is thus not a circular cone as generally used for defining conic sections in two-dimensional geometry.

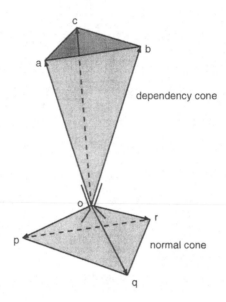

Figure 3.17. Construction of the normal cone. Its faces (opq, oqr and orp) are the normal planes to the vectors a, b and c.

If C_d is a dependency cone in N-space $\mathbb{R}^N$, then the allowed ordering vector cone C_a is obviously defined by:

$$C_a = \{x \in \mathbb{R}^N \mid \forall y \in C_d : x.y \geq 0\}$$

It is clear that this cone is the negative (opposite) of the normal cone of the dependency cone: each dependency vector divides the space in two half-spaces, separated by a hyperplane (the normal plane of the vector). These normal planes define two cones which are point-symmetric to each other: one away from the dependency cone, and one in the same direction as the dependency cone. The first of these is the normal cone (see Fig. 3.17), and the other one is the allowed ordering vector cone (not shown in the figure).

Note that, in special cases like the one described above (a two-dimensional dependency cone in a three-dimensional space), the allowed ordering vector cone will be degenerate. See for example Fig. 3.18.

3.4.1.3 Construction of the cones.   To construct the dependency cone, the extreme dependencies have to be determined. In two dimensions, this can easily be done by computing the angle of all dependencies w.r.t. a fixed axis (using trigonometric functions). The extreme angles correspond to the extreme dependencies, as illustrated in Fig. 3.19. However, computational geometry [401] offers more convenient methods to compute the extreme vectors of a set, without computing the angles themselves.

In three (or more) dimensions, a cone has more than two extreme dependencies; in fact the number of extreme dependencies can even be unlimited. This is illustrated in Fig. 3.20, where six extreme dependencies are present. The computation of the extreme dependencies is in this case a computational geometry problem, similar to computing the convex hull of a set of points. We will use a simple approach, which can reuse existing convex hull methods. It consists of computing the convex hull of all end-points of the dependency vectors, together with the origin. All edges of that convex hull containing the origin, constitute the extremal rays of the dependency cone. It is clear that much faster specific methods can be developed, but this is not crucial in our case, as shown below.

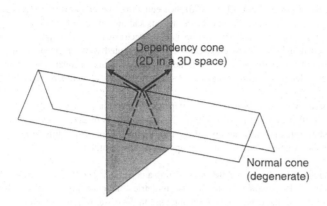

Figure 3.18.   When the dependency cone is of a lower dimension than the common iteration space, the allowed ordering vector cone is degenerate.

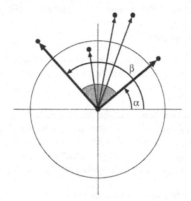

Figure 3.19.   Construction of the dependency cone in two dimensions.

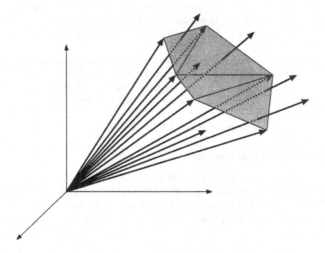

Figure 3.20.   Construction of the dependency cone in three dimensions.

Local dependency cones. For affine recurrence equations (i.e. with affine index functions and affine polytope boundaries), the dependency functions and their domain boundaries are also affine. In that case, it is clear that the extreme dependency vectors also correspond to extremes of the dependency function domain. These extremes can be computed as the intersections of the domain boundaries, and they will always be rather limited in number. Thus it is relatively simple and fast to construct a dependency cone corresponding to a single dependency function between two polytopes.

Global dependency cones. When a global dependency cone has to be constructed as the union of a number of other dependency cones, only the extremal rays of these individual cones have to be considered as potential extremal rays of the global cone. Thus also in this case, the number of vectors of which the convex hull has to be computed, is rather limited.

Allowed ordering vector cones. After constructing a dependency cone, the allowed ordering vector cone can easily be computed as the point-symmetric cone of its normal cone. The boundary hyperplanes of the normal cone are the normal planes to the extremal rays of the dependency cone. It is clear that the extremal rays of the normal cone are also the normal vectors to the boundary hyperplanes of the dependency cone. Thus the normal cone of the normal cone is again the initial cone.

Computational aspects of computing the sharpness of a cone are treated in section 5.2.4 of [127].

3.4.2   Cost functions for locality

In this section, we will discuss locality cost functions, which are related to dependency length (which is not considered at all in the previous section). Two cost functions will be defined: a simple and a refined one.

3.4.2.1   Simple locality cost function. For the potential locality, it seems appropriate to use a rather simple cost function, namely the average dependency length (or equivalently, the sum of all dependency lengths). This means that extreme dependency lengths will not receive special attention, but that is not necessary since one dependency corresponds only to one source signal that has to be stored, and it does not reduce the search space (freedom) for the ordering vector.

3.4.2.2   Refined locality cost function. One refinement can be formulated for this cost function: when several dependencies have the same source signal, it is possible to take only the longest of these dependencies into account.[5] The reason for this is that after all, the source signal has to be stored only once, even if it is consumed multiple times. This is illustrated in Fig. 3.21. Of course, during the placement it is not clear at all whether the longest dependency vector indeed subsumes the time during which the other dependencies related to the same source signal are alive. Geometrically it is a valid heuristic however.

It is clear that this refined locality cost function is already related to data reuse, for which better cost functions will be discussed in the next section. Therefore, it will not be used in the final implementation, which will only use the simple locality cost function and the data reuse cost function. However, it is still discussed here because it is an important building block for the real data reuse cost function in the next section.

Computational aspects are treated in section 5.3.2 of [127].

3.4.3   Cost functions for data reuse

The loop transformation step of the DTSE script is only an enabling step; the intent is that it will lead to improved data reuse, in-place mapping, ... later on in the DTSE script. Consequently, it is strongly desired to have estimations for these aspects during the transformation step. This will provide more direct feedback on the quality of the control flow than only the abstract locality and

---

[5]Of course, this is only valid for the *locality* cost function: when not all dependencies point in the same direction, the shorter ones still have to be taken into account for the dependency cone (regularity).

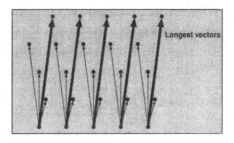

Figure 3.21.   Several dependencies with the same source signal: the refined cost function only uses the longest dependency.

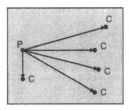
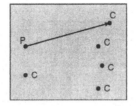
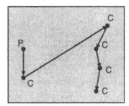

P = production
C = consumption

Figure 3.22.   Example of refined cost function. At the left, all dependencies are shown. In the middle, only the longest dependency is kept. At the right, the refined cost function for data reuse is shown.

regularity measures. Estimations for data reuse are considered the most important because they have the largest impact on the memory hierarchy exploitation; therefore we will thoroughly discuss data reuse in this section. See subsection 3.4.5 for in-place mapping estimations.

Since we have chosen the multi-phase approach of placement and ordering for the loop transformations, it would be preferable to have an estimation for the data reuse in both phases. Of course, the real data reuse can only be estimated when an ordering has been (or is being) defined, but the placement limits the orderings which are possible. So during the placement phase, we need an estimate for the potential data reuse, independent of the ordering.

This can be achieved by adapting the cost functions to the fact that data reuse is *possible*. Indeed, our cost functions up to now optimize the potential locality and regularity, so a logical extension is to make them take into account the potential data reuse too.

Data reuse is essentially possible when several consumptions of the same source signal are close to each other in time. It means that placing a consumption of a signal near to another consumption is as good as placing it close to the production. Indeed, when a signal is consumed it is transferred from a memory to the data path; thus when it is consumed again in the near future we can just as well assume that it has been produced during the last consumption. Of course, the concept of time is not known during the placement, so we will have to place the consumptions *geometrically* close to each other instead.

In subsection 3.4.2, we considered a refined locality cost function which takes only the longest dependency for a certain source signal instance into account. In fact this is already data reuse related, since it assumes that the shorter dependencies can simply reuse the source signal. Here we will look at a better refinement of this cost function, to adapt it to the possibility of data reuse: we will take into account all consumptions, but for each consumption, we consider the distance to the previous consumption, instead of the production. See Fig. 3.22.

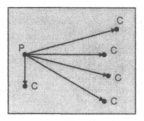 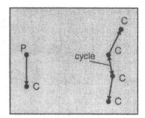

P = production
C = consumption

Figure 3.23.   A problem with the simple refined cost function. At the left, all dependencies are shown. At the right, the dependency to the closest other access is drawn for each consumption.

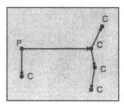

Figure 3.24.   Minimum spanning tree.

Formalizing the cost function.   Developing a formal cost function, requires a little more effort. The problem is again that we do not know the ordering yet, so we do not know which consumption comes first (and thus not which is the "previous" one). Therefore, we propose the following heuristic, which captures all potential data reuse: for each consumption, we choose as dependency length, the shortest length between this consumption and all other accesses.

This would however give a too optimistic measure, as shown in Fig. 3.23: only short dependencies are left (sometimes in a cycle), while it is clear that the source signal will have to be stored in memory at least during the longest original dependency too. What we really need is a graph containing all accesses (one production and n consumptions), which does not contain cycles, and for which the sum of the edges is minimal. Thus our new cost function for data reuse will take the *minimum spanning tree* over all accesses to the same source signal. See Fig. 3.24.

Computational aspects are treated in section 5.4.2 of [127].

3.4.4   Interaction with the ordering phase

3.4.4.1   Cost functions and data reuse.   When the placement phase has been performed, we are left with one common iteration space, in which potential locality, data reuse and regularity have been optimized. The ordering has to exploit these potential properties to optimize the real locality, data reuse and regularity.

In [542], the minimal required storage size for the whole algorithm is optimized. This results in a simple cost function, but it is obviously not the way to go, as it does not consider data reuse, or even the number of background memory accesses.

While the required storage size has to be part of the cost function, also data reuse estimations will be important during the ordering. Indeed, the memory level in which signals are stored has a large influence on the power consumption. E.g. a simplified version of the data reuse tree method [563] could be used during the ordering. This is a topic of current research [512].

We will not elaborate further on the ordering phase, since our main focus in this chapter is on the placement. The ordering is in fact partly similar to conventional space-time mapping techniques, but it must be able to incorporate decisions from the placement phase and to generate partial ordering constraints instead of a fully fixed ordering. This is a topic of current research too [185].

3.4.4.2   Loop tiling. The cost functions of the placement phase optimize the locality and regularity in a geometrical sense, without considering the size of the memories in the memory hierarchy. A result of this is however that it will sometimes not be possible to achieve good locality results for a real memory hierarchy with a linear ordering vector, even for a good placement.

Consider e.g. matrix multiplication. When the three-dimensional polytope of this loop nest is just mapped to the common iteration space in a standard way (without transformation), the locality is quite good. Indeed, the dependencies (certainly the data reuse dependencies) are very short. The regularity is not really that good, because dependencies in three orthogonal directions are present, but this cannot be enhanced anyway.

When a linear ordering vector has to be chosen for this placement, the standard *ijk-, kji-, . . .* loop organizations will result. However, it is known that for large matrices, none of these versions will exhibit a good cache behaviour. Indeed, in addition to the loop interchange transformation, tiling the loop nest is necessary to achieve this.

However, during the placement phase, this is not visible because the size of the memories is not known. Nevertheless, the placement is not faulty: tiling should be included *additionally* in the control flow transformations. In fact, we can express it even stronger: tiling is sometimes necessary, exactly to *exploit* a good placement! Thus it is clear that tiling should also be integrated into our overall control flow transformation methodology. The question is at which point it has to be performed:

1. A first option is to change the placement phase so as to incorporate tiling. Tiling can be performed during the placement, by splitting an N-dimensional polytope over more than N dimensions. However, as we already said, we do not know the size of the memories during the placement, so we do not know the optimal tile size. Furthermore, tiling individual polytopes is conflicting with our "global transformation" strategy, where the global control flow has to be defined first.

2. The second option is to perform tiling in the common iteration space, after the placement, but before the ordering. This is in line with the global transformation strategy, but it leads again to the problem that, when the ordering is not known, also the direction in which we should tile first is not known.

3. The third option is to perform tiling and ordering together. This seems the best alternative, since tiling and ordering cannot be decoupled in an effective way. Also conventional transformation methods perform tiling and space-time mapping together. In this approach, an optimal ordering will normally be determined first, and when this ordering cannot exploit the available data reuse (because the local memories are too small), tiling will be performed.

So, we will choose the third option here, which means that nothing really changes to our strategy and cost functions in the placement phase. The fact that tiling is performed in the common iteration space, allows it to exploit the placement in an optimal way.

3.4.5   Estimations for in-place mapping

In-place mapping is strongly related to the lifetimes of signals: reducing these lifetimes will increase the possibilities for in-place mapping [146, 148]. Since the cost functions used during the placement are exactly designed to lead to shorter lifetimes of the signals, no changes to these cost functions are necessary to optimize the potential in-place mapping.

Apart from that, it would still be useful to have a more direct estimation of the potential in-place mapping for a certain placement. Of course, an exact estimation is not possible at this stage since (again) the ordering has not been fixed yet. However, as shown in chapter 4, lower and upper bounds can be derived on the required memory size (and thus on the in-place mapping) during the placement. The most useful information comes from bounds that take into account *partially fixed* orderings.

3.4.6   Ways to reduce the complexity and increase the freedom of the placement phase

In [127], several ways are proposed to reduce the complexity. These are mainly based on transformations to combine statements to reduce the number of polytopes, unrolling small inner loops to reduce the number of dimensions, polytope splitting and polytope clustering. Several of these actually belong to the pruning and preprocessing step of the DTSE script (see e.g. BG splitting in subsection 1.7.2), but are also relevant here because in some cases they can only be applied after other polytope transformations.

3.5   AUTOMATED CONSTRAINT-BASED POLYTOPE PLACEMENT

3.5.1   Introduction

In the previous section the cost functions for the polytope placement phase have been defined. The next step is to develop a strategy to arrive at the optimal placement w.r.t. these cost functions. In the past, two different methods have already been tried out:

1. Mapping all polytopes at once, namely by writing an ILP (Integer Linear Program) [542]. It is not surprising that these ILPs turn out to be much too complex for real-world algorithms, containing more than a few polytopes.

2. The MASAI-1 tool[6] [541, 542] uses an incremental placement strategy. In a first analysis phase, all polytopes are ordered according to certain characteristics (e.g. their size, meaning that larger polytopes are more important and should thus be placed first), and then they are placed in this order.

   Also this approach does however not lead to good solutions for real-world problems. Only the first few polytopes can be placed in an acceptable amount of search time; for the others, either no good solution is available anymore, or it takes too long to find it.

In this section, we first propose to separate the regularity and locality optimization from each other, by splitting up the placement phase into two steps. From then on, we will mainly focus on the regularity optimization because it is the most complex step.

Next, we propose a constraint-based strategy for the regularity optimization step. In a first phase all dependencies between the polytopes are analyzed, and an optimal relative mapping is determined for each set of polytopes between which a dependency exists. Based on these optimal mappings, constraints are generated to guide the actual placement, which is finally determined by a backtracking-based search algorithm. To prove the feasibility of the concept, we present a prototype tool implementation for this regularity optimization step.

This section is organized as follows: in subsection 3.5.2 some remarks are made w.r.t. the mathematical modeling of polytopes and mappings. In subsection 3.5.3, we propose to split the placement phase into two steps. Next, in subsection 3.5.4, we present the constraint-based optimization approach for the first step. The implementation of the prototype tool is discussed in subsection 3.5.5 and experimental results are presented in subsection 3.5.6. Finally, the second step of the placement approach is briefly discussed in subsection 3.5.7.

3.5.2   Preliminaries

3.5.2.1   Homogeneous coordinates.   For describing index functions, mappings and transformations in this section, we will use homogeneous coordinates [210]. Originally, homogeneous coordinates were used to prove certain problems in N-space. These problems have a corresponding problem in $(N+1)$-space, where a solution can be found more easily. This solution is then projected back to N-space.

---

[6]MASAI is an acronym for Memory-oriented Algorithm Substitution by Automated re-Indexing.

A point $(x_1, x_2, \ldots, x_n) \in \mathbb{R}^n$ can be represented in homogeneous coordinates by the point $(wx_1, wx_2, \ldots, wx_n, w) \in \mathbb{R}^{n+1}$, where $w \in \mathbb{R}_0$ is called the scale factor. We can also project a point $(x_1, x_2, \ldots, x_{n+1})$ in homogeneous coordinates back to affine coordinates by dividing all components by the scale factor x_{n+1}. In this text, we will always choose $w = 1$.

We use homogeneous coordinates in order to transform affine problems into linear problems, such that also translations can be represented as matrix multiplications. E.g., when a point (x, y, z) is translated with (a, b, c), i.e.

$$\begin{bmatrix} x' \\ y' \\ z' \end{bmatrix} = \begin{bmatrix} x \\ y \\ z \end{bmatrix} + \begin{bmatrix} a \\ b \\ c \end{bmatrix} = \begin{bmatrix} x+a \\ x+b \\ x+c \end{bmatrix}$$

This can be expressed in homogeneous coordinates by the following matrix multiplication:

$$\begin{bmatrix} x' \\ y' \\ z' \\ 1 \end{bmatrix} = \begin{bmatrix} 1 & 0 & 0 & a \\ 0 & 1 & 0 & b \\ 0 & 0 & 1 & c \\ 0 & 0 & 0 & 1 \end{bmatrix} \begin{bmatrix} x \\ y \\ z \\ 1 \end{bmatrix} = \begin{bmatrix} x+a \\ y+b \\ z+c \\ 1 \end{bmatrix}$$

Homogeneous mapping functions. Assume we have a mapping function $F(x) = Ax + b : x \in \mathbb{Z}^n \rightarrow x' \in \mathbb{Z}^m$ (thus mapping an n-dimensional polytope to an m-dimensional common iteration space). In homogeneous coordinates, we can represent this as a matrix multiplication $\bar{F}(\bar{x}) = \bar{A}\bar{x} : \bar{x} \in \mathbb{Z}^{n+1} \rightarrow \bar{x}' \in \mathbb{Z}^{m+1}$.

We denote the homogeneous version of a mapping matrix A by $\bar{A}$. It is constructed by appending a column for the translation component, and a unit row. Theoretically we should also use $\bar{F}()$ and $\bar{x}$ for the homogeneous version of the mapping function and the coordinates, but we will not do this since the difference is always clear. We will always use $\bar{A}x$ or $\bar{A}(x)$ to denote the mapping function, and Ax or $A(x)$ to denote its linear part.

From the construction of the homogeneous matrix, it is clear that $rank(\bar{A}) = rank(A) + 1$. So if A is of full rank, also $\bar{A}$ is of full rank, and if A is square and invertible, also $\bar{A}$ is square and invertible.

Homogeneous index expressions. Also index expressions can be represented by homogeneous coordinates, since an index expression is in fact a mapping, namely from n-dimensional coordinates to m-dimensional array subscripts. We will always denote the full index expression by $\bar{J}x$ or $\bar{J}(x)$, and the linear part of it by Jx or $J(x)$.

3.5.2.2 A note on inverse mappings.

The mapping of the iteration domain of a polytope to its definition domain (given by the index function of the signal being defined) is one-to-one and thus invertible. The same is true for the mapping of a polytope to the common iteration space. Nevertheless, the matrix of this mapping is not always directly invertible, e.g. when the common iteration space has more dimensions than the polytope. In that case, the matrix is not square, but it can be extended with unit columns to make it square and invertible.

Some other cases exist in which the mapping matrix is not invertible, but it is always possible to compute an adapted matrix which is invertible. A solution for this problem is discussed in [542], Appendix 5.D (see also [319]). We will not repeat the theory here, but in some examples we will explain how to compute the adapted matrices. In the sequel, we always assume that matrices corresponding to one-to-one mappings are invertible (unless stated otherwise).

3.5.3 Splitting up the placement phase

From section 3.4, we know that we have to optimize the placement for two cost functions which are relatively unrelated, i.e. regularity on the one hand, and locality and data reuse on the other hand.

We will try to separate these two issues, to reduce the complexity of the search space. For this we have the choice between first optimizing for regularity, or first optimizing for locality and data reuse. If we choose to optimize the locality and data reuse first (remember that we are talking about geometrical properties here), it is possible to end up with very large dependency cones, such that the ordering phase will in practice not be able to exploit this (if a valid ordering is still available at all). On the other hand, by optimizing the regularity first, the freedom for locality is usually increased. Thus it is clear that we have to optimize the regularity first; the locality has to be optimized then within the found dependency cone.

The affine mapping of polytopes consists of a linear mapping and a translation component. When we consider two polytopes, we can see that the linear mapping defines mainly the dependency cone (though it has also a small influence on the locality), and the translation defines mainly the locality. Of course this is not true anymore for more than two polytopes, since their relative translation can have a large effect on the dependency cone. Still, the intended complexity reduction can be achieved in the following way:

- First the regularity between sets of polytopes is optimized by fixing their linear mapping.

- Then the polytopes are translated to optimize the locality, while still taking care that the dependency cone does not grow (much) larger (which would undo the effect of the previous step).

It is clear that the first step is the most complex one, since the number of coefficients in the linear mapping is the square of those in the translation component. After the first step, the orientation of all polytopes is fixed, and the complexity of the remaining step is more manageable. The polytopes only have to be shifted around anymore.

3.5.4   Constraint-based regularity optimization

We will now describe how the regularity of dependencies can be optimized by imposing constraints on the relative mapping of polytopes. In the first two subsections we will explain which constraints have to be imposed to make dependencies uniform resp. regular. Next examples will be given for some simple loop transformation. Finally the example of subsection 3.3.2 will be revisited.

3.5.4.1   Constraints to make dependencies uniform.   When imposing constraints to improve the regularity of dependencies, the ideal goal is that all dependencies become uniform, i.e. they have the same direction and the same length. We will now derive the constraints that have to be imposed to accomplish this.

Suppose that an array U is accessed in two polytopes P_1 and P_2 by the index functions $\tilde{J}_1$ and $\tilde{J}_2$:

$$P_1 : \{x \in \mathbb{Z}^{n_1} \mid \tilde{C}_1 x \geq 0\} : U(\tilde{J}_1(x)) = \ldots$$
$$P_2 : \{x \in \mathbb{Z}^{n_2} \mid \tilde{C}_2 x \geq 0\} : \ldots = f(U(\tilde{J}_2(x)))$$

When these polytopes are mapped unchanged (i.e. with the identity function, assuming that this is possible without overlap) to the common iteration space, the dependency function from P_2 to P_1 is:

$$\tilde{D}(x) = \tilde{J}_1^{-1}(\tilde{J}_2(x))$$

The dependency vectors (for those x for which a dependency exists) are thus:

$$x - \tilde{D}x = (\tilde{I} - \tilde{J}_1^{-1}\tilde{J}_2)x$$

These dependency vectors are uniform (independent of x) when the linear part of this expression is zero, thus when

$$(I - J_1^{-1}J_2) = 0$$

or, in terms of the dependency function

$$D = I$$

Another equivalent expression is $J_1 = J_2$, which is indeed immediately clear from the definitions of the polytopes, since uniform accesses can only differ in the offset of the index functions.

When the dependencies are not uniform, we can try to make them uniform by choosing different mappings for the polytopes. We denote the mapping functions by $\tilde{A}_1(x)$ and $\tilde{A}_2(x)$. The recurrence equations in the common iteration space become:

$$P_1' : \{x' \in \mathbb{Z}^{nc} \mid \tilde{C}_1 \tilde{A}_1^{-1} x' \geq 0\} : U(\tilde{J}_1 \tilde{A}_1^{-1} x') = \ldots$$
$$P_2' : \{x' \in \mathbb{Z}^{nc} \mid \tilde{C}_2 \tilde{A}_2^{-1} x' \geq 0\} : \ldots = f(U(\tilde{J}_2 \tilde{A}_2^{-1} x'))$$

The dependency function becomes:

$$\tilde{D}'(x') = \tilde{A}_1 \tilde{J}_1^{-1} \tilde{J}_2 \tilde{A}_2^{-1} x'$$

The new dependency vectors are thus:

$$x' - \tilde{D}' x' = (\tilde{I} - \tilde{A}_1 \tilde{J}_1^{-1} \tilde{J}_2 \tilde{A}_2^{-1}) x'$$

These dependency vectors become uniform when the linear part of this expression is zero, thus when

$$(I - A_1 J_1^{-1} J_2 A_2^{-1}) = 0 \tag{3.2}$$

or, in terms of the new dependency function

$$D' = I$$

The relation with "dependency uniformization" techniques.   Several papers have been written about dependency uniformization techniques [517, 477]. In these techniques, affine dependencies are written as a sum of a limited set (a base) of uniform dependencies. Thus the dependencies are not modified at all, they are just replaced by another set of dependencies (typically a larger set). The new dependencies can be more regularly expressed, which is beneficial e.g. for systolic array mapping.

Since our goal is to modify the dependencies (by transformations) to improve regularity and locality, we cannot make use of these techniques. They have no relation with those described in this section.

3.5.4.2   Constraints to make dependencies regular.   The constraints for uniformity, derived in the previous subsection, are very stringent. Indeed, it means that all instances of a dependency function should have the same direction *and* the same length. The latter requirement is however clearly less important than the former, since the lengths do not affect the dependency cone. Therefore, we will now derive the constraints which are needed to make all dependency instances point in the same direction, while ignoring their length. If this is not fully possible in all dimensions, it should still be maximized in as many dimensions as possible.

We have seen that, after mapping the polytopes to the common iteration space, the dependency vectors have the following form:

$$x' - \tilde{D}' x' = (\tilde{I} - \tilde{D}') x'$$

This is in fact a relation which projects all iteration points onto their associated dependency vector. We will denote the space of the dependency vectors by the *dependency vector space*. All dependency vectors are the same (uniform) when this projection is constant, i.e. when its linear part $(I - D') = 0$. Note that another way to express this is:

$$\text{rank}(I - D') = 0$$

This means that the dimension of the dependency vector space is zero when only its linear part is considered. An example is shown in Fig. 3.25.

When we just want the *direction* of all dependency vectors to be uniform, this projection should map all iteration points onto a subspace of dimension one. An example where this is the case is shown in Fig. 3.26. An example where this is not the case, is shown in Fig. 3.27.

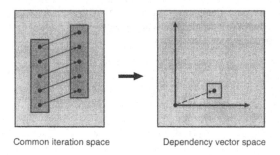

Common iteration space   Dependency vector space

Figure 3.25.   An example of a uniform dependency vector space.

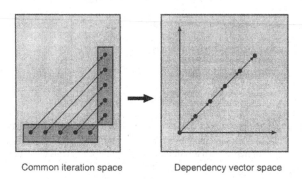

Common iteration space   Dependency vector space

Figure 3.26.   An example of a one-dimensional dependency vector space.

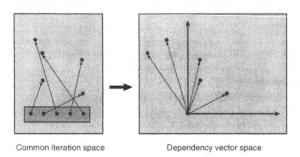

Common iteration space   Dependency vector space

Figure 3.27.   An example of an irregular dependency vector space.

Thus the dependency direction is uniform when

$$\mathrm{img}(x' - \bar{D}'x') \qquad (3.3)$$

is minimally enclosed in a one-dimensional vector-space. This constraint can also be formulated in terms of the rank of the associated matrix:

$$\mathrm{rank}(\bar{I} - \bar{D}') = 1$$

However care should be taken when extended (invertible) matrices have been used in the process of computing $\bar{D}'$. Therefore, we will directly use the vector-space representation here. Contrary to the uniform case, this constraint cannot directly be formulated in terms of the rank of the linear matrix. Still, it is clear that the rank of the linear matrix cannot be higher than the rank of the affine (homogeneous) matrix. Thus, the following is a necessary condition for the linear mapping:

$$\text{rank}(I - D') = 1$$

This method can be extended to a more general case: if we cannot make the dimensionality of the dependency cone one or zero, we can still use it as an optimization criterion, i.e. we can try to minimize this dimension. This is not an exact method however, since in N-space even a two-dimensional dependency cone can give rise to an empty allowed ordering vector cone in the extreme case. Nevertheless this criterion can be used as a good heuristic: if the dependency cone is not full-dimensional, total ordering freedom is available in the remaining dimensions.

3.5.4.3   Examples for individual loop transformations.

Loop Interchange.   Consider the following code fragment:

```
1: (i: 1 .. N)::
     (j: 1 .. M)::
       a[i][j] = ...;

2: (j: 1 .. M)::
     (i: 1 .. N)::
       ... = f(a[i][j]);
```

In this example, we have a two-dimensional array $a[][]$, which is produced in polytope 1 by the index function $\tilde{J}_1(x)$, and consumed in polytope 2 by the index function $\tilde{J}_2(x)$, given by:

$$\tilde{J}_1 = \begin{bmatrix} 1 & 0 & 0 \\ 0 & 1 & 0 \\ 0 & 0 & 1 \end{bmatrix}$$

$$\tilde{J}_2 = \begin{bmatrix} 0 & 1 & 0 \\ 1 & 0 & 0 \\ 0 & 0 & 1 \end{bmatrix}$$

Thus the dependency function is:

$$\tilde{D} = \tilde{J}_1^{-1}\tilde{J}_2 = \begin{bmatrix} 0 & 1 & 0 \\ 1 & 0 & 0 \\ 0 & 0 & 1 \end{bmatrix}$$

We know that we can make this dependency uniform by applying a loop interchange transformation. In PDG terms, the dependency function can be made uniform by mapping the polytopes to the common iteration space with mappings satisfying equation (3.2). When mapping to a two-dimensional space, a possible set of mappings is:

$$\tilde{A}_1 = \begin{bmatrix} 1 & 0 & a_1 \\ 0 & 1 & b_1 \\ 0 & 0 & 1 \end{bmatrix}$$

$$\tilde{A}_2 = \begin{bmatrix} 0 & 1 & a_2 \\ 1 & 0 & b_2 \\ 0 & 0 & 1 \end{bmatrix}$$

Computing the transformed dependency function $\tilde{D}'$ shows indeed that its linear part D' is the identity matrix:

$$\tilde{D}' = \tilde{A}_1\tilde{J}_1^{-1}\tilde{J}_2\tilde{A}_2^{-1} = \begin{bmatrix} 1 & 0 & a_1 - a_2 \\ 0 & 1 & b_1 - b_2 \\ 0 & 0 & 1 \end{bmatrix}$$

We have left the translation unspecified because this does not influence the regularity of the dependencies. In any case, overlap between the polytopes is not allowed, and as a result the dependencies will be quite long when mapping to a two-dimensional space. This can be avoided by mapping the polytopes to a three-dimensional space, e.g. with the following mapping functions:

$$\tilde{A}_1 = \begin{bmatrix} 1 & 0 & 0 \\ 0 & 1 & 0 \\ 0 & 0 & 0 \\ 0 & 0 & 1 \end{bmatrix}$$

$$\tilde{A}_2 = \begin{bmatrix} 0 & 1 & 0 \\ 1 & 0 & 0 \\ 0 & 0 & 1 \\ 0 & 0 & 1 \end{bmatrix}$$

To make these matrices invertible, we extend them with a unit column:

$$\tilde{A}_1^* = \begin{bmatrix} 1 & 0 & 0 & 0 \\ 0 & 1 & 0 & 0 \\ 0 & 0 & 1 & 0 \\ 0 & 0 & 0 & 1 \end{bmatrix}$$

$$\tilde{A}_2^* = \begin{bmatrix} 0 & 1 & 0 & 0 \\ 1 & 0 & 0 & 0 \\ 0 & 0 & 1 & 1 \\ 0 & 0 & 0 & 1 \end{bmatrix}$$

The index functions $\tilde{J}_1$ and $\tilde{J}_2$ have to be extended with a column and row to accommodate the increased dimensionality of the common iteration space. For $\tilde{J}_1$, the extended matrix should be invertible; for $\tilde{J}_2$, this is not necessary, but it is still convenient since $\tilde{J}_2$ is already invertible anyway:

$$\tilde{J}_1^* = \begin{bmatrix} 1 & 0 & 0 & 0 \\ 0 & 1 & 0 & 0 \\ 0 & 0 & 1 & 0 \\ 0 & 0 & 0 & 1 \end{bmatrix}$$

$$\tilde{J}_2^* = \begin{bmatrix} 0 & 1 & 0 & 0 \\ 1 & 0 & 0 & 0 \\ 0 & 0 & 1 & 0 \\ 0 & 0 & 0 & 1 \end{bmatrix}$$

The fact that these matrix extensions do not change the behaviour of the algorithm is illustrated by the following equivalent code, which has still completely the same behaviour as the initial code:

```
1: (i: 1 .. N)::
      (j: 1 .. M)::
         (k: 0 .. 0)::
            a[i][j][k] = input();

2: (j: 1 .. M)::
      (i: 1 .. N)::
         (k: 0 .. 0)::
            output = f(a[i][j][k]);
```

Now we can see that the linear part of the transformed dependency function becomes the identity matrix again:

$$\tilde{D}' = \tilde{A}_1^* \tilde{J}_1^{*-1} \tilde{J}_2^* \tilde{A}_2^{*-1} = \begin{bmatrix} 1 & 0 & 0 & 0 \\ 0 & 1 & 0 & 0 \\ 0 & 0 & 1 & -1 \\ 0 & 0 & 0 & 1 \end{bmatrix}$$

Loop Skewing.   Consider the following code fragment:

```
1: (i: 1 .. N)::
       a[i]= ...;

2: (i: 1 .. N)::
     (j: 0 .. i-1)::
       ... = f(a[i-j]);
```

Contrary to the previous example, reuse of data is present here, which means that dependencies with the same source signal cannot be made identical (since overlapping mappings are not allowed). Thus it will be impossible to get fully uniform dependencies; the best we can achieve is that all dependencies have the same direction. In this example, we have a one-dimensional array $a[]$, produced and consumed respectively by the index functions:

$$\tilde{J}_1 = \begin{bmatrix} 1 & 0 \\ 0 & 1 \end{bmatrix}$$

$$\tilde{J}_2 = \begin{bmatrix} 1 & -1 & 0 \\ 0 & 0 & 1 \end{bmatrix}$$

These index functions can be extended to: (note that we cannot make $\tilde{J}_2^*$ invertible in this case, because of the reuse)

$$\tilde{J}_1^* = \begin{bmatrix} 1 & 0 & 0 \\ 0 & 1 & 0 \\ 0 & 0 & 1 \end{bmatrix}$$

$$\tilde{J}_2^* = \begin{bmatrix} 1 & -1 & 0 \\ 0 & 0 & 0 \\ 0 & 0 & 1 \end{bmatrix}$$

Thus the dependency function is

$$\bar{D} = \tilde{J}_1^{-1}\tilde{J}_2 = \begin{bmatrix} 1 & -1 & 0 \\ 0 & 0 & 0 \\ 0 & 0 & 1 \end{bmatrix}$$

To get regular dependencies in the common iteration space, we should make (3.3) a one-dimensional vector space. This can be satisfied by choosing, for example, the following mappings for the two polytopes:

$$\bar{A}_1^* = \begin{bmatrix} 1 & 0 & 0 \\ 0 & 1 & 0 \\ 0 & 0 & 1 \end{bmatrix}$$

$$\bar{A}_2^* = \begin{bmatrix} 1 & -1 & 0 \\ 0 & 1 & 1 \\ 0 & 0 & 1 \end{bmatrix}$$

The dependency vectors are now given by:

$$x' - \bar{D}'x' = \begin{bmatrix} 0 & 0 & 0 \\ 0 & 1 & 0 \\ 0 & 0 & 0 \end{bmatrix} x'$$

The rank of this matrix is one, which shows that all dependencies are enclosed in a one-dimensional vector-space.

3.5.4.4 Example of placing an algorithm. We will now revisit the example of subsection 3.3.2. The initial code is the following:

```
A: (i: 1..N)::
      (j: 1..N-i+1)::
         a[i][j] = in[i][j] + a[i-1][j];

B: (q: 1..1)::
      (p: 1..N)::
         b[p][1] = f( a[N-p+1][p], a[N-p][p] );

C: (k: 1..N)::
      (l: 1..k)::
         b[k][l+1] = g( b[k][l] );
```

Note that loop nest B is in fact one-dimensional, but an additional iterator q has been inserted into the code, to make the index and mapping functions square (and invertible). Normally, this would be done by computing extended matrices J^* and A^*.

Initial Placement. A trivial placement of the polytopes in the common iteration space is shown in Fig. 3.28. For this trivial placement, the mapping functions are the identity functions plus an offset (to avoid overlap), e.g. for C: (in the figures, x_1 is the vertical direction)

$$\begin{bmatrix} x'_{C,1} \\ x'_{C,2} \\ 1 \end{bmatrix} = \bar{A}_C \begin{pmatrix} k \\ l \\ 1 \end{pmatrix} = \begin{bmatrix} 1 & 0 & N+1 \\ 0 & 1 & N \\ 0 & 0 & 1 \end{bmatrix} \begin{bmatrix} k \\ l \\ 1 \end{bmatrix}$$

The dependency cones can be computed as described in the previous section. E.g. for the dependency between B and C (through array $b[][]$), the index functions are the following:

$$\bar{J}_{p(roduction)} \begin{pmatrix} q \\ p \\ 1 \end{pmatrix} = \begin{bmatrix} 0 & 1 & 0 \\ 1 & 0 & 0 \\ 0 & 0 & 1 \end{bmatrix} \begin{bmatrix} q \\ p \\ 1 \end{bmatrix}$$

$$\bar{J}_{c(onsumption)} \begin{pmatrix} k \\ l \\ 1 \end{pmatrix} = \begin{bmatrix} 1 & 0 & 0 \\ 0 & 1 & 0 \\ 0 & 0 & 1 \end{bmatrix} \begin{bmatrix} k \\ l \\ 1 \end{bmatrix}$$

The transformed dependency vectors can now be computed from the transformed dependency function $\bar{D}'(x')$:

$$x' - \bar{D}'(x') = x' - \bar{A}_B(\bar{J}_p^{-1}(\bar{J}_c(\bar{A}_C^{-1}(x'))))$$

$$= x' - \left(\begin{bmatrix} 0 & 1 & 0 \\ 1 & 0 & -N-1 \\ 0 & 0 & 1 \end{bmatrix} \begin{bmatrix} x'_1 \\ x'_2 \\ 1 \end{bmatrix} \right)$$

$$= \begin{bmatrix} 1 & -1 & 0 \\ -1 & 1 & N+1 \\ 0 & 0 & 0 \end{bmatrix} \begin{bmatrix} x'_1 \\ x'_2 \\ 1 \end{bmatrix}$$

Substituting $x'_2 = N+1$ (the part of polytope C for which the dependency exists), we get $x' - \bar{D}'(x') = (x'_1 - N - 1, -x'_1 + 2N + 2, 0)$. So the dependency cone between B and C is two-dimensional, as can be seen in Fig. 3.28. In this figure, also the evaluations of the cost functions for regularity, locality and data reuse defined in section 3.4 are given:

- The total dependency cone can easily be deducted from the figure. It measures $\pi/2$.

- The locality cost function is simply the sum of all dependency lengths in the figure (according to the Manhattan distance). E.g. for the dependency between B and C, the distances are

$$(N+1) + ((N-1)+2) + \cdots + (1+N) = \frac{N(N+1)}{2}$$

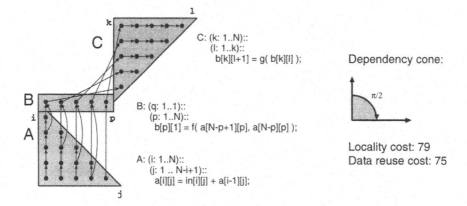

C: (k: 1..N)::
  (l: 1..k)::
    b[k][l+1] = g(b[k][l]);

B: (q: 1..1)::
  (p: 1..N)::
    b[p][1] = f(a[N-p+1][p], a[N-p][p]);

A: (i: 1..N)::
  (j: 1 .. N-i+1)::
    a[i][j] = in[i][j] + a[i-1][j];

Dependency cone:

$\pi/2$

Locality cost: 79
Data reuse cost: 75

Figure 3.28.   Initial placement.

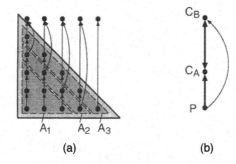

(a)   (b)

Figure 3.29.   Basic groups for polytope A and extended dependency graph for basic set A_2.

When the lengths are added for all dependencies, we obtain $3N^2 + N - 1$. E.g. for $N = 5$ (as in the figure), this expression evaluates to 79. This result is also immediately obtained by the algorithm of section 5.3.2 in [127].

■ For the data reuse cost function, we observe that the values defined in polytope A can be split into three basic groups A_1, A_2 and A_3 (see Fig. 3.29(a)). The values of A_1 and A_3 are consumed only once, thus the data reuse cost is the same as the locality cost for these sets. For the values of A_2, which are consumed twice, an extended dependency graph is constructed in Fig. 3.29(b). The dependency lengths and pseudo-dependency lengths in this graph can be obtained by summing the corresponding Manhattan distances at the left of the figure, or by applying algorithms of section 5.3.2 and 5.4.2 in [127]. E.g. for $N = 5$, we obtain

$$Loc(D_{P \to C_A}) = 4$$
$$Loc(D_{P \to C_B}) = 14$$
$$Loc(D_{C_A \leftrightarrow C_B}) = 10$$

It is clear that the minimum spanning tree consists of $D_{P \to C_A}$ and $D_{C_A \leftrightarrow C_B}$, as indicated by the thick arrows in the figure. In this way, we finally get the value of 75 for the total data reuse cost of the initial placement.

Partially Optimized Placement.   In Fig. 3.30, a mapping is shown in which all dependency directions are constant for the individual cones, but not for the overall cone. The mapping function for C

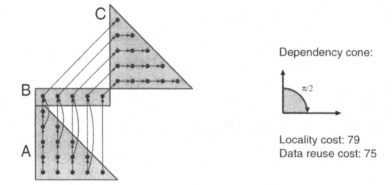

Figure 3.30.   Intermediate placement.

is now:

$$\begin{bmatrix} x'_{C,1} \\ x'_{C,2} \\ 1 \end{bmatrix} = \tilde{A}_C \begin{pmatrix} k \\ l \\ 1 \end{pmatrix} = \begin{bmatrix} -1 & 0 & 2N+2 \\ 0 & 1 & N \\ 0 & 0 & 1 \end{bmatrix} \begin{bmatrix} k \\ l \\ 1 \end{bmatrix}$$

For the dependency between B and C, the transformed dependency vectors are now:

$$\begin{aligned} x' - \bar{D}'(x') &= x' - \left(\begin{bmatrix} 0 & 1 & 0 \\ -1 & 0 & 2N+2 \\ 0 & 0 & 1 \end{bmatrix} \begin{bmatrix} x'_1 \\ x'_2 \\ 1 \end{bmatrix} \right) \\ &= \begin{bmatrix} 1 & -1 & 0 \\ 1 & 1 & -2N-2 \\ 0 & 0 & 0 \end{bmatrix} \begin{bmatrix} x'_1 \\ x'_2 \\ 1 \end{bmatrix} \end{aligned}$$

Again after substituting $x'_2 = N + 1$, we get $x' - \bar{D}'(x') = (x'_1 - N - 1, x'_1 - N - 1, 0)$. So this dependency cone is one-dimensional, i.e. the dependency direction is constant. The size of the total dependency cone is however still $\pi/2$, as depicted in the figure. Also the cost functions for locality and data reuse, which have been computed as described above, have not changed compared to the initial placement.

Fully Optimized Placement.   For this simple example, it is possible to make all dependency functions uniform, and to make the angle of the total dependency cone zero, as shown in Fig. 3.31. This has been achieved by imposing the constraints for uniform dependencies. E.g., the mappings for B and C are now:

$$\begin{bmatrix} x'_{B,1} \\ x'_{B,2} \\ 1 \end{bmatrix} = \tilde{A}_B \begin{pmatrix} q \\ p \\ 1 \end{pmatrix} = \begin{bmatrix} 1 & -1 & N+1 \\ 0 & 1 & 0 \\ 0 & 0 & 1 \end{bmatrix} \begin{bmatrix} q \\ p \\ 1 \end{bmatrix}$$

$$\begin{bmatrix} x'_{C,1} \\ x'_{C,2} \\ 1 \end{bmatrix} = \tilde{A}_C \begin{pmatrix} k \\ l \\ 1 \end{pmatrix} = \begin{bmatrix} -1 & 1 & N+2 \\ 1 & 0 & 0 \\ 0 & 0 & 1 \end{bmatrix} \begin{bmatrix} k \\ l \\ 1 \end{bmatrix}$$

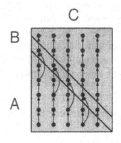

Dependency cone:

Locality cost: 38
Data reuse cost: 34

Figure 3.31.   Final placement.

For the dependency between B and C, the transformed dependency vectors are now:

$$x' - \tilde{D}'(x') = x' - \left(\begin{bmatrix} 1 & 0 & -1 \\ 0 & 1 & 0 \\ 0 & 0 & 1 \end{bmatrix} \begin{bmatrix} x'_1 \\ x'_2 \\ 1 \end{bmatrix} \right)$$

$$= \begin{bmatrix} 0 & 0 & 1 \\ 0 & 0 & 0 \\ 0 & 0 & 0 \end{bmatrix} \begin{bmatrix} x'_1 \\ x'_2 \\ 1 \end{bmatrix}$$

This shows that the dependencies are indeed uniform for this cone. Computing the other cones shows that this is true for all dependency functions, and that even the overall dependency direction is constant.

By optimizing the offsets, the potential locality and data reuse have been improved too. The values obtained by evaluating the corresponding cost functions are given in the figure.

3.5.5 Automated exploration strategy

Based on the constraints developed in the previous section, we have developed an exploration strategy for traversing the placement search space. A prototype tool has been implemented to prove the feasibility of the concept. This section explains the exploration strategy, as well as the basics of the tool implementation.

3.5.5.1 Basics of the strategy.   The goal is to find linear polytope mappings which minimize the rank of the vector-space of as many as possible dependency functions. Since the constraints can be quite stringent, this approach should only be applied to clusters of polytopes which are strongly connected. Indeed, even when a dependency function between two polytopes has only one instance (one signal), it still has a rank. Optimizing this rank may impose strong constraints which are not necessary at all. Such polytopes should be separated from each other in the clustering phase.

We will first present an outline of the strategy, and then some necessary refinements to make it more usable.

First Outline.   The envisaged strategy consists of three phases:

1. In a first phase, we will check the relative mappings of the polytopes involved in each dependency, to determine the rank of the mapped dependency function. This is done with an exhaustive search, which is possible since only two polytopes are involved in each dependency. All possible relative mappings of these polytopes are checked, and the associated ranks are stored.

2. In a second phase, the mappings which have been generated in the previous phase, are combined per polytope. For all dependencies in which a polytope is involved, the optimal rank is determined, and the mappings which potentially lead to the optimal rank for all the dependencies, are determined.

3. Finally, a backtracking algorithm is started in which the complete search space is traversed, and where a global mapping (which satisfies all rank constraints) is determined. The backtracking level is the polytope being mapped. Mappings for each polytope are selected from the combined set generated in the previous phase. Every time a mapping is chosen, all dependencies in which the polytope is involved, are checked to see if the rank is optimal. If this is the case, the algorithm proceeds to the next polytope. If not, another mapping is selected. If no mappings are available anymore, backtracking occurs.

Refinements.   Based on initial experiments, the strategy defined above had to be refined in two ways:

- The order in which the backtracking procedure traverses the search space of the polytopes, is very important. When a conflict between two polytopes occurs too many times at a very deep level in the search tree, the exploration can take very long. By moving those polytopes to the front of the tree, the conflict is detected much faster, and backtracking happens earlier in the tree.

  For this reason, the polytopes between which the most potential conflicts occur, should come first in the search space. However, it is difficult to know this in advance. The polytopes can be sorted according to the number of dependency functions in which they are involved, but in practice this has proven to be insufficient. Therefore, we have implemented a dynamic scheme, in which the search tree is reordered whenever too many conflicts occur at a level too deep in the tree.

- The third phase as defined above is too stringent, as it searches immediately for the optimal ranks, which is not always feasible. In the latter case, the whole search space has to be explored, which can take a very long time. Moreover, the only result of the exploration in that case is that no solution exists. The exploration can be repeated with looser constraints until a solution is found, but each time (except the last one) the whole search space has to be explored.

  On the other hand, when a solution does indeed exist, it can also take a long time to find it, which may cause the designer to stop the exploration. Also in that case, the only available result is that no solution has been found yet.

  Therefore, an incremental approach has been implemented, where the constraints are first selected to be very loose, and as long as a solution is found, they are gradually strengthened. Usually, the first solutions are found very quickly, and it takes more time as the constraints get stronger. The advantage of this approach is that also intermediate results are available. Thus when the exploration starts taking too much time, the designer can stop it and choose the last generated solution.

  Moreover, this approach has proven to be faster even when the optimal solution does exist. The reason lies in the combination with the first refinement above, which causes the polytopes to be reordered as the constraints are strengthened. As a result, they are almost in an optimal order when the final exploration is started, such that the optimal solution is found very fast.

3.5.5.2 Selection of the starting set of possible mappings.

An important issue that has not been discussed yet, is the set of possible mappings which should initially be present in the search tree. It is clear that this set can become very large, since a linear mapping matrix in n-dimensional space contains n^2 coefficients. When each coefficient has k possible values, the total number of possible mappings is:

$$k^{(n^2)}$$

To control the complexity, we have to restrict the set of possible mappings. Since most polytopes are n-simplices, rectangles or triangles and their higher-dimensional equivalents, a straightforward choice is to limit the possible coefficients in the mapping matrices to -1, 0 and 1. This is enough to make such polytopes fit onto each other. In the prototype tool we have developed, we have also excluded the -1 coefficient. This limitation should be removed in the near future, since it does not allow loop reversal transformations.

Furthermore the starting set has been restricted by removing all singular mappings.

```
function GenerateRankConstraints
AllMappings = GenerateStartingSet ;
ID = eye (Dimension) ;  % Identity matrix
for dep = 1 : length (Dependency)
    p1 = Dependency[dep].source_polytope ;
    p2 = Dependency[dep].dest_polytope ;
    c = Dependency[dep].dest_consumption ;
    Constraint[p1].mappings[dep].rank[:] = Dimension ;   % Init. with max. possible rank
    Constraint[p2].mappings[dep].rank[:] = Dimension ;
    Ddl = Dependency[dep].Ddl ;   % Linear part of dependency function
    for k1 = 1 : length (AllMappings)
        A1 = AllMappings[k1] ;
        for k2 = 1 : length (AllMappings)
            A2 = AllMappings[k2] ;
            Dl_transformed = A1 x Ddl x A2⁻¹ ;
            DV = ID − Dl_transformed ;
            r = rank (DV) ;
            Constraint[p1].mappings[dep].rank[k1] =
                min (Constraint[p1].mappings[dep].rank[k1] , r) ;
            Constraint[p2].mappings[dep].rank[k2] =
                min (Constraint[p2].mappings[dep].rank[k2] , r) ;
        end
    end
end
```

Figure 3.32.   Code for generating rank constraints per polytope.

3.5.5.3 Implementation.   The prototype tool (called MASAI-2) has been implemented in the programming language of the MATLAB package[7]. MATLAB is very convenient and efficient for matrix-based computations. It is much less efficient for executing loops and processing integer data types however. Since the backtracking algorithm contains a lot of while-loops, the performance is not very high. We estimate that an implementation in C++ would be up to 100 times faster (based on some small benchmarks).

In the figures in this section, we will use pseudo-code based on MATLAB notation to illustrate the implementation.

Phase 1.   The first phase of the tool, i.e. determining the rank of all possible mapped dependency functions, is shown in Fig. 3.32. The code iterates over all dependencies (which are stored in the array Dependency). For the two polytopes involved in a dependency, all possible relative mappings are tried, and for each one the rank of the mapped dependency function is computed. These ranks are then stored in an array Constraint.

Phase 2.   The pseudo-code for the second phase is shown in Fig. 3.33. The outer loop iterates over all polytopes (which are stored in the array Polytope). For each polytope, a variable combmappings is initialized to contain the full starting set of mappings. Then the optimal rank is determined (and stored in TargetRank[].opt) for all dependencies in which the polytope is involved. The mappings which can certainly not lead to this optimal rank, are removed from combmappings.

Phase 3.   Finally, the pseudo code for the actual search space exploration is shown in Figs. 3.34–3.35. The dependency ranks for which a global mapping is being searched, are stored in TargetRank[].current. Before the first iteration of the outer while loop, these target ranks are initialized to the dimension of the common iteration space, i.e. the highest possible (and thus most easily achievable) ranks. The goal of the algorithm is that the current target ranks evolve towards the optimal target ranks. If the backtracking algorithm finishes without finding a solution, the search space exploration is not stopped, but the optimal target ranks are adapted (and stored in TargetRank[].best) to reflect this fact.

---

[7]by The MathWorks, Inc.

```
function CombineConstraints
for  p = 1 : length (Polytope)
      Constraint[p].combmappings = AllMappings ;   % Initialize with starting set of mappings
      for  dep = 1 : length (Dependency)
           if  not isempty (Constraints[p].mappings[dep])   % If polytope is involved in dependency
               found = 0 ;
               targetrank = -1 ;
               while  found == 0   % Look for lowest occurring rank
                   targetrank = targetrank + 1 ;
                   possible_mappings = Constraint[p].mappings[dep].rank[:] ≤ targetrank ;
                        % possible_mappings now contains indices of all mappings with rank ≤ targetrank
                   found = any(possible_mappings) ;
               end
               TargetRank[dep].opt = targetrank ;
               Remove mappings which are not in possible_mappings
                   from Constraint[p].combmappings ;
           end
      end
end
```

Figure 3.33. Code for combining the constraints per polytope and determining the optimal rank per dependency.

The second level while loop is the actual backtracking loop. The polytope for which a mapping is searched at this level is called plevel. In the loop, a new mapping is selected from the combined set of mappings generated in the previous phase for polytope plevel. Then the dependencies between this polytope and the ones which have already been mapped (those which come before plevel in the search tree) are checked to see whether the target rank constraint is satisfied. If this is the case, we store the mapping in Mapping[plevel].A and proceed to the next polytope. If not, we go back to the start of the while loop. When all mappings have been tried without finding a solution, backtracking occurs.

When a solution has been found, the target ranks are updated to find a more optimal solution. On the other hand, when no solution has been found, TargetRank[].best is updated to reflect this. The algorithm is terminated when a solution has been found for TargetRank[].best, i.e. for the best achievable target ranks.

The search tree reordering algorithm works by monitoring the number of   conflicts in which each polytope is involved. A conflict occurs whenever a polytope cannot be mapped anymore due to dependencies with other polytopes that have already been mapped. Basically, a maximum number of iterations is set at the start of the algorithm. During the backtracking loop, the number of conflicts in which each polytope is involved, is stored in the array Conflict. When no solution has been found after the maximum number of iterations, the search tree is reordered such that the polytopes which have the most conflicts, are at the beginning of the tree. The procedure is then repeated with an increased maximum number of iterations, such that more time is available during the exploration of the new search tree.

3.5.6 Experiments on automated polytope placement

The prototype tool has been applied to a number of applications to prove the validity of our approach. In this section, we will first apply the tool to the example of subsection 3.5.4.4. Next we will apply it to two realistic test algorithms: the algebraic path problem and an updating singular value decomposition algorithm. For each test case, different initial descriptions have been used, in order to verify whether the final result generated by the tool is the same. This is a very good measure for its effectiveness.

3.5.6.1 Simple example.   We revisit the example which has first been described in subsection 3.3.2 and which has manually been optimized with our placement approach in subsection 3.5.4.4. The PDG of this example contains three polytopes and five dependencies.

```
function SearchSpaceExploration
for i = 1 : length (Dependency)
    TargetRank[i].best = TargetRank[i].opt ;
    TargetRank[i].current = Dimension ;   % Start with loose rank constraints (rank = Dimension)
end
org_max_its = 1000 ;   max_its = org_max_its ; optimal = 0 ;
Conflict = zeros (size (Polytope)) ;   thisdp = length (Dependency);   % Last changed targetrank
while not optimal
    iteration = 0 ;
    plevel = 1 ;  % polytope number = backtracking level
    mappings_seen[plevel] = 0 ;   solution_found = 0 ;
    while (plevel > 0) & (not solution_found) & (iteration < max_its)
        if mappings_seen[plevel] == length (Constraint[plevel].combmappings)
            plevel = plevel − 1 ;
        else
            mappings_seen[plevel] = mappings_seen[plevel] + 1 ;   iteration = iteration + 1 ;
            curmapping = Constraint[plevel].combmappings[mappings_seen[plevel]] ;
            ok = 1 ;
            pother = 1 ;
            while ok & (pother < plevel)
                % Check dependencies with all other polytopes which have already been mapped
                c = 1 ;
                while ok & (c ≤ NumDep (plevel, pother))
                    dp = LookupDep (plevel, pother, c) ;   Ddl = Dependency[dp].Ddl ;
                    Dl_trafo = curmapping.A x Ddl x Mapping(pother).A⁻¹ ;
                    r = rank (ID − Dl_trafo) ;
                    ok = (r ≤ TargetRank[dp].current) ;
                    if not ok
                        Conflict[plevel] = Conflict[plevel] + 1 ;
                        Conflict[pother] = Conflict[pother] + 1 ;
                    end
                    c = c+1 ;
                end
                c = 1 ;
                while ok & (c ≤ NumDep (pother, plevel))
                    ...   % Analogous to previous while loop but for reverse dependencies
                end
                pother = pother + 1 ;
            end
            if ok
                Mapping[plevel].A = curmapping.A ;
                if plevel == length (Polytope)
                    solution_found = 1 ;
                else
                    plevel = plevel + 1 ;
                    mappings_seen[plevel] = 0 ;
                end
            end
        end
    end
end
... (continued in Fig. 3.35)
```

Figure 3.34.   Backtracking algorithm for traversing the search space.

The optimal ranks determined after the first two phases of the tool are all zero. Thus all dependencies can be made uniform when considered in isolation from the other polytopes. The third phase of the tool starts with finding a mapping in which all dependency ranks are 2. Next these ranks are decremented one by one, and each time a corresponding mapping is found. This procedure is repeated until all ranks are optimal (zero). After 37 iterations (in total) of the backtracking loop, the following mapping matrices are finally generated by the tool:

$$A_A = \begin{bmatrix} 1 & 1 \\ 0 & 1 \end{bmatrix} \qquad A_B = \begin{bmatrix} 1 & 0 \\ 0 & 1 \end{bmatrix} \qquad A_C = \begin{bmatrix} 0 & 1 \\ 1 & 0 \end{bmatrix}$$

```
... (continued from Fig. 3.34)
if solution_found
   % Improve target if solution found
      max_its = org_max_its ;
      thisdp = mod (thisdp, length (Dependency)) + 1 ;
      orgthisdp = thisdp ;
      while (TargetRank[thisdp].current == TargetRank[thisdp].best) & (not optimal)
         thisdp = mod (thisdp, length (Dependency)) + 1 ;
         if thisdp == orgthisdp
            optimal = 1 ;
         end
      end
      if not optimal
         TargetRank[thisdp].current = TargetRank[thisdp].current - 1 ;
      end
   elseif iteration == max_its
      % Reorder polytope search tree based on Conflicts (details are left out)
      max_its = max_its x 2 ;
   else
      % No solution found (plevel==0): adapt best achievable solution
      TargetRank[thisdp].current = TargetRank[thisdp].current + 1 ;
      TargetRank[thisdp].best = TargetRank[thisdp].current ;
      max_its = org_max_its ;
      Conflict = zeros (size (Polytope)) ;
   end
end
```

Figure 3.35.   Backtracking algorithm for traversing the search space – continued.

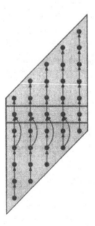

Figure 3.36.   Placement generated by prototype tool for simple example.

Fig. 3.36 gives a graphical representation of this mapping, where we have also selected a good translation for clarity. The mapping is not the same as the one found manually in Fig. 3.31, but it is a complete affine transformation of it. Therefore, it represents exactly the same placement and it will give rise to the same choices for the ordering phase.

Next we have modified the initial description by applying a loop interchange transformation to the first loop. Also in this case, an optimal solution (with all ranks zero) has been easily found. The final mapping is the following:

$$A_A = \begin{bmatrix} 1 & 1 \\ 1 & 0 \end{bmatrix} \qquad A_B = \begin{bmatrix} 1 & 0 \\ 0 & 1 \end{bmatrix} \qquad A_C = \begin{bmatrix} 0 & 1 \\ 1 & 0 \end{bmatrix}$$

It is clear that this mapping represents the same solution as in Fig. 3.36.

3.5.6.2 Algebraic path problem. The Algebraic Path Problem is a general framework unifying several algorithms into a single form, such as the determination of the inverse of a real matrix, shortest distance in a weighted graph, etc. Its optimization in the PDG model has been thoroughly discussed in [542], where it is one of the few examples for which the MASAI-1 tool produces a result. The PDG of this example contains 10 polytopes and 35 dependencies.

Our tool needs about 900 iterations to generate a mapping which satisfies all constraints. The final mapping matrices are the following:

$$A_i = \begin{bmatrix} 0 & 0 & 1 \\ 0 & 1 & 0 \\ 1 & 0 & 0 \end{bmatrix} \quad \forall i = 1 \ldots 10$$

Since the mapping matrices are the same for all polytopes, the solution is simply a complete affine transformation (in this case a loop permutation) of the initial description. Indeed, when we take a close look at the transformations carried out in [542] for the APP, it is clear that the polytopes are only translated. This is probably the reason why the MASAI-1 tool could produce a result for this application. As motivated in subsection 3.5.3, the search space for the translation is much smaller than for the linear mapping.

To check whether our approach would generate the same result also when the initial description is modified such that actual linear transformations of the polytopes are indeed required, we have tried a number of alternative initial versions of the APP. In a first version, we have applied a loop interchange transformation to the polytope which comes first in the search tree of the backtracking scheme. Since this polytope is mapped first, a different mapping is required for all other polytopes (unless the search tree would be reordered, which is not the case here). Indeed, the following solution is generated:

$$A_1 = \begin{bmatrix} 0 & 0 & 1 \\ 0 & 1 & 0 \\ 1 & 0 & 0 \end{bmatrix}$$

$$A_i = \begin{bmatrix} 0 & 0 & 1 \\ 1 & 0 & 0 \\ 0 & 1 & 0 \end{bmatrix} \quad \forall i = 2 \ldots 10$$

About 12000 iterations are required to find this solution, which is much more than the 900 for the original description, where no transformation was really performed.

In a second version, we have applied a loop permutation and a double loop skewing transformation to another polytope (polytope 3). The following solution is generated in about 15000 iterations:

$$A_1 = \begin{bmatrix} 0 & 0 & 1 \\ 0 & 1 & 0 \\ 1 & 0 & 0 \end{bmatrix}$$

$$A_3 = \begin{bmatrix} 1 & 0 & 0 \\ 1 & 1 & 1 \\ 1 & 1 & 0 \end{bmatrix}$$

$$A_i = \begin{bmatrix} 0 & 0 & 1 \\ 1 & 0 & 0 \\ 0 & 1 & 0 \end{bmatrix} \quad \forall i = 2, 4, 5 \ldots 10$$

Two more versions have been tried. In the last one, almost all polytopes have been transformed in the initial description. Also in this case, the final generated solution is the same. The number of search iterations required to find the solution does not increase further; instead it decreases again to about 6000 iterations for the last version (see Table 3.1).

On this final placement result, the ordering algorithm that we have developed in co-operation with INSA Lyon [186] has been applied. Our prototype tool indeed produces the best ordering of the axes that minimizes the sum of the lengths of the dependencies. As far as we know that outcome is also the optimal one in terms of access locality in a DTS context. This demonstrates the effectiveness of our overall methodology, the techniques used, and our prototype tool implementations.

3.5.6.3 Updating singular value decomposition.   The Updating Singular Value Decomposition (USVD) algorithm is frequently used in signal processing applications. The PDG of this example contains 26 polytopes and 87 dependencies.

Our tool generates a global mapping which satisfies all constraints after about 45000 iterations. The mapping matrices generated are quite diverse, showing that the initial description was not optimal at all.

Also for this example, we have tried a modified initial description, where a number of polytopes have been transformed. The final solution generated by the tool is the same again. It requires 43000 iterations to be found.

3.5.6.4 MPEG-4 video motion estimation kernel.   For the video motion estimation (ME) kernel (see subsection 4.5.1), we have started from an executable code but many different loop organizations (and hence orderings) are feasible initially. As shown in chapter 4 and [276] at least 6 of these are relevant to study in more depth. Starting from all of these different initial codes (and orderings), the ordering algorithm that we have developed in co-operation with INSA Lyon [186], automatically arrives at the globally best ordering that minimizes the storage requirements, as identified in extensive manual design studies [67] and as confirmed by the estimation technique presented in chapter 4.

3.5.6.5 Discussion.   In each of the above examples, the solution generated by the tool is independent of the initial description. This shows that the optimization approach really works.

Table 3.1 summarizes the CPU times and the number of search iterations which are needed for each of the test cases. For the simple example, the CPU times are very low, as could be expected. For the APP algorithm, the tool takes about 5 minutes to complete, which is quite reasonable. The CPU time is almost independent of the initial description.

The USVD example requires about 11 minutes to complete. However, the table shows that, while the time required for the first two phases increases more or less linearly with the number of polytopes, for the third phase it increases exponentially. The reason is that within the first phases the polytopes are considered individually or per dependency, while in the third phase a global search is performed, with the number of polytopes being the depth of the search tree.

Therefore, we feel that the 26 polytopes and 87 dependencies of the USVD is about the limit for this prototype implementation. Of course, an efficient implementation in C++ would be much faster and should be able to handle a larger number of polytopes. Moreover, a clustering approach will allow a hierarchical tackling of much larger applications.

The examples discussed in this section had a two-dimensional (for the simple example) or three-dimensional (for the APP and USVD) common iteration space. It is clear that the complexity of the search space increases exponentially with this dimension. For the three-dimensional case, it is still feasible to handle all possible relative mappings with a limited number of coefficients. In four or more dimensions, this is not feasible anymore (e.g. about 22000 non-singular 4×4 matrices with coefficients 0 or 1 can be generated). Therefore, an important issue in higher dimensions is to reduce the starting set of mappings further. This can be done e.g. by limiting the number of loop skewing transformations which are performed at the same time (thus limiting the number of 1's on a row of column). This is a topic of current research.

3.5.7 Optimization of the translation component

Even when the linear mapping of all polytopes has been fixed, optimizing the actual position of the polytopes in the common iteration space is still a major task. This task has not been investigated in detail yet, but currently we foresee an approach where the translation component would be optimized in two phases:

Application	Version	CPU time (s)			Search iterations
		Phase 1	Phase 2	Phase 3	
Simple example	Initial	0.030	0.001	0.034	37
	Modified	0.035	0.002	0.048	37
APP	Initial	286	0.036	3.56	910
	Modified 1	240	0.037	14.3	11828
	Modified 2	225	0.038	20.2	14866
	Modified 3	285	0.037	12.7	8580
	Modified 4	261	0.037	10.6	6006
Updating SVD	Initial	428	0.025	119	45355
	Modified	537	0.017	134	42948

Table 3.1.   CPU times and search iterations for tool experiments on HP-PA 8000 workstation.

1. In a first phase, the polytopes are approximated by their bounding boxes. This makes it easy to build a construction with these boxes without having to consider the actual shapes of the polytopes. In this way, a crude exploration of the search space can be performed while concentrating on the global dependency cone and the locality.

2. In a second phase, the actual shapes of the polytopes are considered, and starting from the solution generated in the first phase, they are further shifted towards each other, in order to make the common iteration space as compact as possible.

Sometimes the shapes will not fit well onto each other in the second phase, and holes will be present between the polytopes. These holes can affect the global dependency cone and the locality (+ data reuse):

■ When the dependency cone is affected such that the allowed ordering vector cone would be limited too severely, the first phase is redone with some extra constraints in order to find a better mapping.

■ When only the locality and data reuse are affected, the holes between the polytopes do not present a problem. Indeed, they will be eliminated during the ordering and code generation phases (see subsection 3.3.1.1).

3.6 SUMMARY

In this chapter, we have presented the model we use for representing algorithms (i.e. iterations and dependencies between them), namely the Polyhedral Dependency Graph (PDG) model. We have also shown how we perform loop transformations in this model, namely with a multi-phase approach consisting of several placement phases and an ordering phase. The placement phases are purely geometrical and it has been the main focus of this chapter.

The placement optimization itself is split into two subphases: in the first subphase the linear mapping of the polytopes is determined to improve regularity, and in the second subphase the translation is optimized for a global locality optimisation. Cost functions have been defined to steer each of these subphases.

For the first subphase, we have developed a constraint-based optimization approach, where the placement of the polytopes is guided by constraints put on the dependencies between them. We have presented a prototype implementation of this approach, which generates very good results in acceptable runtimes.

4 SYSTEM-LEVEL STORAGE ESTIMATION WITH PARTIALLY FIXED EXECUTION ORDERING

Estimators at the system-level are crucial to help the designer in making global design decisions and trade-offs. For data-dominant applications in the multi-media and telecom domains, the system-level description is typically characterized by large multi-dimensional loop nests and arrays. A major aspect of system cost related to such codes is due to the data transfer and storage (DTS) aspects, as motivated in chapter 1. Cost models for the amount of area per memory cell or the energy consumption per access for a given memory plane size can be obtained from vendors. However, in order to identify the amount of memory accesses or data transfers and the required memory size, automatable estimation techniques are required. Effective approaches for this are described in this chapter.

4.1 CONTEXT AND MOTIVATION

In the ATOMIUM tool-set at IMEC [61], efficient techniques have been incorporated to count the read and write accesses for individual arrays at different scopes of the program. Up till recently, no accurate method was available however for the expected storage requirement of the arrays, that is the amount of memory required to store the arrays during program execution. The real life-times of the arrays and the possibilities for intra- and inter-signal in-place mapping must be taken into account as discussed below.

A structural code example is given in Fig. 4.1 a). A straightforward way of estimating the storage requirement of such a code is to find the size of each array by multiplying the size of each dimension, and then to add together the different arrays. Fig. 4.1 b) shows an example of such memory usage for the *A* and *B* arrays of the code in Fig. 4.1 a). This will normally result in a huge overestimate however, since not all the arrays, and possibly not all parts of one array, are alive at the same time. In this context, an array element is alive from the moment it is written, or produced, and until it is read for the last time. This last read is said to consume the element. To achieve a more accurate estimate, we have to take into account these non-overlapping lifetimes and their resulting opportunity for mapping arrays and parts of arrays at the same locations in memory. The process of optimizing the reuse of these locations, or places, is denoted the *in-place mapping* problem [519], as also discussed in previous and subsequent chapters of this boook. To what degree it is possible to perform in-place mapping, depends heavily on the order in which the elements in the arrays are produced and consumed. This is mainly determined by the execution ordering of the loop nests surrounding the arrays. Depending on the ordering of the loops and loop nests, the arrays can to a greater or lesser extent be mapped to the same physical locations in memory as shown in Fig. 4.1 c). Since not all parts of one array are alive simultaneously, even the memory size required by each array can be greatly reduced through in-place mapping if the loops and loop nests are organized optimally, as indicated in Fig. 4.1 d).

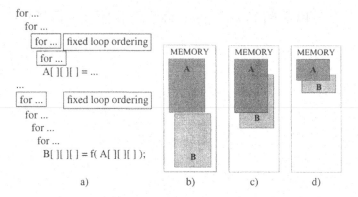

Figure 4.1.   Code structure and memory usage for data intensive applications.

At the beginning of the design process, no information about the execution order is known, except what is given from the data dependencies between the statements in the code. As the process progresses, the designer makes decisions that gradually fix the ordering, as indicated as fixed loop ordering in Fig. 4.1a). In the end, the full execution ordering is known. To steer this process, estimates of the upper and lower bounds on the storage requirement are needed at each step, given the partially fixed execution ordering.

In the DTSE script that we propose (see section 1.7) these estimations are especially useful at the start and in the early steps. This is particularly true for the global loop transformation (see chapter 3) and data reuse decision (see chapter 5) steps where explicit size estimates and guidelines on loop reorganization to reduce the estimated size are crucial inputs.

4.2 PREVIOUS WORK

In addition to the DTSE design methodology, there exist a number of academic and commercial system level design environments. Examples are SpecSyn [190, 191], COSMOS [241, 346], Ptolemy [109], COSSAP [494], and SPW [76]. Even higher (pre-specification) design environments have recently been proposed [53]. Most of these are optimized towards and work well for certain application domains, such as control dominated systems, communicating systems, and digital signal processing systems. Their focus tend to be on data path and communication design and is as such complimentary to the DTSE methodology. They may still benefit from storage requirement estimation and optimization techniques though, and the rest of this section reviews previous work in the estimation area.

4.2.1 Scalar-based estimation

By far the major part of all previous work on storage requirement has been scalar-based. The number of scalars, also called signals or variables, is then limited, and if arrays are treated, they are flattened and each array element is considered a separate scalar. Using scheduling techniques like the left-edge algorithm, the lifetime of each scalar is found so that scalars with non-overlapping lifetimes can be mapped to the same storage unit [301]. Techniques such as clique partitioning are also exploited to group variables that can be mapped together [510]. In [393], a lower bound for the register count is found without the fixation of a schedule, through the use of As-Soon-As-Possible and As-Late-As-

Possible constraints on the operations. A lower bound on the register cost can also be found at any stage of the scheduling process using Force-Directed scheduling [426]. Integer Linear Programming techniques are used in [195] to find the optimal number of memory locations during a simultaneous scheduling and allocation of functional units, registers and busses. A thorough introduction to the scalar-based storage unit estimation can be found in [189]. Common to all scalar based techniques is that they break down when used for large multi-dimensional arrays. The problem is NP-hard and its complexity grows exponentially with the number of scalars. When the multi-dimensional arrays present in the applications of our target domain are flattened, they result in many thousands or even millions of scalars.

4.2.2 Estimation with fixed execution ordering

To overcome the shortcomings of the scalar-based techniques, several research teams have tried to split the arrays into suitable units before or as a part of the estimation. Typically, each instance of array element accessing in the code is treated separately. Due to the code's loop structure, large parts of an array can be produced or consumed by the same code instance. This reduces the number of elements the estimator must handle compared to the scalar approach.

Most previous work in this domain assumes a fully fixed execution ordering. The input code is then executed sequentially in a *procedural* form. A C description is for instance usually interpreted in such a way by both designers and compilers. This is then in contrast to *applicative* or *nonprocedural* languages, such as Silage [227], where operations can be executed in any order, as long as data dependency relations in the code are not violated. All data must be produced before they are consumed.

In [523], a production time axis is created for each array. This models the relative production and consumption time, or date, of the individual array accesses. The difference between these two dates equals the number of array elements produced between them. The maximum difference found for any two depending instances gives the storage requirement for this array. The total storage requirement is the sum of the requirements for each array. An Integer Linear Programming approach is used to find the date differences. To be able to generate the production time axis and production and consumption dates, the execution ordering has to be fully fixed. Also, since each array is treated separately, only in-place mapping internally to an array (intra-array in-place) is consider, not the possibility of mapping arrays in-place of each other (inter-array in-place).

Another approach is taken in [213]. Assuming procedural execution of the code, the data-dependency relations between the array references in the code are used to find the number of array elements produced or consumed by each assignment. The storage requirement at the end of a loop equals the storage requirement at the beginning of the loop, plus the number of elements produced within the loop, minus the number of elements consumed within the loop. The upper bound for the occupied memory size within a loop is computed by producing as many array elements as possible before any elements are consumed. The lower bound is found with the opposite reasoning. From this, a memory trace of bounding rectangles as a function of time is found. The total storage requirement equals the peak bounding rectangle. If the difference between the upper and lower bounds for this critical rectangle is too large, better estimates can be achieved by splitting the corresponding loop into two loops and rerunning the estimation. In the worst-case situation, a full loop-unrolling is necessary to achieve a satisfactory estimate.

[572] describes a methodology for so-called exact memory size estimation for array computation. It is based on live variable analysis and integer point counting for intersection/union of mappings of parameterized polytopes. In this context, as well as in our methodology, a polytope is the intersection of a finite set of half-spaces and may be specified as the set of solutions to a system of linear inequalities. It is shown that it is only necessary to find the number of live variables for one statement in each innermost loop nest to get the minimum memory size estimate. The live variable analysis is performed for each iteration of the loops however, which makes it computationally hard for large multi-dimensional loop nests.

In [443], a reference window is used for each array in a perfectly nested loop. At any point during execution, the window contains array elements that have already been referenced and will also be referenced in the future. These elements are hence stored in local memory. The maximal window size found gives the memory requirement for the array. The technique assumes a fully fixed execution ordering. If multiple arrays exists, the maximum reference window size equals the sum of the windows for individual arrays. Inter-array in-place is consequently not considered.

At the early steps of the design trajectory, no or only a partial execution ordering is determined. Using an estimation methodology that requires a fully fixed ordering necessitates a full exploration of all alternatives for the unfixed parts. The complexity of investigating for example all loop interchange solutions for a loop nest without any constraints is N!, where N is the number of dimensions in the loop nest. For realistic examples, N can be as large as six even if the dimensionalities of the individual arrays are smaller. Since it is the number of loop dimensions that determine the complexity, a full exploration is very time consuming and hence not feasible when the designer needs fast feedback regarding an application's storage requirement when the execution ordering is not fully fixed.

4.2.3 Estimation without execution ordering

In contrast to the methods in the previous section, the storage requirement estimation technique presented by Balasa et al. in [38] and [39] does not require the execution ordering to be fixed. On the contrary, it does not take execution ordering into account at all, except what is given from data dependencies in the code. The first step towards estimation is an extended data-dependency analysis where not only dependencies between array accesses in the code are taken into account, but also *which parts* of an array produced by one statement that are read by another (or possibly the same) statement. For each statement in the code, a definition domain is extracted, containing each array element produced by the statement. Similarly, operand domains are extracted for the parts of arrays read by the statement. Through an analytical partitioning of the arrays involving, among other steps, intersection of these domains, the end product is a number of non-overlapping *basic sets* (in general called basic groups to include also non-array data types) and the dependencies between them. Array elements common to a given set of domains and only to them constitute a basic set. The domains and basic sets are described as polytopes, using linearly bounded lattices (LBLs) of the form

$$x = T \bullet i + u | A \bullet i \geq b$$

where $x \in Z^m$ is the coordinate vector of an m-dimensional array, and $i \in Z^n$ is the vector of n loop iterators. The array index function is characterized by $T \in Z^{m \times n}$ and $u \in Z^m$, while the polytope defining the set of iterator vectors is characterized by $A \in Z^{2n \times n}$ and $b \in Z^{2n}$. The basic set sizes, and the sizes of the dependencies, are found using an efficient lattice point counting technique. The dependency size is the number of elements from one basic set that is read while producing the depending basic set. This information is used to generate a data-flow graph where the basic sets are the nodes and the dependencies between them are the branches. The total storage requirement for the application is found through a traversal of this graph, where basic sets are selected for production by a greedy algorithm. A basic set is ready for production when all basic sets it depends on have been produced, and a basic set is consumed when the last basic set depending on it has been produced. As the application is executed, the size of the currently alive basic sets changes over time as shown in Fig. 4.2. The maximal combined size of simultaneously alive basic sets gives the storage requirement.

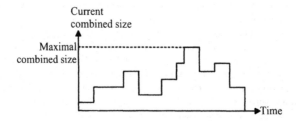

Figure 4.2.   Maximal combined size of simultaneously alive basic sets.

Fig. 4.3 presents a simple illustrative example written in single-assignment form. Note that very small loop bounds are used here to allow a graphical representation, but the mathematics used in the techniques and tools result in a run time complexity that is nearly independent of the actual loop bounds. Let us focus on the accesses to array B[u][v][w] in Fig. 4.3. The array elements are produced by statements S.1 and S.2, and read (and finally consumed) by statements S.2, S.3, and S.4. Fig. 4.4 gives examples of the LBL descriptions of the production and reading of array B[u][v][w] as found using the methodology from [39].

```
L.1 for (j=0; j<=3; j++)
        for (k=0; k<=2; k++)
S.1     B[0][j][k] = f1( A[j][k] );
L.2 for (i=1; i<=5; i++)
        for (j=0; j<=3; j++)
        for (k=0; k<=2; k++) {
S.2     B[i][j][k] = f2( B[i-1][j][k] );
S.3     if (j >= 2) C[i][j][k] = f3( B[i][j-2][k] );
S.4     if (i >= 5) D[j][k] = f4( B[5][j][k] );
        }
```

Figure 4.3.   Code example.

The part of array B[u][v][w] that is produced by statement S.1 is read and only read, and thus also consumed, by statement S.2. It therefore results in one basic set only, B(0). The part of array B[u][v][w] produced by statement S.2, however, is read by statement S.2, S.3, and/or S.4. Through intersection of the different definition and operand domains, this part of array B[u][v][w] is split into several basic sets depending on how they interrelate with each other. Some of the elements are for instance read by S.2 and S.3 but not S.4. These elements constituting basic set B(2). Fig. 4.5 gives a graphical description of the basic sets for the B[u][v][w] and C[u][v][w] arrays. To avoid unnecessarily complex figures, two-dimensional rectangles are used to indicate the basic sets. All nodes within a rectangle are part of the corresponding basic set. Fig. 4.6 gives an example of the LBL description for one of the basic sets, B(2). For more complex basic sets the representation of one basic set may consist of a collection of LBLs. Refer to [39] for a mathematical description and algorithm of how this partitioning is performed.

$$
a) \quad B\begin{bmatrix} u \\ v \\ w \end{bmatrix} = \begin{bmatrix} 1 & 0 & 0 \\ 0 & 1 & 0 \\ 0 & 0 & 1 \end{bmatrix} \begin{bmatrix} i \\ j \\ k \end{bmatrix} \quad \begin{bmatrix} 1 & 0 & 0 \\ -1 & 0 & 0 \\ 0 & 1 & 0 \\ 0 & -1 & 0 \\ 0 & 0 & 1 \\ 0 & 0 & -1 \end{bmatrix} \begin{bmatrix} i \\ j \\ k \end{bmatrix} \geq \begin{bmatrix} 1 \\ -5 \\ 0 \\ -3 \\ 0 \\ -2 \end{bmatrix}
$$

$$
b) \quad B\begin{bmatrix} u \\ v \\ w \end{bmatrix} = \begin{bmatrix} 1 & 0 & 0 \\ 0 & 1 & 0 \\ 0 & 0 & 1 \end{bmatrix} \begin{bmatrix} i \\ j \\ k \end{bmatrix} + \begin{bmatrix} 0 \\ -2 \\ 0 \end{bmatrix} \quad \begin{bmatrix} 1 & 0 & 0 \\ -1 & 0 & 0 \\ 0 & 1 & 0 \\ 0 & -1 & 0 \\ 0 & 0 & 1 \\ 0 & 0 & -1 \end{bmatrix} \begin{bmatrix} i \\ j \\ k \end{bmatrix} \geq \begin{bmatrix} 1 \\ -5 \\ 2 \\ -3 \\ 0 \\ -2 \end{bmatrix}
$$

Figure 4.4.   LBL descriptions for a) definition domain (elements produced) and b) operand domains (elements read).

When the basic sets and the dependencies between them for all arrays in the code have been derived, the data-flow graph of Fig. 4.7 is generated. The annotations to the nodes are the names and sizes of the corresponding basic sets, while the annotations to the branches are the dependency sizes. During the traversal of this graph, as explained above, it is detected that the maximal current storage requirement is reached when basic sets C(0), B(1) and B(3) are alive simultaneously. Together they require 30+6+6=42 storage locations for the code example in Fig. 4.3. Note that B(4) only requires six locations due to its self-dependency and to the fact that only six of its elements are required by other basic sets. B(2) on the other hand needs 24 storage locations, since all of its elements are required by C(0). Observe also that basic sets B(2) and C(0) can not be alive together since all children of basic set B(2) must be produced before or as C(0) is produced.

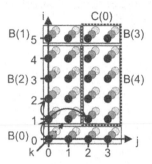

Figure 4.5.   Iteration space for array B[u][v][w] and C[u][v][w] in the code example in Fig. 4.3. Dependency shown for B[1][0][0].

In summary, all of the previous work on storage requirement entails a fully fixed execution ordering to be determined prior to the estimation. The only exception is the methodology in this section, which allows any ordering not prohibited by data dependencies. None of the approaches permits the designer to specify partial ordering constraints. As mentioned earlier and discussed in detail in the next section, this is essential during the early exploration of the code transformations.

$$
B(2): B \begin{bmatrix} u \\ v \\ w \end{bmatrix} = \begin{bmatrix} 1 & 0 & 0 \\ 0 & 1 & 0 \\ 0 & 0 & 1 \end{bmatrix} \begin{bmatrix} i \\ j \\ k \end{bmatrix} \begin{bmatrix} 1 & 0 & 0 \\ -1 & 0 & 0 \\ 0 & 1 & 0 \\ 0 & -1 & 0 \\ 0 & 0 & 1 \\ 0 & 0 & -1 \end{bmatrix} \begin{bmatrix} i \\ j \\ k \end{bmatrix} \geq \begin{bmatrix} 1 \\ -4 \\ 0 \\ -1 \\ 0 \\ -2 \end{bmatrix}
$$

Figure 4.6.   LBL description for basic set B(2) of array B[u][v][w].

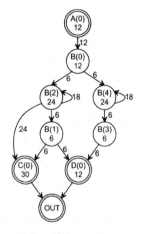

Figure 4.7.   Data-flow graph for the code example in Fig. 4.3.

4.3 ESTIMATION WITH PARTIALLY FIXED EXECUTION ORDERING

4.3.1 Motivation and context

Section 4.2.3 has described an estimation methodology that on purpose ignores the execution ordering. It will now be shown how this affects the estimation results and how a more accurate estimate can be achieved by taking available partially fixed execution ordering into account. Consider basic sets B(2) and C(0) in Fig. 4.7 and the dependency between them. As the data-flow graph shows, the size of basic set B(2) is 24. This corresponds to its worst-case storage requirement if the complete set B(2) is produced before the first element of C(0) is produced. As can be seen from Fig. 4.5, this will happen if for example the k dimension is fixed at the *innermost nest level* , followed by i, and with j at the *outermost nest level*. This signifies that the k-iterator runs from 0 to 2 between each increment of the i-iterator, which again runs from 0 to 5 between each increment of the j-iterator. As Fig. 4.5 also shows, the best-case storage requirement is two, achieved for example if j and k are interchanged in the above ordering. Then only elements B[1][0][0] and B[1][1][0] are produced before element C[1][2][0], which can potentially be mapped in-place with B[1][0][0]. Assume now that early in the design trajectory, the designer wants to explore the consequences of having k at the outermost nest level. Through inspection of Fig. 4.5 it can be seen that the dependency between B(2) and C(0) does not cross the k dimension. Accordingly all elements produced for one value of k are also consumed for this value of k. Consequently the worst-case and best-case storage requirements are now 8 and 2 respectively, indicating that it is indeed advantageous to place k outermost. In practice, it is not this simple however, since there can be conflicting dependencies in other parts of the code. Obviously there exists a need for an automated estimation and optimization tool to be able to take all dependencies into account. This also motivates the need for estimates using a partially fixed execution ordering. With estimation tools requiring the execution ordering to be fully specified, every alternative ordering with k at the outermost nest level would have to be explored to find the bounds on the storage requirement. It also shows the shortcomings of not considering the execution ordering at all, as in [39], since this results in a large overestimate due to the assumption that the whole basic set is produced before any of its elements are consumed. The methodology in [39] may also cause an underestimate because the data-flow graph traversal is somewhat optimistic and always selects the least costly basic set. Through the traversal a partial execution ordering is assumed, and the estimate may not be met if the final implementation follows another ordering. The new technique described here will overcome the overestimation problems by taking the partially fixed execution ordering into account when the basic set and dependency sizes are being calculated. Instead of using a graph traversal to detect simultaneously alive basic sets, a *common iteration space* is generated for all statements in the code. This space is then inspected to find the total storage requirement of the application, taking any partially fixed execution ordering into account.

The new methodology is useful for a large class of applications. There are certain restrictions on the code that can be handled in the present version however. The main requirements are that the code is single assignment and has affine array indexes. This is achievable by a thorough array data flow analysis preprocessing, see [171] and [436]. Also the resulting Dependency Parts, see section 4.3.2.1, must be orthogonal, or made orthogonal, as described in section 4.3.2.3.

When in the sequel the words upper and lower bounds are used, the bounds after approximations are meant. The lower bounds may not be factual, since some approximations tend to result in overestimates so that realizations with lower storage requirements may exist. For a large class of applications, where the Dependency Parts are already orthogonal, and when the generation of the best-case and worst-case common iteration space is possible, both bounds are correct.

A prototype CAD tool (named STOREQ for STOrage REQuirement estimation and optimization) has been developed to prove the feasibility and usability of the techniques. It includes major parts of the storage requirement estimation and optimization methodology. Run time experiments have shown an overall time complexity in the order of seconds, even for large applications. This corresponds well to the theoretical time and space complexities which are linear with the number of dependencies in the code. See [275] for details regarding the tool and the complexities.

4.3.2 Concept definitions

The *iteration space* of a loop nest reflects the values over which the iterators run as shown in Fig. 4.5. Even though each dimension in the iteration space corresponds to a given loop in the loop nest, the order in which the dimensions are traversed is not given from the iteration space. It is first when the execution ordering is (partially) fixed that dimensions are associated with given nest levels in the (partially) fixed loop nest. At each node in the iteration space, denoted an *iteration node*, the statements within the loop nest is executed. If the production of data by the statement is constrained, for example by an if-clause, data will only be produced at a subset of the iteration nodes. This subset is then defined as the *iteration domain* (ID) of the statement [171].

The starting point for the estimation methodology that is presented in this chapter is the basic set and dependency information as has been described in section 4.2.3. The main principles can be used on any complete polyhedral description of sets of signals and their dependencies though. The Polyhedral Dependency Graph model discussed in section 3.3 is one example. In the methodology presented below, the detailed splitting of arrays into basic sets that is used in [39] is not necessarily required. Iteration domains for individual statements in the code can be used directly. It is in both cases the iteration nodes at which array elements are produced and read that are of interest. Which individual array elements that are actually being produced or read is not important after the dependency analysis is finished. The focus is thus somewhat different from most other work in this field, since it is on the iteration nodes where array elements of individual dependencies are produced and read, not on the actual array elements produced and read by individual statements. To avoid the use of too many terms, the concept of iteration domain is used in the sequel, independent of how the dependency analysis has been performed. A dependency is also said to exist between two iteration domains even though it actually runs between the array elements produced at those iteration domains.

To overcome the problem of overestimation found in [39], information about the partially fixed execution ordering is exploited to calculate upper and lower bounds on the size of the dependencies between iteration domains. This section defines some important concepts used by the estimation methodology while the next section gives a general introduction to the main principles of the methodology. The concepts described in this section were first presented in [274].

To illustrate the following concept definitions, a simple example code is given in Fig. 4.8 with the corresponding iteration space with iteration domains and data-flow graph in Fig. 4.9. Again, two-dimensional rectangles are used to indicate the iteration domains. All nodes within a rectangle are part of the corresponding iteration domains. Note that the partitioning algorithm of [39] will split the array elements produced by statement S.1 into two basic sets, A(0) and A(1). This is not done here. Instead A(0) and B(0) are denoted iteration domains as justified above.

```
for i=0 to 5 {
  for j=0 to 5 {
    for k=0 to 2 {
S.1    A[i][j][k] = f1( input );
S.2    if(i>=1)&(j>=2) B[i][j][k] = f2( A[i-1][j-2][k] );
}}}
```

Figure 4.8.   Code example for concept definitions.

4.3.2.1 Dependency part. Each branch in the data-flow graph represents a data-dependency from one iteration domain to another. Often only a subset of the array elements produced in one iteration domain is read by the statement producing elements within the depending iteration domain. This has led to the definition of a *Dependency Part* (DP) containing the iteration nodes where array elements are produced that are read within the depending iteration domain. This concept differs from a normal operand domain. An operand domain defines which array elements that are read by a statement, but these elements may have been produced by multiple other statements giving rise to multiple dependencies. The DP on the other hand, focuses on the elements carried by a single

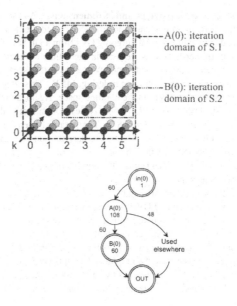

Figure 4.9.   Iteration space and data-flow graph of example in Fig. 4.8.

dependency. The DP can be described using a Linearly Bounded Lattice (LBL). The iteration nodes in the depending iteration domain where the depending statement reads array elements produced within the DP are denoted *depending iteration nodes*.

The *length* of the DP in a dimension d_i is denoted $|DP_{d_i}|$ and equals the number of iteration nodes covered by a projection of the DP onto this dimension. Both the start and end node of a dimension is hence included in its length.

When array elements are being produced within the iteration domain of S.2 in Fig. 4.9, a subset of the array elements produced within the iteration domain of S.1 is read. The corresponding DP is given in Fig. 4.10. In this case, all iteration nodes in the iteration domain of S.2 are depending iteration nodes. The length of the DP in each dimension is also indicated.

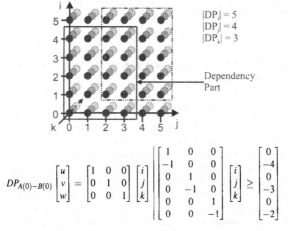

Figure 4.10.   Dependency Part of A(0) with respect to B(0) of example in Fig. 4.8.

4.3.2.2 Dependency vector polytope and its dimensions. A *Dependency Vector* (DV) runs from an iteration node where an array element is produced and to the iteration node where it is read. A dependency is said to be uniform if all outgoing DVs from a DP have the same direction and length. When DVs of uniform dependencies are used in the estimation methodology, it is only necessary to consider one of them. For simplicity, the DV is chosen that starts at the iteration node with the lowest position following a lexicographic ordering, [453]. This single dependency vector spans a *Dependency Vector Polytope* (DVP) in the iteration space. Again, an LBL description can be used to describe the DVP. The DVP covers all or (more typically) a subset of the dimensions in the iteration space, defined as *Spanning Dimensions* (SD). The iterator dimensions not covered by the DV are defined as *Nonspanning Dimensions* (ND). The largest value for each dimension in the DVP is denoted *Spanning Value* (SV) of that dimension. Similar to the DP, the length of the DVP in a dimension d_i is denoted $|DVP_{d_i}|$ and equals the number of iteration nodes covered by a projection of the DVP onto this dimension.

For the dependency between A(0) and B(0) in Fig. 4.9, the DV starts at (i,j,k)-node (0,0,0) and end at (1,2,0). The corresponding DVP is given in Fig. 4.11 together with its SVs and the lengths of its dimensions. Only iteration nodes fully encircled by the rectangle are part of the DVP. When the rectangle is drawn on top of an iteration node, it indicates that this node and all nodes behind it is not a part of the DVP. (i,j,k)-node (1,1,0) is thus a part of the DVP while (1,1,1) and (1,1,2) are not. Note that for comparison with the graphical description and the DP a three-dimensional LBL is used, even though the DVP really only has two dimensions, the i and j dimensions. The SDs of this DVP are the i and j dimensions, while the k dimension is the only ND.

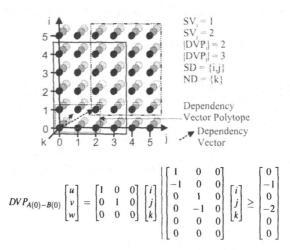

$$DVP_{A(0)-B(0)} \begin{bmatrix} u \\ v \\ w \end{bmatrix} = \begin{bmatrix} 1 & 0 & 0 \\ 0 & 1 & 0 \\ 0 & 0 & 0 \end{bmatrix} \begin{bmatrix} i \\ j \\ k \end{bmatrix} \left| \begin{bmatrix} 1 & 0 & 0 \\ -1 & 0 & 0 \\ 0 & 1 & 0 \\ 0 & -1 & 0 \\ 0 & 0 & 0 \\ 0 & 0 & 0 \end{bmatrix} \begin{bmatrix} i \\ j \\ k \end{bmatrix} \geq \begin{bmatrix} 0 \\ -1 \\ 0 \\ -2 \\ 0 \\ 0 \end{bmatrix} \right.$$

Figure 4.11. Dependency Vector and Dependency Vector Polytope for the dependency between A(0) and B(0) of example in Fig. 4.8.

In many cases, the end point of the DV falls outside the DP. This happens when no overlap exists between the two depending iteration domains. For now, the DVP is then intersected with the DP. Note however that this definition of the DVP will change somewhat in section 4.4, where a more detailed and accurate mathematical estimation methodology is presented.

When a dependency is not uniform, the DVs for different iteration nodes within its DP can differ in both length and direction. The DVP is then generated using the extreme DVs [127]. Fig. 4.12 shows a code example with non-uniform array indexes. Fig. 4.13 shows all dependency vectors for the example. For readability, the figure is split into three, one for each value of the k-iterator. The extreme DVs are drawn with thicker lines. Again denoting the iteration domains of S.1 and S.2 A(0) and B(0) respectively, the corresponding DVP is given in Fig. 4.14. Due to orthogonalization, see section 4.3.2.3, both iteration domains cover all iteration nodes of the iteration space.

```
for i=0 to 4 {
  for j=0 to 5 {
    for k=0 to 2 {
S.1    A[i][j][k] = f1( input );
S.2    if(i>=k)&(j>=i-k) B[i][j][k] = f2( A[i-k][j-i+k][k] );
}}}
```

Figure 4.12.   Code example with non-uniform indexes.

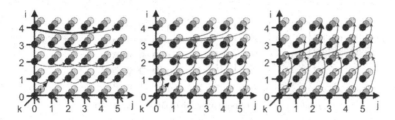

Figure 4.13.   Dependency Vectors for each iteration node of the example in Fig. 4.12.

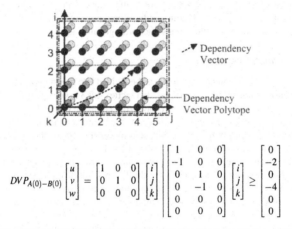

$$DVP_{A(0)-B(0)} \begin{bmatrix} u \\ v \\ w \end{bmatrix} = \begin{bmatrix} 1 & 0 & 0 \\ 0 & 1 & 0 \\ 0 & 0 & 0 \end{bmatrix} \begin{bmatrix} i \\ j \\ k \end{bmatrix} \begin{bmatrix} 1 & 0 & 0 \\ -1 & 0 & 0 \\ 0 & 1 & 0 \\ 0 & -1 & 0 \\ 0 & 0 & 0 \\ 0 & 0 & 0 \end{bmatrix} \begin{bmatrix} i \\ j \\ k \end{bmatrix} \geq \begin{bmatrix} 0 \\ -2 \\ 0 \\ -4 \\ 0 \\ 0 \end{bmatrix}$$

Figure 4.14.   Dependency Vector and Dependency Vector Polytope for the dependency between A(0) and B(0) of example in Fig. 4.12.

Use of the extreme DVPs is a simple approach to uniformization suitable for estimation when a short run time is required to give fast feedback to the designer. Alternative techniques for uniformization of affine dependency algorithms are presented in [477]. The complexity of these procedures is however much larger.

4.3.2.3   Orthogonalization.   The DPs may have complex shapes that do not lend themselves easily to the type of calculations included in the estimation techniques. It is therefore necessary to *orthogonalize* the DP. Each dimension is extended as needed until all dimensions are orthogonal with respect to each other. Fig. 4.15 provides an example of two-dimensional orthogonalization.

The orthogonalization induces a certain estimation error. For many applications, this is not a problem, since most DPs are already orthogonal. The final implementation may also require that the storage order is linear to avoid an overly complex address generation [148]. Even non-orthogonal DPs will then require an orthogonal physical storage area, so the approximation will yield correct estimates. If DP shapes that are more complex appear in the code, it may also be a viable solution

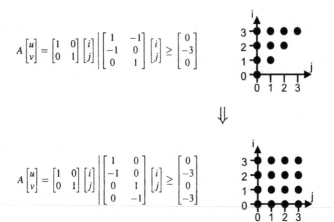

Figure 4.15.   Orthogonalization of Dependency Part.

to split the dependency so that each DP has a more regular shape. This will reduce the estimation error at the cost of a higher computational complexity. When in the sequel an *Upper Bound* (UB) or *Lower Bound* (LB) is used, these are the bounds after orthogonalization.

4.3.3   Overall estimation strategy

The storage requirement estimation methodology can be divided into four steps as shown in Fig. 4.16. The first step uses the technique from [39] presented in section 4.2.3 to generate a data-flow graph reflecting the data dependencies in the application code. Any complete polyhedral dependency model can be used however, and work is in progress to integrate the estimation methodology with the ATOMIUM tool, [30], [61], [386], [389]. The second step places the DPs of the iteration domains in a common iteration space to enable the global view of the fourth step. Best-case and worst-case placements may be used to achieve lower and upper bounds respectively. Chapter 3 describes work that can be utilized for this step, [124, 123]. The third step focuses on the size of individual dependencies between the iteration domains in the common iteration space. The final step searches for simultaneously alive dependencies in the common iteration space, and calculates their maximal combined size.

The main emphasis of the research presented in this chapter is on the two last steps. After a short description of a simplified technique for generation of a common iteration space, the principles of these steps are described in the next sections. The principles for estimation of individual dependencies that is presented in section 4.3.5 is a rough sketch of the main ideas. A more detailed and accurate methodology is given in section 4.4. Section 4.3.6 presents the main techniques for detection of simultaneously alive dependencies.

4.3.4   Generation of common iteration space

In the original application code, an algorithm is typically described as a set of imperfectly nested loops. At different design steps, parts of the execution ordering of these loops are fixed. The execution ordering may include both the sequence of separate loop nests, and the order and direction of the loops within nests. To perform global estimation of the storage requirement, taking into account the overall limitations and opportunities given by the partial execution ordering, a common iteration space for the code is needed. A common iteration space can be regarded as representing one loop nest surrounding the entire, or a given part of the code. This is similar to the global loop reorganization described in [123]. Fig. 4.17 shows a simple example of the steps required for generation of such a common iteration space. Note that even though it here includes rewriting the application

Generate data-flow graph reflecting data dependencies in the application code

↓

Position DPs in a common iteration space according to their dependencies and the partially fixed execution ordering

↓

Estimate upper and lower bounds on the size of individual dependencies based on the partially fixed execution ordering

↓

Find simultaneously alive dependencies and their maximal combined size
⇒
Bounds on the application's storage requirement

Figure 4.16.   Overall storage requirement estimation strategy.

```
for (i=0; i<=5; i++)
   A[i] = ...                       Original code
for (j=0; j<=3; j++)
   B[j] = f( A[j] );
```

```
for (t=0; t<=1; t++)
   for (i=0; i<=5; i++)             Add outer pseudo-dimension
      if (t==0) A[i] = ...          Add/enlarge dimensions and add if-
for (t=0; t<=1; t++)                clauses to enable merging
   for (i=0; i<=5; i++)
      if (t==1) & (i<=3) B[i] = f (A[i]);
```

```
                                    Merge loops
for (t=0; t<=1; t++)
   for (i=0; i<=5; i++) {
      if (t==0) A[i] = ...
      if (t==1) & (i<=3) B[i]= f( A[i] );
   }
```

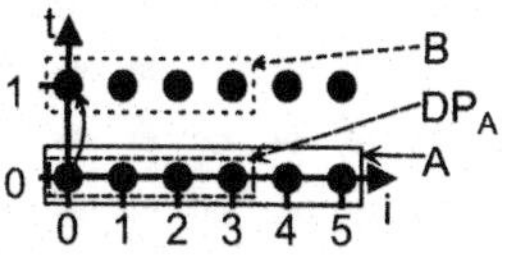

```
                                    Optimize placement
for (t=0; t<=1; t++)
   for (i=0; i<=5; i++) {
      if (t==0) A[i] = ...
      if (t==0) & (i<=3) B[i]= f( A[i] );
   }
```

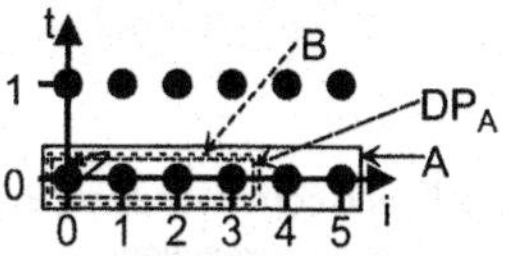

Figure 4.17.   Generation of Common Iteration Space.

code, this is not necessary to perform the estimation. The common iteration space can be generated directly using the PDG model of [93].

The introduction of the common iteration space opens for an aggressive in- place mapping which may not be used in the final implementation. The placement of the DPs in the common iteration space entails certain restrictions on the possible execution ordering of the still unfixed parts of the code. The sizes of the individual dependencies will hence be influenced by the placement of their DPs and depending DPs in the common iteration space. To have realistic upper and lower bounds it is therefore necessary to generate two common iteration spaces, one with a worst-case and one with a best-case placement, both taking into account the available execution ordering and other realistic design constraints. Chapter 3 presents techniques for placement that enable the designer to organize the data accessing in such a way that aggressive in-place optimization can be employed. The worst-case placement can be based on a full loop body split, so that each statement is assigned individual iterator values in the t dimension. For the rest of this chapter, optimal iteration spaces are assumed, but the methodologies presented work equally well on alternative organizations of the iteration space.

4.3.5   Estimation of individual dependencies

The following definition is used for the size of a dependency.

Definition 4.1 *The size of a dependency between two iteration domains equals the number of iteration nodes visited in the DP before the first depending iteration node is visited.*

Since one array element is produced at each iteration node in the DP, this size equals the number of array elements produced before the first depending array element is produced that potentially can be mapped in-place of the first array element. Only elements produced by one statement and read by the depending statement are counted. Following Definition 4.1, the size of a dependency may vary greatly with the chosen execution ordering. With no execution ordering fixed, the upper and lower bounds on the storage requirement can be estimated from the size of the original DP and DVP respectively. No matter what execution order is chosen in the end, the iteration nodes of the DVP will always be visited before the first depending iteration node. Similarly, no more than all iteration nodes, that is the full DP, can be visited before the first depending iteration node. If overlap exists between the DP/DVP and the depending iteration nodes, not all of their elements will be produced before the first depending element that can potentially be mapped in-place of the first DP/DVP element. The overlap can therefore be removed from the DP and DVP. The DP and DVP in Fig. 4.10 and Fig. 4.11 thus gives the following estimates with no execution ordering fixed:

$$UB = size(DP) - size(DP_{overlap}) = 60 - 24 = 36$$
$$LB = size(DVP) - size(DVP_{overlap}) = 6 - 1 = 5$$

The UB and LB are shown graphical in Fig. 4.18.

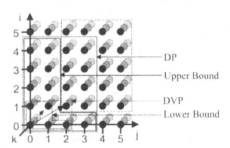

Figure 4.18.   DP and DVP and corresponding UB and LB for the dependency between A(0) and B(0) of example in Fig. 4.8 without any execution ordering fixed.

Unless no overlap exists between the DP and the depending iteration nodes, the bounds found using these simple principles cannot be reached. The detailed description of a more accurate estimation technique, but which is still based on the same principles, is given in section 4.4.

The execution ordering can be fixed through a fixation of dimensions starting with the innermost or outermost nest levels. SDs and NDs must be treated differently. Table 4.1 summarizes the estimation principles for each of these categories. The main effects on the DVP and DP are written in bold. If an ND is fixed outermost, only the DP changes, giving rise to a reduced upper bound and an unchanged lower bound. A fixation of an ND innermost has the opposite effect, an unchanged upper bound and an increased lower bound. If an SD is fixed outermost, both DVP and DP changes,

	Fixation starts outermost	Fixation starts innermost
SD	**Expanded DVP, reduced DP:** Expand all unfixed SDs of the DVP to the boundary of the DP. Add all unfixed NDs to the DVP. *If* (no SD fixed so far): remove from the DP elements that for the current dimension have values larger than its SV. *Else*: remove elements that will not be visited from the DP	**Expanded DVP, reduced DP:** *If* (not last SD): expand the current dimension of the DVP to the boundary of the DP. Remove from the DP elements that for all unfixed SDs have values larger than their SV.
ND	**Unchanged DVP, reduced DP:** *If* (no SD fixed so far): remove the dimension from the DP *Else*: remove elements that will not be visited from the DP	**Expanded DVP, unchanged DP:** *If* (unfixed SDs still exist): add the dimension to the DVP

Table 4.1.   Main principles for treatment of fixed dimensions.

giving rise to a reduced upper bound and an increased lower bound. This is also the result if an SD is fixed innermost. In this last case however, the change in the DVP only covers the fixed SDs, while the fixation of an SD outermost effects all unfixed dimensions. The increase in the lower bound is therefore in most cases much smaller if the SD is fixed innermost. Note that there are many special cases that Table 4.1 does not explain fully, partly covered by the *Else* clauses. They will be handled in section 4.4. The methodology presented there will also remove the requirement of fixation starting at the innermost or outermost nest level.

The main principles will now be demonstrated using the example code from Fig. 4.8. Assume that SD j has been fixed at the innermost nest level. The j dimension of the DVP is then extended to the boundary of the DP since these iteration nodes will now certainly be visited before the first depending iteration node. All iteration nodes of the DP with values in the i dimension larger then the SV_i (that is > 1) can be removed, since they are now certainly not visited before the first depending iteration node. The resulting DP and DVP are given in Fig. 4.19. For enhanced readability the corresponding UB and LB are not drawn. As before, they are found through a removal of the overlap with the depending iteration nodes. The estimated bounds are now:

$$UB = size(DP) - size(DP_{overlap}) = 24 - 6 = 18$$
$$LB = size(DVP) - size(DVP_{overlap}) = 8 - 2 = 6$$

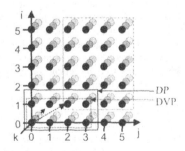

Figure 4.19.   DP and DVP of the dependency between A(0) and B(0) of example in Fig. 4.8 with SD dimension j fixed innermost.

When one dimension is fixed either innermost or outermost the estimation can continue similarly for any additionally fixed dimension. Assume for instance that ND k is first fixed outermost, followed by SD i second outermost. The resulting DP and DVP are shown in Fig. 4.20 and the estimated bounds are now:

$$UB = \text{size}(DP) - \text{size}(DP_{overlap}) = 8 - 2 = 6$$
$$LB = \text{size}(DVP) - \text{size}(DVP_{overlap}) = 8 - 2 = 6$$

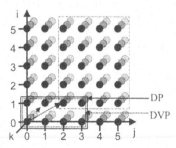

Figure 4.20.   DP and DVP for the dependency between A(0) and B(0) of example in Fig. 4.8 with dimension k fixed outermost and i fixed second outermost.

For the execution orderings used in Fig. 4.20, the simple estimation principles outlined in Table 4.1 end up with converging and true upper and lower bounds. This is however not always the case, so more complex, but still not computationally hard, calculation techniques are required. These will be discussed in section 4.4.

4.3.6   Estimation of global storage requirement

After having found the upper and lower bounds on the size of each dependency in the common iteration space, it is necessary to detect which of the dependencies that are alive simultaneously. Their combined size gives the current storage requirement at any point during the execution of the application.

Definition 4.2 *The maximal combined size of simultaneously alive dependencies over the lifetime of an application, gives the total storage requirement of the application.*

When multiple dependencies cover the same array elements, this is taken into account during the calculation of the combined size of the dependencies found to be alive simultaneously. This part of the estimation methodology is presented in [276].

The methodology that is presented in [39] does not use a common iteration space. Instead, a traversal of the data-flow graph is performed to identify the maximum size of simultaneously alive basic sets. As [39] states, the search for this maximum is an NP-complete problem, so a full search is not feasible for real applications. Even for the very simple graph of Fig. 4.7, there are 31 alternative traversals. The number grows exponentially with the number of basic sets that can be selected as the next basic set during the traversal. As a consequence, a traversal heuristic is needed that at a low computational cost gives a reasonable estimate. [39] keeps continuous track of the current storage requirement throughout the traversal, and selects for production the basic set that gives the least increase or largest decrease in current storage requirement. The chosen traversal will in itself define a partial execution order, and the resulting upper bound on the storage requirement only holds if this order is followed. The extreme worst-case execution ordering may give rise to a much larger storage requirement. A possible heuristic to find this requirement may be to always select for production the basic set that results in the largest increase or least decrease in current storage requirement. This may however give a very unrealistic upper bound since it is not probable that this execution ordering will be selected for the final implementation. Research on a traversal heuristic that finds realistic upper and lower bounds on the storage requirements is an interesting path for future work.

The estimation methodology that is presented in this chapter avoids the NP-hard graph traversal through use of the common iteration space. To find the best-case and worst-case placements in the common iteration space are on the other hand also NP-complete problems. This is however covered by other work, as described in chapter 3. Currently a prototype tool is available that demonstrates that the automatic production of an optimized common iteration is feasible. In the near future a more generally applicable version is expected to become available for more elaborate experiments. For the rest of this section a given common iteration space is assumed.

4.3.6.1 Simultaneous aliveness for two dependencies. Two dependencies with DPs placed at given positions in the common iteration space, may or may not be partially alive simultaneously. The deciding factor is the way the SDs and NDs of the DPs overlap. In this context, a dimension is an SD if it is an SD for any of the two DPs. For the inspection of overlap, the DP is extended in all SDs with the length of the DV in that dimension. This is necessary since the dependency is alive until all of its elements have been read, that is until all iteration nodes within this *Extended DP* (EDP) have been visited. In Fig. 4.21, DP A is extended (dotted line) from iterator value 3 to iterator value 5 in the i dimension. Overlap exists in a dimension if for one or more iterator values of that dimension, the two EDPs both have elements. EDP A and EDP B are thus overlapping in the i and k dimensions, but not in the j dimension.

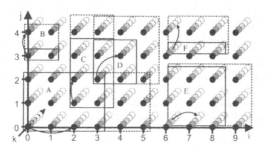

Figure 4.21.   Overlap between EDPs.

If all dimensions, SDs and NDs, are partially overlapping, that is the EDPs cover some of the same nodes in iteration space, the dependencies will be alive at least partially simultaneously, regardless of the chosen execution ordering. Elements are produced and/or consumed at the same point in time. EDP A and C have this kind of overlap in Fig. 4.21. If none of the EDPs' SDs are overlapping, the dependencies can not be alive simultaneously. If some of the NDs are overlapping, the EDPs may however alternate in being alive. EDP A and F can never be alive simultaneously since no dimensions are overlapping. EDP A and E can not be alive simultaneously either, since none of the SDs are overlapping. They are overlapping in ND j, however, so if this dimension is placed outermost, the two dependencies will alternate in being alive.

The most complex form of overlap occurs when two EDPs are overlapping in one or more dimensions, including at least one SD, but still not in all dimensions. Whether the two dependencies are alive simultaneously will then depend on the execution ordering. Simultaneous aliveness is avoided if at least one dimension without overlap (SD or ND) is fixed outside all overlapping SDs. EDP A and B are overlapping in all dimensions except the j dimension, which must hence be placed outside i to avoid simultaneous aliveness. An interesting consequence of this is that the lower bound on the size of the B dependency is doubled. The cost of avoiding simultaneous aliveness must therefore be balanced against the increased dependency size. EDP B and F are only overlapping in the j dimension, so ND i can be placed outermost, which is optimal for both dependencies. Table 4.2 summarizes the overlap of spanning and nonspanning dimensions for the EDPs of Fig. 4.21.

4.3.6.2 Simultaneous aliveness for multiple dependencies. More than two dependencies may be alive simultaneously. The detection of this is not straightforward, since one dependency may be alive simultaneously with for example two others, but not necessarily at the same point in time. In Fig. 4.21, EDP A is overlapping with both EDP B and E, but EDP B and E do not overlap. It is therefore impossible for all three of them to be alive simultaneously. As before, the most complex situation occurs when only a subset of the dimensions is overlapping. EDP D, E and F

	B	C	D	E	F
A	SD=i ND=k	SD=i,j ND=k	SD=i,j ND=k	ND=j	
B		SD=j ND=k	SD=j ND=k		SD=j
C			SD=i,j ND=k	SD=j,k	SD=j,k
D				SD=j,k	SD=j,k
E					SD=i,k

Table 4.2.   Dimensions with overlap for EDPs in Fig. 4.21.

Val.	0	1	2	3	4
	A_s E_s	C_s	D_s A_e E_e	B_s F_s	B_e C_e D_e F_e

Table 4.3.   List of start and end points for EDPs in j dimension.

are all overlapping with each other but depending on the chosen execution ordering, only one, any combination of two, or all three will be alive partially simultaneously.

```
for each iterator value of pseudo-dimension t {
    for each dimension dᵢ {
    generate list ( length = |dᵢ| )
    for each EDP
        place start and end points at corresponding list location
    traverse list and group EDPs overlapping in dimension dᵢ
    }
    sort groups of EDPs according to their number of EDPs
    for each group starting with the largest group {
    intersect the group with all smaller groups
    remove groups that are fully covered by the current group
} }
return groups of possibly partially simultaneously alive EDPs
```

Figure 4.22.   Simultaneous aliveness for multiple dependencies.

To overcome the complexity of deciding whether three or more dependencies really are alive simultaneously the task is divided into two consecutive steps. First dependencies that may be alive simultaneously are grouped, followed by an inspection to reveal if they can really be alive simultaneously for a given partially fixed execution ordering. An algorithm that groups EDPs is given in Fig. 4.22. It assumes that a pseudo t dimension is used to generate a common iteration space from multiple loop nests. As will be shown shortly, it can be ignored if it is not present. Use of the algorithm will now be demonstrated on the iteration space of Fig. 4.21. In this case, the pseudo dimension t has only one iterator value and is therefore ignored. Table 4.3 displays the list for one of the three dimensions, the j dimension, after insertion of each EDP's start (X_s) and end (X_e) points in that dimension. The traversal of the lists detects overlapping EDPs by adding EDPs to a group as long as start points are encountered. When one or more end points are encountered, the current group is terminated and a new one is started without the EDPs that have ended. For the j dimension in Table 4.3 EDPs A, E, C, and D are added to the first group at iterator values 0, 1, and 2. At iterator value 2 the end points of EDP A and E are encountered, so the first group is finished and a second one is started containing C and D. At iteration node 3, EDPs B and F are added to this second group. The groups for this and the other dimensions are summarized in Table 4.4. It also shows the sorting of all groups from all dimensions and the removal of covered groups. Finally it lists the combined upper bounds on the size for all DPs in each group.

The next step is to perform a check of the possible simultaneous size of the different groups, starting with the biggest group. The global upper and lower bounds are found when the lower bound

	In each dimension	Sorted	Covered by ABCD	Covered by CDEF	Combined size
i	AB, ACD, EF	ABCD	ABCD	ABCD	41
j	AECD, CDBF	AECD	AECD	AECD	49
k	ABCD, CDEF	CDBF	CDBF	CDBF	35
		CDEF	CDEF	CDEF	43
		ACD	~~ACD~~	~~EF~~	
		AB	~~AB~~		
		EF	EF		

Table 4.4.   Groups of overlapping EDPs.

Outermost	Simultaneously alive	UB	LB
i	C and D or E and F	20	12
j	C and D and E or C and D and F	31	27
k	C and D and E and F	23	18

Table 4.5.   Actual simultaneous aliveness for group CDEF.

for one group is bigger than the combined size of the next group. Starting with group AECD it can be seen from Table 4.2 that EDPs A and E only overlap in ND j. The group is hence split into two new groups, ACD and ECD, both of which are covered by larger groups. CDEF is now the largest group. C and D are overlapping in all dimensions, and will consequently be alive simultaneously independent of the chosen execution ordering. E and F are overlapping in SDs i and k and will thus be alive simultaneously unless the dimension without overlap, j, is placed outermost. Both C and D are overlapping with E and F in SDs j and k, so these pairs will be alive simultaneously unless i is placed outermost. For all four to be alive simultaneously k must consequently be placed outermost. The actually simultaneously alive EDPs for three partially fixed execution orderings, and their upper and lower bounds are given in Table 4.5. The upper and lower bounds of groups of EDPs are found through addition of the bounds of the individual DPs. Since these may require conflicting orderings they may not be reachable simultaneously. Even closer inspections of the groups are therefore needed to ensure that the bounds are reachable. In this case no conflicting orderings exists and it can be concluded that for group CDEF, the lowest storage requirement can be reached if i is placed outermost. Note that the ordering that results in the largest number of simultaneously alive dependencies is not the ordering with the largest storage requirement. This is because this ordering results in smaller individual dependencies.

The discussion above covers simultaneously alive dependencies. Multiple dependencies may however stem from the same, or partly the same, array elements. This must be taken into account in such a way that only the dependency with the longest lifetime for a given partial ordering is counted.

4.3.7 Guiding principles

Table 4.1 shows the consequences of fixing SDs and NDs innermost and outermost. This knowledge can be used by the designer to guide the fixation of the execution ordering if the main goal is to reduce the storage requirement through optimization of the in-place mapping opportunity. The size of a dependency is minimized if the execution ordering is fixed so that its NDs and SDs are fixed at the outermost and innermost nest levels respectively. The proof of this and the other guiding principles discussed here can be found in [275]. The order of the NDs is of no consequence as long as they are all fixed outermost. If one or more SDs are ordered in between the NDs, it is however better to fix the *shortest* NDs inside the SDs. The length of ND d_i is determined by the length of the DP in this dimension as defined in section 4.3.2.1. Rewriting statement S.2 of Fig. 4.8 as follows:

S.2 if (j >= 2) B[i][j][k] = f2(A[i][j-2][k]);

gives two NDs, i and k, and one SD, j. As can be seen from Fig. 4.23, $|DP_i| = 5$ while $|DP_k| = 3$. Note again that both the starting and ending node is counted when the lengths are calculated. Assuming that j is fixed second innermost, the k dimension should be fixed innermost since $|DP_k| < |DP_i|$. This results in a dependency size of six while the alternative ordering, with the i dimension innermost, results in a dependency size of ten.

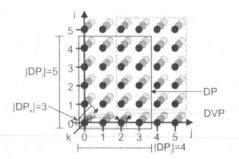

Figure 4.23.   Length of DPs for the code example of Fig. 4.8 with modified S.2.

The ordering of SDs is important even if all of them are fixed innermost. The SD with the largest *Length Ratio* (LR) should be ordered innermost. Details regarding the derivation and proof of the use of the LR can be found in [275]. Its definition is given by

$$LR_{d_i} = \frac{|DVP_{d_i}| - 1}{|DP_{d_i}| - 1}$$

or the length of the DVP in dimension d_i minus one divided by the length of the DP in dimension d_i minus one. Returning to the original code of Fig. 4.8 with two SDs, the calculation of the LRs for the i and j dimensions are illustrated in Fig. 4.24. According to the guiding principles, the j dimension should be fixed innermost giving a dependency size of six as shown in Fig. 4.20. The alternative ordering, with the i dimension innermost, would give a dependency size of eleven.

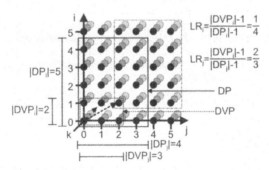

Figure 4.24.   Calculation of LRs for the code example of Fig. 4.8.

The guiding principles can be used directly in the cases where one or a few major dependencies determine the total storage requirement of a loop nest. For more general cases with multiple and possibly contending optimal orderings, an automated tool is needed. A prototype tool has been developed to show the feasibility of this approach. This tool makes use of feedback from the estimation tool to find the cost of not following the optimal ordering for a given dependency. In any case, dimensions that are NDs for all dependencies should be fixed outermost to optimize the in-place mapping opportunity. Early ideas for the optimization methodology are presented in [272].

4.4   SIZE ESTIMATION OF INDIVIDUAL DEPENDENCIES

A most crucial step of the storage requirement estimation methodology is the estimation of the size of individual data dependencies in the code. This information can be used directly to investigate the largest and thus globally determining dependencies, or it can be used as input to the subsequent search for the maximal size of simultaneously alive dependencies. Important parts of the techniques covered in this chapter were first presented in [273].

4.4.1   General principles

The estimation technique for individual dependencies presented in this section is a mathematically based and more accurate version of the rough sketch that is introduced in section 4.3.5. The main enhancement is the utilization of the guiding principles from section 4.3.7. They are used to find a best-case and worst-case ordering for the dependency. Using these two orderings, the exact upper and lower bounds on the storage requirement are calculated. Any dimensions fixed at given nest levels in the common loop nest are taken into account. Furthermore, the restriction that dimensions should be fixed starting with the innermost or outermost nest level is lifted. Even partial ordering among dimensions is allowed.

For reasons to become apparent later in this chapter, the definition of the DVP must be changed somewhat. It is extended with one in all dimensions for which the DV falls outside the DP. This will typically happen when the DP and the depending iteration nodes do not overlap. Rewriting statement S.2 of Fig. 4.8 as follows:

$$\text{if } (i >= 1) \ \& \ (j >= 2) \ \& \ (j <= 4) \ B[i][j][k] = f2(A[i-1][j-2][k]);$$

reduces the DP in the j dimension as shown in Fig. 4.25 so that it only runs from 0 to 1. The DVP is still the same though, since it has been extended with one in the j dimension after the intersection with the DP.

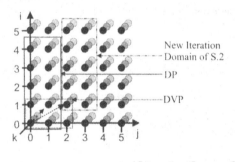

$$DVP_{A(0)-B(0)} \begin{bmatrix} u \\ v \\ w \end{bmatrix} = \begin{bmatrix} 1 & 0 & 0 \\ 0 & 1 & 0 \\ 0 & 0 & 0 \end{bmatrix} \begin{bmatrix} i \\ j \\ k \end{bmatrix} \begin{vmatrix} \begin{bmatrix} 1 & 0 & 0 \\ -1 & 0 & 0 \\ 0 & 1 & 0 \\ 0 & -1 & 0 \\ 0 & 0 & 0 \\ 0 & 0 & 0 \end{bmatrix} \begin{bmatrix} i \\ j \\ k \end{bmatrix} \geq \begin{bmatrix} 0 \\ -1 \\ 0 \\ -2 \\ 0 \\ 0 \end{bmatrix} \end{vmatrix}$$

Figure 4.25.   Dependency Vector and Dependency Vector Polytope for reduced Dependency Part.

4.4.2   Automated estimation with a partially fixed ordering

Fig. 4.26 describes an algorithm for automated estimation of the Upper Bound (UB) and Lower Bound (LB) on the storage requirement for individual dependencies. The algorithm also outputs the best case ordering of the unfixed dimensions. The input to the algorithm consists of the LBL descriptions of the DP and DVP in addition to the ordered set of *Fixed Dimensions* (FD) starting with the outermost nest level. The unfixed dimensions of the FD set are represented by X. A five dimensional iteration space with dimension d_3 fixed second outermost and d_4 fixed innermost is thus presented as FD=$\{X, d_3, X, d_4\}$. As defined in section 4.3.2 the use of $|DVP_{d_i}|$ in the algorithm denotes the length of the DVP projected onto the d_i dimension while $|DP_{d_i}|$ denotes the length of the DP projected onto the d_i dimension. Note also that for the increased readability the result of a multiplication operation over an empty set is assumed to return 1 ($\prod_{\forall d_i \in S} |A_{d_i}| = 1 \ if \ S = \emptyset$) and thus can be ignored.

EstimateDependencySize(DependencyPart (DP), DependencyVectorPolytope (DVP),

FixedDimensions (FD))

    define set AllDimensions (AD) = (all dimensions in DP)
    define set SpanningDimensions (SD) = (all dimensions in DVP)
    define set NonspanningDimensions (ND) = AD - SD
    define set LowerboundFixedSpanningDimensions (LFSD) = SD ∩ FD
    define ordered set LowerboundUnfixedSpanningDimensions (LUSD)=

SD - LFSD

    define set LowerboundFixedNonspanningDimensions (LFND) = ND ∩ FD
    define ordered set LowerboundUnfixedNonspanningDimensions (LUND) =

ND - LFND

    order $\forall d_i \in LUSD \left| \left(d_i \prec d_{i+1} \Leftrightarrow \frac{|DVP_{d_i}-1|}{|DP_{d_i}-1|} < \frac{|DVP_{d_{i+1}}-1|}{|DP_{d_{i+1}}-1|} \right) \right.$ /* Incr. LR */
    order $\forall d_i \in LUND \left| \left(d_i \prec d_{i+1} \Leftrightarrow |DP_{d_i}| > |DP_{d_{i+1}}| \right) \right.$ /* Decr. $|DP_{d_i}|$ */
    define set UpperboundFixedSpanningDimensions (UFSD) = LFSD
    define ordered set UpperboundUnfixedSpanningDimensions (UUSD) =

LUSD in opposite order

    define set UpperboundFixedNonspanningDimensions (UFND) = LFND
    define ordered set UpperboundUnfixedNonspanningDimensions (UUND) =

LUND in opposite order

    define ordered set FixedDimensionsBestCase (FDBC) = [ ]
    define value LowerBound (LB) = 0
    define value UpperBound (UB) = 0
    define value LBFinished = 0
    define value UBFinished = 0
    if GuidingPrinciplesOK(DP, DVP, FD, SD, LUSD) {
      for each element $c_i \in FD$ {
        if FD_{c_i} == X { /* The current dimension is unfixed */
          if LBFinished $\neq$ 1 { /* No non-overlapping SDs so far used in LB calc. */
            if LUND $\neq \emptyset$ /* Remove unfixed nonspanning dimension from DP */
              LUND = LUND - (next dimension in LUND)
            else { /* Add unfixed spanning dimension to DVP */
              d_i = (next dimension in LUSD)
              $FDBC_{c_i} = d_i$
              if $LR_{d_i} > 1$ /* The dimension is non-overlapping */
                LBFinished = 1
              LUSD = LUSD - d_i
              $LB = LB + \left(|DVP_{d_i}| - 1 \right) * \prod_{\forall d_j \in LUSD} |DP_{d_j}| *$

$$\prod_{\forall d_j \in LFSD} |DP_{d_j}| * \prod_{\forall d_j \in LFND} |DP_{d_j}|$$

      } }
      else
        $FDBC_{c_i} = 0$

Figure 4.26.   Pseudo-code for dependency size estimation algorithm (cont'd in Fig. 4.27).

A very important part of the estimation algorithm is the utilization of the guiding principles. They are used to sort the unfixed dimensions in a best-case and worst-case order before the calculation of the upper and lower bounds on the dependency size.

if UBFinished $\neq 1$ { /* No non-overlapping SDs so far used in UB calc. */
   if UUSD $\neq \emptyset$ { /* Add unfixed spanning dimension to DVP */
     d_i = (next dimension in UUSD)
     if $LR_{d_i} > 1$ /* The dimension is non-overlapping */
      UBFinished = 1
     UUSD = UUSD - d_i

$$UB = UB + \left(\left|DVP_{d_i}\right| - 1\right) * \prod_{\forall d_j \in UUSD}\left|DP_{d_j}\right| *$$
$$\prod_{\forall d_j \in UUND}\left|DP_{d_j}\right| * \prod_{\forall d_j \in UFSD}\left|DP_{d_j}\right| * \prod_{\forall d_j \in UFND}\left|DP_{d_j}\right|$$

     }
   else /* Remove unfixed nonspanning dimension from DP */
     UUND = UUND - (next dimension in UUND)
}}
else /* The current dimension is fixed */
   $d_i = FD_{c_i}$
   $FDBC_{c_i} = d_i$
   if $d_i \in ND$ { /* Remove fixed nonspanning dimension from DP */
     LFND = LFND - d_i
     UFND = UFND - d_i
   }
   else { /* Add fixed spanning dimension to DVP */
     LFSD = LFSD - d_i
     UFSD = UFSD - d_i
     if LBFinished $\neq 1$ /* No non-overlapping SDs so far used in LB calc. */

$$LB = LB + \left(\left|DVP_{d_i}\right| - 1\right) * \prod_{\forall d_j \in LUSD}\left|DP_{d_j}\right| *$$
$$\prod_{\forall d_j \in LUND}\left|DP_{d_j}\right| * \prod_{\forall d_j \in LFSD}\left|DP_{d_j}\right| * \prod_{\forall d_j \in LFND}\left|DP_{d_j}\right|$$

     if UBFinished $\neq 1$ /* No non-overlapping SDs so far used in UB calc. */

$$UB = UB + \left(\left|DVP_{d_i}\right| - 1\right) * \prod_{\forall d_j \in UUSD}\left|DP_{d_j}\right| *$$
$$\prod_{\forall d_j \in UUND}\left|DP_{d_j}\right| * \prod_{\forall d_j \in UFSD}\left|DP_{d_j}\right| * \prod_{\forall d_j \in UFND}\left|DP_{d_j}\right|$$

     if $LR_{d_i} > 1$ /* The current dimension is non-overlapping */
      UBFinished = LBFinished = 1
}}}
return LB, UB, FDBC

Figure 4.27.   (cont'd from Fig. 4.26) Pseudo-code for dependency size estimation algorithm.

The algorithm starts by defining a number of sets that are used during the calculation of the LB and UB. They divide the dimensions into fixed and unfixed Spanning Dimensions (SDs) and Nonspanning Dimensions (NDs). The unfixed SDs are ordered according to increasing LR for the LB calculation and decreasing LR for the UB calculation. The unfixed NDs are ordered according to decreasing $|DP_{d_i}|$ for the LB calculation and increasing $|DP_{d_i}|$ for the UB calculation. The reasoning behind this ordering is based on the guiding principles.

Starting with the outermost nest level of the FD set, the contributions of each dimension to the LB and UB are calculated. If the current FD set element, c_i, is unfixed, represented by an X, the sets containing unfixed SDs and NDs are used for the contribution calculation. As long as unfixed NDs exist, nothing is added to the LB. The only action taken is the removal of the next dimension from the set of unfixed NDs. If there are no more unfixed NDs left, the next unfixed SD is removed from the LB set of unfixed SDs and used for the LB contribution calculation. Its $|DVP_{d_i}|$ minus 1 is multiplied with the $|DP_{d_i}|$ of all other remaining dimensions. This is demonstrated in Fig. 4.28 for the code example of Fig. 4.8. The contribution of the j dimension if placed outermost is indicated. The number of iteration nodes within the contribution rectangle equals $\left(\left|DVP_j\right| - 1\right) * |DP_k| * |DP_i| = (3-1) * 3 * 5 = 30$.

Note that this is not the ordering that would have been used for the LB calculation, but it demonstrates nicely how the contribution is calculated.

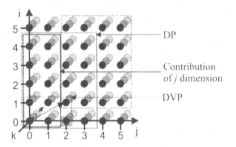

Figure 4.28.   Contribution of *j* dimension when placed at the outermost nest level.

In some cases, the LR of a dimension is larger than one, indicating that no overlap exists between the DP and the depending iteration nodes. This is the case for the *j* dimension in Fig. 4.25. The calculated contribution of this dimension will then include all elements of the remaining dimensions within the DP, as can be seen if Fig. 4.25 and Fig. 4.28 are compared. Consequently, there will be no additional contribution of the remaining dimensions, and the LB calculation can be terminated.

The calculation of the contribution of c_i to the UB is performed in a similar way as the LB calculation, except that unfixed SDs are placed outside the unfixed NDs since this corresponds to the worst case ordering. Since the SD with the largest LR is used first, any unfixed non-overlapping dimensions will immediately cause all elements in the DP to be included in its contribution, and hence the UB calculation can be terminated. Note that the LB calculation must continue, so the function does not terminate.

If the current FD set element c_i is fixed, represented by the dimension fixed at this position, the dimension itself is used for the contribution calculation. If it is an ND, it is simply removed from the sets containing the fixed NDs for the UB and LB calculations. If it is an SD, its $|DVP_{c_i}|$ minus 1 is multiplied with the $|DP_{d_i}|$ of all other remaining dimensions, as demonstrated in Fig. 4.28. This is done for both the UB and LB contribution calculation but unless the current dimension is fixed outermost, the content of the sets used for the calculation of the two contributions may be different. The resulting UB and LB contributions may then also be different. The contributions to both the UB and LB are terminated as soon as the contribution of a non-overlapping SD is calculated.

In addition to returning the upper and lower bounds on the storage requirement, the algorithm also returns an ordered set, *Fixed Dimensions Best Case* (FDBC), with the best case ordering of the unfixed dimensions. For each element c_i in the FD set, the corresponding element in FDBC is the dimension used for calculating the contribution to the LB. If the current element in FD is fixed, its dimension is directly inserted at the same location in FDBC. If the current element in FD is unfixed, the same dimension selected for LB calculation is inserted as the current element in FDBC. The FDBC can guide the designer towards an ordering of dimensions that is optimal for one or for a number of dependencies.

In a few rare situations, the guiding principles do not result in a guaranteed optimal ordering. This may happen if there exist internally fixed dimensions. In the following FD set, $FD = \{X, d_3, X, X, d_4\}$, dimension d_3 is fixed internally since it has an X on both sides. In most cases the guiding principles can be used even with internally fixed dimensions. The function *GuidingPrinciplesOK( )* in Fig. 4.26 checks for certain properties of the internally fixed dimensions, and returns FALSE if the guiding principles can not be used with a guaranteed result. See [275] for a detailed discussion of the situations where this is the case.

4.4.3 *Estimation with partial ordering among dimensions*

In section 4.2.2 it is assumed that dimensions are fixed at a given position in the FD set and thus at certain nest levels in the loop nest. It is however also interesting to be able to estimate upper and lower bounds on the dependency size when the only information available is a partial ordering among some of the dimensions. The constraint may for instance be that one dimension is not fixed outermost among (a subset of) the dimensions in the iteration space. This may for example happen when a dependency has three SDs, and one of them is negatively directed through use of the following array index in the consuming array access: [i+1]. If this dimension is fixed outermost among the SDs, the whole dependency will become negative, which is illegal for any flow dependency [290]. The designer still has full freedom in placing the three dimensions in the FD set and also regarding which one of the others (or both) to place outside the negative dimension. The main difference between a partially fixed ordering and a partial ordering among dimensions is thus that the partial ordering does not include fixation of dimensions at given nest levels in the FD set.

It is possible to use a different algorithm for the dependency size estimation when these kinds of constraints are given. An alternative approach using the same algorithm has been chosen here since it is usually only necessary to activate the existing algorithm twice with two different FD sets to obtain the results, one for the upper bound and one for the lower bound. The typical situation is that the partial ordering prohibits one dimension to be fixed innermost or outermost among a group of dimensions. Currently these dimensions have to follow each other in the best case and worst case ordering. The extension to allow other dimensions to have an LR that is larger than some and smaller than other of the LRs of the group is subject to future work. For the typical situation, where the partial ordering is among the dependency's entire set of SDs, these dimensions will anyway follow each other in the best case and worst case orderings. To find which FD sets to use, the LRs of the dimensions in the partial ordering is compared with the best-case and worst-case orderings. Depending on the result of this comparison, FDs are generated as follows:

1. If the partial ordering is in accordance with both the best case and worst case ordering, the estimation algorithm is simply activated with a fully unfixed FD (FD={...,X,X,...}). The resulting LB and UB returned from the algorithm is then immediately correct for this partial ordering.

2. If the dimension d_i with the largest LR among a group of dimensions is not to be ordered outermost among these dimensions, then the LB is found through an activation of the estimation algorithm with a fully unfixed FD (FD={...,X,X,X,...}). If there are n dimensions with larger LR than d_i, then the UB is found through an activation of the estimation algorithm with an FD where d_i is preceded by n+1 X's (FD={X,X,d_i,X...} if n=1).

3. If the dimension d_i with the largest LR among a group of dimensions is not to be ordered innermost among these dimensions, then the UB is found through an activation of the estimation algorithm with a fully unfixed FD (FD={...,X,X,X,...}). If there are n dimensions with larger LR than d_i, then the LB is found through an activation of the estimation algorithm with an FD where d_i is followed by n+1 X's (FD= {...X,d_i,X,X} if n=1).

4. If the dimension d_i with the smallest LR among a group of dimensions is not to be ordered outermost among these dimensions, then the UB is found through an activation of the estimation algorithm with a fully unfixed FD (FD={...,X,X,X,...}). If there are n dimensions with smaller LR than d_i, then the LB is found through an activation of the estimation algorithm with an FD where d_i is preceded by n+1 X's (FD= {X,X,d_i,X...} if n=1).

5. If the dimension d_i with the smallest LR among a group of dimensions is not to be ordered innermost among these dimensions, then the LB is found through an activation of the estimation algorithm with a fully unfixed FD (FD={...,X,X,X,...}). If there are n dimensions with larger LR than d_i, then the UB is found through an activation of the estimation algorithm with an FD where d_i is followed by n+1 X's (FD= {...X,d_i,X,X} if n=1).

Table 4.6 summarizes the use of FDs for the four situations where the partial ordering is not in accordance with both the best case and worst case ordering.

The reasoning behind the use of these FD sets is based on the guiding principles. When the partial order is in accordance with both the best case and worst case ordering a single fully unfixed FD can be used to find both the UB and LB since both the worst case and best case ordering used by the estimation algorithm is legal. For the other cases, the partial ordering is in accordance with

	d_i not outermost in group	d_i not innermost in group
d_i has largest LR in group	$FD_{LB} = \{...,X,X,X,...\}$ $FD_{UB} = \{X,X,d_i,X,...\}$ if n=1	$FD_{LB} = \{...,X,d_i,X,X\}$ if n=1 $FD_{UB} = \{...,X,X,X,...\}$
d_i has smallest LR in group	$FD_{LB} = \{X,X,d_i,X,...\}$ if n=1 $FD_{UB} = \{...,X,X,X,...\}$	$FD_{LB} = \{...,X,X,X,...\}$ $FD_{UB} = \{...,X,d_i,X,X\}$ if n=1

Table 4.6. FDs used for dependency size estimation with partial ordering among dimensions.

either the best case or the worst case ordering. The fully unfixed FD can then be used to find the UB or LB respectively since the ordering used by this part of the estimation algorithm is legal. The other ordering used by the estimation algorithm places dimension d_i illegally outermost or innermost among the dimensions in the group. The legal best case or worst case ordering can be reached if d_i is swapped with the dimension in the group ordered next to it. This is achieved through a fixation of d_i at this dimension's position in the worst case or best case ordering. The resulting FDs are shown in Table 4.6 for the cases where d_i is normally placed second innermost or second outermost (n=1) in the ordered set of all dimensions in the Dependency Part.

Given that the dimensions are already sorted according to their LR, it would also be possible to use fully fixed FDs to find the UB and LB. This would however prohibit estimation with a combination of the partially fixed ordering methodologies presented in section 4.4.2 and the partial ordering among dimensions presented here. To have the freedom of this combination for future work, the FDs are kept as unfixed as possible also for estimations with a partial ordering among dimensions.

4.4.4   Automated estimation on three dimensional code examples

In this section, estimates are performed on the three dimensional example presented in Fig. 4.8. It is quite simple, so the full details of the estimation algorithm can be illustrated. Estimations using larger and realistic design examples can be found in section 4.5.

Let us go back to the example code in Fig. 4.8. If no execution ordering is specified, FD={X,X,X}, the algorithm in Fig. 4.26 orders all SDs according to their LR. $LR_i = (2-1)/(5-1) = 1/4$ and $LR_j = (3-1)/(4-1) = 2/3$, so LUSD={i,j} for the LB calculation and UUSD={j,i} for the UB calculation (see Fig. 4.26 for an explanation of the set abbreviations). Since k is the only ND, the orderings of the LUND and UUND sets are trivial. With $c_1 = X$ the LB calculation simply removes the unfixed ND k from the LUND set. The UB calculation selects the first UUSD dimension, j. The UFSD and UFND sets are empty, so together with the remaining UUSD element, i, and the UUND element, k, the resulting temporary UB value is $UB_{tmp} = UB_{tmp} + (|DVP_j| - 1) * |DP_i| * |DP_k| = 0 + 2*5*3 = 30$. With $c_2 = X$, the LUND set is empty, and the LB calculation selects the first LUSD dimension, i. The only nonempty set is for the LB calculation is LUSD resulting in a temporary LB value of $LB_{tmp} = LB_{tmp} + (|DVP_i| - 1) * |DP_i| = 0 + 1*4 = 4$. The UB calculation selects the next UUSD dimension, i, resulting in a temporary UB value of $UB_{tmp} = UB_{tmp} + (|DVP_i| - 1) * |DP_k| = 30 + 1*3 = 33$. With $c_3 = X$ the LB calculation selects the next LUSD dimension, j, resulting in a final LB value of $LB = LB_{tmp} + (|DVP_j| - 1) = 4 + 2 = 6$. The UB calculation simply removes the unfixed ND k from the UUND set leaving the previously calculated UB as the final UB. A graphical description of the A-array elements that make up the upper and lower bounds is given in Fig. 4.29.

Assume now that the k dimension is specified as the outermost dimension while the ordering between the i and j dimensions are still unresolved, FD={k,X,X}. The LUSD and UUSD sets are then initially ordered the same way as before while the k dimension is moved from the LUND and UUND to the LFND and UFND sets respectively. With $c_1 = k$, the LB and UB calculations simply removes k from the LFND and UFND sets. During the subsequent calculation of the UB, all but the UFSD set is empty and the calculation is therefore reduced to $UB_{tmp} = UB_{tmp} + (|DVP_i| - 1) * |DP_i| = 0 + 2*5 = 10$, and $UB = UB_{tmp} + (|DVP_i| - 1) = 10 + 1 = 11 = 11$. The LB calculation already assumed NDs to be placed outermost. Consequently, placing the k dimension outermost gives an unchanged LB and a decreased UB.

For the next partial fixation, the j dimension is placed innermost and the k dimension second innermost, FD={X,k,j}. The resulting sets are UFSD = LFSD = {j}, UFND = LFND = k, UUSD =

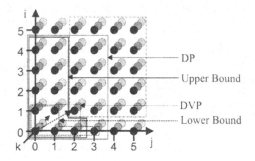

Figure 4.29.   UB and LB with no execution ordering fixed.

$LUSD = \{i\}$, and UUND = LUND = $\emptyset$. With $c_1 = X$ and LUND = $\emptyset$ the LB calculation is $LB_{tmp} = LB_{tmp} + (|DVP_i| - 1) * |DP_j| * |DP_k| = 0 + 1 * 4 * 3 = 12$. Since UUSD $\neq \emptyset$ the UB calculation is $UB_{tmp} = UB_{tmp} + (|DVP_i| - 1) * |DP_j| * |DP_k| = 0 + 1 * 4 * 3 = 12$. $c_2 = k$ results in removal of k from the LFND and UFND sets. Finally $c_2 = j$ gives $LB = LB_{tmp} + (|DVP_j| - 1) = 12 + 2 = 14$ and $UB = UB_{tmp} + (|DVP_j| - 1) = 12 + 2 = 14$. Note that given what is actually a fully fixed ordering since only one dimension is unfixed, the UB and LB calculations are identical.

Consider the situation where the j dimension is fixed internally, FD= $\{X,j,X\}$. The resulting sets are UFSD = LFSD = $\{j\}$, UUSD = LUSD = $\{i\}$, UUND = LUND = $\{k\}$, and UFND = LFND = $\emptyset$. With $c_1 = X$ the LB calculation simply removes the k dimension from the LUND set. The UB calculation uses the i dimension from the UUSD set to find $UB_{tmp} = UB_{tmp} + (|DVP_i| - 1) * |DP_j| * |DP_k| = 0 + 1 * 4 * 3 = 12$. With $c_2 = j$ the LB calculation is $LB_{tmp} = LB_{tmp} + (|DVP_j| - 1) * |DP_k| = 0 + 2 * 5 = 10$ and the UB calculation is $UB_{tmp} = UB_{tmp} + (|DVP_j| - 1) * |DP_k| = 12 + 2 * 3 = 18$. Finally with $c_3 = X$ the LB calculation uses the i dimension from the LUSD set to find $LB_{tmp} = LB_{tmp} + (|DVP_i| - 1) = 10 + 1 = 11$. The UB calculation simply removes the k dimension from the UUND set leaving the previously calculated UB as the final UB.

A partial ordering is now given forcing the j dimension not to be placed innermost among the j dimension and i dimension. Since $LR_j > LR_i$ this is not in accordance with the best case ordering and FDs are generated as described in clause 3 of section 4.4.3. No dimensions with LR larger than j exist, so n=0 giving FDLB = $\{X,j,X\}$ and FDUB = $\{X,X,X\}$. Both of these have been detailed above, and the resulting LB found using FDLB = $\{X,j,X\}$ is 11, while the UB found using FDUB = $\{X,X,X\}$ is 33.

Table 4.7 summarizes the estimation results for the different execution orderings and methodologies; upper and lower bounds from the new methodology described here, the methodology presented in [39], and manually calculated exact worst-case (WC) and best-case (BC) numbers. Since no orthogonalization is needed for this example, these numbers are identical with the estimates. For larger real-life code, the exact numbers can of course not be computed manually any longer since they require a full exploration of all alternative orderings. The table also includes the estimation results of a stepwise fixation of the optimal ordering; j innermost, i second innermost, and k outermost. Note the difference between the estimates found here and the more inaccurate results in section 4.3.5. For the partially fixed ordering with j innermost, the estimated upper bound is reduced from 18 in section 4.3.5 to 14 here. Fig. 4.30 shows the iteration nodes that are included in this accurate estimate compared to the iteration nodes included in Fig. 4.19. In Fig. 4.30 it is only for the i-value 0 that iteration nodes with k-values different than 0 is included.

Assume now that the example code of Fig. 4.8 is extended with a third statement:

S.3 if $(i > 1)$ C[i][j][k] = h(A[i-2][j][k]);

The new dependency has only one SD, the i dimension. As was discussed in section 4.3.7, the lowest dependency size is reached when the SDs are placed innermost. Table 4.8 gives the estimation results of the dependency between S.1 and S.3 with no execution ordering fixed, with i innermost, and with j innermost. It also shows the size estimates of the dependency between S.1 and S.2 with i fixed innermost. The size penalty of fixing i innermost for this dependency is smaller (11-6=5) then

Fixed Dimension(s)	LB	UB	[39]	Exact BC/WC
FD = {X,X,X}	6	33	60	6/33
FD = {k,X,X}	6	11	60	6/11
FD = {X,k,j}	14	14	60	14/14
FD = {X,j,X}	11	18	60	11/18
j not innermost among i and j	11	33	60	11/33
FD = {X,X,j}	6	14	60	6/14
FD = {X,i,j}	6	6	60	6/6
FD = {k,i,j}	6	6	60	6/6

Table 4.7.   Dependency size estimates of code example in Fig. 4.8.

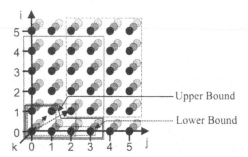

Figure 4.30.   UB and LB with SD *j* fixed innermost.

Dependency	Fixed innermost	LB	UB	[39]	Exact BC/WC
S.1 - S.3	None	2	36	90	2/36
S.1 - S.3	i	2	2	90	2/2
S.1 - S.3	j	12	36	90	12/36
S.1 - S.2	i	11	31	90	11/31

Table 4.8.   Dependency size estimates of extended code example of Fig. 4.8.

that of fixing j innermost for the S.1-S.3 dependency (12-2=10). Using the new dependency size estimation algorithm, it is therefore already with such small amount of information available possible to concluded that the i dimension should be ordered innermost. As can be seen from column [39] in Table 4.7 and Table 4.8, not taking the execution ordering into account results in large overestimates. If a technique requiring a fully fixed execution ordering is used, N! alternatives have to be inspected where N is the number of unfixed dimensions. N can be large (>4) for real life examples. Note that N is not the number of dimensions in the arrays, but the number of dimensions in the loop nests surrounding them. Even algorithms with only two- and three-dimensional arrays often contain nested loops up to six levels deep.

4.5   ESTIMATION ON REAL-LIFE APPLICATION DEMONSTRATORS

In this section the usefulness and feasibility of the estimation methodology is demonstrated on several real life applications (test-vehicles).

4.5.1   MPEG-4 motion estimation kernel

4.5.1.1   Code description and external constraints.   MPEG-4 is a standard from the *Moving Picture Experts Group* for the format of multi-media data-streams in which audio and video objects

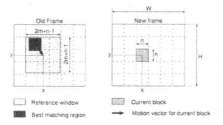

Figure 4.31. Motion estimation kernel: principle operation.

can be used and presented in a highly flexible manner [380, 482]. An important part of the coding of this data-stream is the motion estimation of moving objects (Fig. 4.31). See [70] for a more detailed description of the motion estimation part of the standard. This real life application will now be used to show how storage requirement estimation can be used during the design trajectory. A part of the code where major arrays are produced and consumed is given in Fig. 4.32.

The sad-array is the only one both produced and consumed within the boundaries of the loop nest. It is produced by instruction S.1, S.2, and S.3, and consumed by instructions S.2, S.3, and S.4. For this example, the dependency analysis technique presented in [39] is utilized. This results in a number of basic sets and their dependencies as shown in the corresponding data-flow graph in Fig. 4.33. These basic sets can then be placed in a common iteration space.

```
for (y_s=0; y_s<=31; y_s++)
  for (x_s=0; x_s<=31; x_s++)
    for (y_p=0; y_p<=15; y_p++)
      for (x_p=0; x_p<=15; x_p++)
S.1   if ((x_p == 0)&(y_p == 0)) sad[y_s][x_s][y_p][x_p]=
          f1(curr[y_p][x_p], prev[y_s+y_p][x_s+x_p]);
S.2   else if ((x_p == 0)&(y_p != 0)) sad[y_s][x_s][y_p][x_p]=
          f2(sad[y_s][x_s][y_p-1][15], curr[y_p][x_p], prev[y_s+y_p][x_s+x_p]);
S.2   else sad[y_s][x_s][y_p][x_p]=
          f3(sad[y_s][x_s][y_p][x_p-1], curr[y_p][x_p], prev[y_s+y_p][x_s+x_p]);
for (y_s=0; y_s<=31; y_s++)
  for (x_s=0; x_s<=31; x_s++)
S.4   result[y_s][x_s] = f4(sad[y_s][x_s][15][15]);
```

Figure 4.32. MPEG-4 motion estimation kernel.

For the remainder of the MPEG-4 example, some typical examples of external constraints are assumed. The curr[y_p][x_p] pixels are presented sequentially and row first (curr[0][0], curr[0][1], ... curr[0][15], curr[1] [0] ...) at the input. The prev-array is already stored in local memory from the previous calculation and does not need to be taken into account. There are no output constraints, so the result can be presented at the output in any order.

The loops in the upper nest of the code in Fig. 4.32 can be interchanged in 4!=24 ways. Because of the symmetry in the loops, only six of these have to be investigated. Interchange between y_s & x_s OR y_p & x_p leads to the same data in-place mapping opportunity. An interchange between y_p and x_p would also require additional transformations, since there exists two negative dependencies in the current code, implying a partial ordering between y_p and x_p with y_p outside x_p. According to these restrictions, the six interchange alternatives are shown in Table 4.9. The detection of these symmetry properties is not straightforward however, so the typical search space for these kind of applications is actually N!, where N is the number of loop nests.

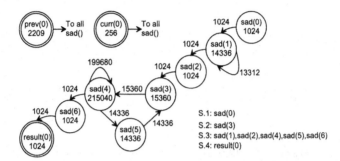

S.1: sad(0)
S.2: sad(3)
S.3: sad(1),sad(2),sad(4),sad(5),sad(6)
S.4: result(0)

Figure 4.33.   Data-flow graph for MPEG-4 motion estimation kernel.

↓ Outermost			Innermost ↓
1) y_s	x_s	y_p	x_p
2) y_p	x_p	y_s	x_s
3) y_s	y_p	x_s	x_p
4) y_p	y_s	x_p	x_s
5) y_s	y_p	x_p	x_s
6) y_p	y_s	x_s	x_p

Table 4.9.   Interchange alternatives.

4.5.1.2   Individual dependencies.   Due to the input constraint, the storage requirement for the curr-array can be made as small as possible using interchange alternative 2) from Table 4.9. Indeed, all calculations for each pixel can be completed before the next pixel arrives, and only one storage location is needed for the curr-array. In any of the other implementations, the curr-array elements must be buffered, since they are needed in bursts. This will require 16x16=256 storage locations. As a first design step, it is therefore natural to investigate the storage requirement of the sad-array dependencies given that the execution order is optimized for the curr-array. y_p is then first fixed at the outermost nest level. The corresponding estimation results are listed in Table 4.10 together with the estimation results without any execution ordering fixed. The table also contains the Spanning Dimensions (SDs) for each dependency. For the two dependencies with y_p and x_p as SDs, y_p is an SD due to boundary conditions causing the negative dependencies mentioned in the previous section. The last dependency does not have an SD since sad(6) and result(0) are placed in such a way in the common iteration space that elements are produced and consumed at the same iteration nodes. The array elements are therefore consumed immediately after production, and the storage requirement is one independently of the chosen execution ordering.

The maximal increase in the Lower Bound (LB) on the storage requirement for a dependencies with y_p as the outermost loop (1024 - 1 = 1023), exceeds the storage requirement for the entire curr-array (256). Consequently, it is possible to rule out any interchange alternative with y_p as the outermost loop, since the lower bound storage requirement with no ordering constraints indicates that a better solution exists. This leaves interchange method 1), 3), and 5), each with y_s as the outermost loop.

y_s is not a Spanning Dimension (SD) for any of the dependencies, so with a fixation of y_s at the outermost nest level the LB on the storage requirements will stay the same as for the unordered situation. The Upper Bound (UB) requirements are reduced however, by a factor 32 everywhere (except for sad(6) → result(0)) since y_s can be removed from every Dependency Part (DP).

The guiding principles can now be utilized to find the ordering of the remaining dimensions. NDs reduce the UB without altering the LB if they are placed outermost, while SDs with similar reasoning should be placed innermost. In this case, the guiding principles indicate that the lowest storage requirement for the dependencies can be achieved when y_p and x_p are ordered as the innermost loops, as in interchange alternative 1). Table 4.11 shows the change of estimated UB and

	SD	No ordering		y_p outermost	
		LB	UB	LB	UB
sad(0) → sad(1)	x_p	1	1024	1	1024
sad(1) → sad(1)	x_p	1	1024	1	1024
sad(1) → sad(2)	x_p	1	1024	1	1024
sad(2) → sad(3)	y_p, x_p	1	1024	1024	1024
sad(3) → sad(4)	x_p	1	15360	1	1024
sad(4) → sad(4)	x_p	1	15360	1	1024
sad(4) → sad(5)	x_p	1	14336	1	1024
sad(5) → sad(3)	y_p, x_p	1	1024	1024	1024
sad(4) → sad(6)	x_p	1	1024	1	1024
sad(6) → result(0)	-	1	1	1	1
curr(0)		1	256	1	256

Table 4.10. Estimation results for the sad-array and curr- array of the MPEG-4 motion estimation kernel.

LB as the execution order is gradually fixed. The UB and LB gradually converge and finally meet when the ordering is fully specified.

4.5.1.3 Simultaneously alive dependencies. As discussed in section 4.3.6, multiple dependencies may be alive simultaneously during the execution of an application. The total storage requirement of the application is then decided by the maximally combined size of simultaneously alive dependencies at any point in time during the execution. The technique presented in section 4.3.6 will now be employed on the MPEG-4 Motion Estimation kernel to find the application's total storage requirement.

Table 4.12 lists the start and end points for the Extended Dependency Parts (EDPs) of the dependencies in the four dimensions. sad0_1_s indicates the starting point for the EDP of the dependency from sad(0) to sad(1) while sad0_1_e indicates the end point.

All EDPs overlap in the y_s and x_s dimensions. Since they are NDs for all the dependencies however, they have no influence on the degree of simultaneous aliveness. As discussed in section 4.3.6, NDs can at worst cause dependencies to alternate in being alive. These dimensions are therefore ignored in the sequel. Using the algorithm in Fig. 4.22, Table 4.12 is inspected to reveal groups of possibly simultaneously alive dependencies. The outcome of the removal of fully covered groups and a sorting in accordance with the combined upper bound size of their dependencies is shown in Table 4.13.

The final step is now to inspect each group to find the dependencies that are really alive simultaneously for a given partial ordering among dimensions. The negative dependencies force x_p to be fixed inside y_p. Apart form this constraint, the ordering of the dimensions, including y_s and x_s, can be chosen arbitrarily. Table 4.14 summarizes in which dimensions the dependencies of the largest group are overlapping. Certain dependencies are not counted as possibly simultaneously alive, even if their EDPs are overlapping. This is the case if one of the dependencies is directly depending on the other and the overlapping between them only covers the extension of the EDP for one dependency and the iteration nodes where the dependency has not reach its full size for the other. Direct in-place mapping can be performed, and the two dependencies can directly be considered as not simultaneously alive. In Table 4.14, this is indicated with Not sim. Most of the remaining overlap either occurs only in NDs, e.g. between sad4_4 and sad5_3, or in an SD that, due to the specified partial ordering, is known to be placed inside a non-overlapping dimension, e.g. between sad2_3 and sad4_5. These overlaps will not cause simultaneous aliveness between the dependencies. The remaining overlap to be investigated further is thus the one between sad2_3 and sad4_4. y_p is here only an SD because of the boundary condition of the negative dependency sad2_3. It can be ignored when x_p is fixed inside y_p. The conclusion is thus that no dependencies are alive simultaneously in this group. Similar reasoning reveals that this is indeed the case for the other groups as well. This

	SD	y_s outermost		x_s 2nd outermost	
		LB	UB	LB	UB
sad(0) → sad(1)	x_p	1	32	1	1
sad(1) → sad(1)	x_p	1	32	1	1
sad(1) → sad(2)	x_p	1	32	1	1
sad(2) → sad(3)	y_p, x_p	1	32	1	1
sad(3) → sad(4)	x_p	1	480	1	15
sad(4) → sad(4)	x_p	1	480	1	15
sad(4) → sad(5)	x_p	1	448	1	14
sad(5) → sad(3)	y_p, x_p	1	32	1	1
sad(4) → sad(6)	x_p	1	32	1	1
sad(6) → result(0)	-	1	1	1	1
curr(0)		256	256	256	256

	SD	y_p 3rd outermost		x_p 4th outermost	
		LB	UB	LB	UB
sad(0) → sad(1)	x_p	1	1	1	1
sad(1) → sad(1)	x_p	1	1	1	1
sad(1) → sad(2)	x_p	1	1	1	1
sad(2) → sad(3)	y_p, x_p	1	1	1	1
sad(3) → sad(4)	x_p	1	1	1	1
sad(4) → sad(4)	x_p	1	1	1	1
sad(4) → sad(5)	x_p	1	1	1	1
sad(5) → sad(3)	y_p, x_p	1	1	1	1
sad(4) → sad(6)	x_p	1	1	1	1
sad(6) → result(0)	-	1	1	1	1
curr(0)		256	256	256	256

Table 4.11.   Estimation results for sad-array and curr-array of the MPEG-4 motion estimation kernel with stepwise fixation of execution ordering.

Dim	0	1	...	14	15
y_s	All_s				All_e
x_s	All_s				All_e
y_p	sad0_1_s	sad3_4_s		sad4_5_e	sad4_6_s
	sad1_1_s	sad4_4_s			sad6_res0_s
	sad1_2_s	sad5_5_s			sad5_3_e
	sad2_3_s	sad5_3_s			sad3_4_e
	sad0_1_e	sad2_3_e			sad4_4_e
	sad1_1_e				sad4_6_e
	sad1_2_e				sad6_res0_e
x_p	sad0_1_s	sad1_1_s		sad1_2_s	sad2_3_s
	sad3_4_s	sad4_4_s		sad4_5_s	sad5_3_s
	sad2_3_e	sad0_1_e		sad4_6_s	sad6_res0_s
	sad5_3_e	sad3_4_e		sad1_1_e	sad1_2_e
				sad4_4_e	sad4_5_e
					sad4_6_e
					sad6_res0_e

Table 4.12.   Start and end points for EDPs in all dimensions.

Groups	Size
sad2_3,sad3_4,sad4_4,sad4_5,sad5_3	47104
sad3_4,sad4_4,sad5_3,sad4_6,sad6_res0	32769
sad1_1,sad4_4,sad1_2,sad4_5,sad4_6	32768
sad0_1,sad3_4,sad1_1,sad4_4	32768
sad1_2,sad4_5,sad4_6,sad2_3,sad5_3,sad6_res0	18433
sad0_1,sad3_4,sad2_3,sad5_3	18432
sad0_1,sad1_1,sad1_2,sad2_3	4096

Table 4.13.   Groups of overlapping EDPs after sorting.

	sad3_4	sad4_4	sad4_5	sad5_3
sad2_3	Not sim.	SD=y_p	SD=x_p	SD=x_p
sad3_4		Not sim.	Not sim.	SD=y_p,x_p
sad4_4			Not sim.	ND=y_p
sad4_5				Not sim.

Table 4.14.   Overlapping dimensions for the largest group.

is however very important information for the designer, since the focus can then be solely on the task of minimizing the size of individual dependencies. Since this conclusion can be drawn at a very early stage of the design process, decisions regarding whether to use a certain algorithm can be taken without having to go to a full detailed implementation.

Fig. 4.34 summarizes estimation results using different techniques and partially fixed execution ordering. The leftmost column shows the combined declared size of the sad-array and curr-array. This is the memory size (262400) the designer would have to assume if no estimation and optimization tools were available. The next column shows the result found using the estimation technique described in [39] (45296). This result will be the same independently of the execution ordering. Finally the columns marked a) through e) show the size estimates found for a number of partially fixed execution orderings using the methodology presented in this chapter. Both results for the combined size of individual dependencies and the size of simultaneously alive dependencies are reported. With no execution ordering fixed, the UB on the combined size of individual dependencies is 51457 while the simultaneously alive dependencies have an UB on the combined size of 15616. This shows the importance of both steps in the estimation methodology. Using the guiding principles and feedback from the estimation tool, the designer is finally able to reach a storage requirement of 257 when the full execution ordering is fixed.

In this example, the prev-array has not been taken into account. It is a large array however (2209 memory locations), and it may very well be that not all of it should be placed in local memory at the same time. Another important design issue would then come into play, i.e. the possibility for data reuse. How is it possible to avoid repeated copying of the same data from background memory and still keep the local memory as small as possible? This step in the overall design methodology is described in chapter 5. It requires estimates of the size of all intermediate data copies to be able to perform an efficient exploration of the search space. A fast estimation tool is hence also here indispensable.

4.5.2 SVD updating algorithm

The results from the storage requirement estimation can also be used for feedback during interactive or tool driven global loop reorganization, [123]. This will now be demonstrated using the *Singular Value Decomposition* algorithm [376] for instance required in beamforming for antenna systems. The algorithm continuously updates matrix decompositions as new rows are appended to a matrix. Fig. 4.35 shows the two major arrays, R and V, and the important loop nest and statements for their

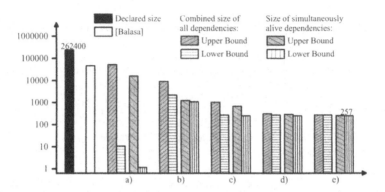

Figure 4.34.   Estimation results for MPEG-4 motion estimation using various techniques and partially fixed execution orderings. a) No ordering, b) y_p outermost, c) y_s outermost, d) x_s second outermost, e) y_p third outermost (fully fixed).

production and consumption. Two smaller arrays, *phi* and *theta*, used during the production of the *R* and *V* arrays, are also shown.

```
    for (k=0; k<=n-2; k++){
S.1   theta[k] = f1( R[...][...][2*k] );
S.2   phi[k] = f2( R[...][...][2*k], theta[k] ) ;
        for (i=0; i<=n-1; i++)
          for (j=0; j<=n-1; j++)
          ...
S.3      else R[i][j][2*k+1] = f3( R[i][j][2*k], theta[k] );
        for(i=0;i<=n-1;i++)
          for(j= 0;j<=n-1;j++) {
          ...
S.4      else R[i][j][2*k+2] = f6( R[i][j][2*k+1], phi[k] );
          ...
S.5      else V[i][j][k+1] = f9( V[i][j][k], phi[k] );
        }}
```

Figure 4.35.   USVD algorithm, diagonalization loop nest

Table 4.15 a) shows estimation results for a number of the dependencies in the code with no execution ordering fixed (n=10). The k dimension is the only SD for all the dependencies, and according to the guiding principles, this dimensions should therefore be fixed at the innermost nest level to minimize the storage requirement. However, due to data dependencies not explicitly shown in Fig. 4.35, the k dimension must be placed outermost during the production of the R-array. Table 4.15 b) gives the estimation results for this partial ordering. The upper and lower bounds of the three largest dependencies converge at their previous upper bound of 100. An investigation of the global storage requirement reveals that the maximal combined size occurs while dependencies S.1 → S.3, S.2 → S.5, S.4 → S.3, and S.5 → S.5 are alive simultaneously. This gives rise to a storage requirement of 202 memory locations, as shown in Fig. 4.36.

The fixation of the k dimension at the outermost nest level is not required for the production of the V-array, as long as the necessary phi-values are available. A comparison of the resulting storage requirement for an alternative loop organization, where a loop body splitting places the V-array in a separate loop nest, is therefore needed. Table 4.15 c) shows estimation results after this reorganization and with the partial ordering that k is not to be fixed outermost in the new V-array

	S.1 → S.3	S.2 → S.5	S.3 → S.4	S.4 → S.3	S.5 → S.5
a)	1/9	1/9	1/100	1/100	1/100
b)	1/1	1/1	100/100	100/100	100/100
c)	1/1	9/9	100/100	100/100	1/10
d)	1/1	9/9	100/100	100/100	10/10

Table 4.15.   LB/UB for dependencies in the USVD algorithm with a) no execution ordering fixed, b) k fixed outermost without loop body split, c) k fixed outermost in R-loop nest and k not fixed outermost in V-loop nest (with loop body split), d) fully fixed ordering for both loop bodies.

nest. Finally Table 4.15 d) shows the estimation results for a fully fixed legal ordering with loop body split.

Due to the loop body split, dependencies S.4 → S.3 and S.5 → S.5 are not alive simultaneously anymore. The global storage requirement is hence approximately halved to 110 memory locations as shown in Fig. 4.36. The V-array is placed alongside the R-array, since they are not alive simultaneously. Information regarding the global storage requirement is vital for the designer when comparing different implementation alternatives. The data dependency size estimates thus guides the designer through the early steps of the design trajectory towards a solution with low storage requirements.

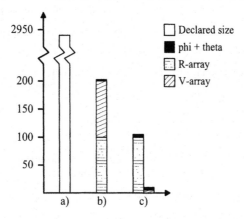

Figure 4.36.   Estimated storage requirement for the USVD algorithm: a) Declared size, b) without loop body split, c) with loop body split (n=10).

4.5.3   Cavity detection algorithm

4.5.3.1   Code description.   The estimation methodology has been applied to a cavity detection algorithm used for detection of cavity perimeters in medical imaging. Fig. 4.37 shows the important parts of the code after pruning. A pseudo t dimension is inserted to generate of a common iteration space. As a starting point, the pseudo t dimension is defined so that statements placed in separate loop nests in the original code are executed sequentially. The corresponding data flow graph is given in Fig. 4.38. Each node depicts the *iteration domain* (ID) of a statement, while the edges represent dependencies between the IDs. The dependencies are enumerated to ease the following discussion. The t dimension opens for transformations such as loop merging. The storage requirement of the transformed code with alternative placements and orderings can thus be compared to the storage requirement of the original ordering. In this example, the focus is on parts with major impact on the storage requirement, so boundary conditions are ignored. It is assumed that the input image can be presented at the input of the application in any order.

The STOREQ CAD tool was used extensively during the exploration of the cavity detection algorithm. As allowed by the tool, some of the dependencies in the code are non-uniform. Up to

```
#define GB 1
#define SIZE_TOT (2*GB)+1
#define NB 8
#define N 480
#define M 640
int x_offset[NB]={1,1,1,0,0,-1,-1,-1};
int y_offset[NB]={1,0,-1,1,-1,1,0,-1};
for (t=0; t<=5; t++)/* Dimension 1 */
  for (x=0; x<=N-1; x++) /* Dimension 2 */
    for (y=0; y<=M-1; y++) /* Dimension 3 */
      for (k=0; k<=NB-1; k++) { /* Dimension 4 */
        if (t==0 & x>=GB & x<=N-1-GB & y>=GB & y<=M-1-GB &
                                                       k<=SIZE_TOT-1)
S.1     gauss_x_compute[x][y][k+1] = f1( gauss_x_compute[x][y][k],
                                              image_in[x+k-1][y] );
        if (t==0 & k==SIZE_TOT-1 & x>=GB & x<=N-1-GB & y>=GB
                                                       & y<=M-1-GB)
S.2     gauss_x_image[x][y] = f2( gauss_x_compute[x][y][SIZE_TOT] );
        if (t==1 & x>=GB & x<=N-1-GB & y>=GB & y<=M-1-GB &
                                                       k<=SIZE_TOT-1)
S.3     gauss_xy_compute[x][y][k+1] = f3( gauss_xy_compute[x][y][k],
                                          gauss_x_image[x][y+k-1] );
        if (t==1 & k==SIZE_TOT-1 & x>=GB & x<=N-1-GB & y>=GB
                                                       & y<=M-1-GB)
S.4     gauss_xy_image[x][y] = f4( gauss_xy_compute[x][y][SIZE_TOT]);
        if (t==2 & x>=GB & x<=N-1-GB & y>=GB & y<=M-1-GB)
S.5     maxdiff_compute[x][y][k+1] = f5( gauss_xy_image[x+x_offset[k]]
                  [y+y_offset[k]], gauss_xy_image[x][y], maxdiff_compute[x][y][k] );
        if (t==2 & k==NB-1 & x>=GB & x<=N-1-GB & y>=GB &
                                                       y<=M-1-GB)
S.6     comp_edge_image[x][y] = f6( maxdiff_compute[x][y][NB] );
        if (t==3 & x>=GB & x<=N-1-GB & y>=GB & y<=M-1-GB)
S.7     out_compute[x][y][k+1] = f7( comp_edge_image[x+x_offset[k]]
                  [y+y_offset[k]], comp_edge_image[x][y], out_compute[x][y][k] );
        if (t==3 & k==NB-1 & x>=GB & x<=N-1-GB & y>=GB &
                                                       y<=M-1-GB)
S.8     image_out[x][y] = f8( out_compute[x][y][NB] );
      }
```

Figure 4.37.   Code for Cavity Detection example.

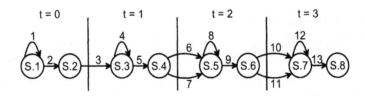

Figure 4.38.   Data-flow graph for Cavity Detection example.

four extreme DVs are needed for their description as was discussed in section 4.3.2.2. Statement S.3 has for example a non-uniform accessing of array gauss_x_image. Depending on the k value, an array element is accessed that was produced at an iteration point with a y value smaller, equal, or larger than the current y value. This gives rise to two extreme DVs, one in each direction along the y dimension and with different lengths in the k dimension. Furthermore, a number of negative dependencies exist in the code. Table 4.16 summarizes the negative dimensions, the SDs, and the

Dep.	DV1	DV2	DV3	DV4
1	-t -x -y k			
2	-t -x -y -k			
3	t -x y k̲	t -x y̲ -k		
4	-t -x -y k			
5	-t -x -y -k			
6	t -x -y k̲	t -x -y k		
7	t x̲ y̲ k	t x̲ y -k	t x y̲ k	t x y k
8	-t -x -y k			
9	-t -x -y -k			
10	t -x -y k̲			
11	t x̲ y̲ k	t x̲ y k̲	t x y̲ k̲	t x y -k
12	-t -x -y k			
13	-t -x -y -k			

Table 4.16. NDs (indicated with -), SDs, and negative dimensions (underlined) for each DV of the dependencies in the cavity detection code.

NDs for each DV of the dependencies. This information is automatically dumped from the STOREQ tool. The dependency enumeration is in accordance with Fig. 4.38.

4.5.3.2 Storage requirement estimation and optimization. It will now be shown how the STOREQ tool can be used to rearrange the common loop nest and to determine the execution ordering that gives an optimized storage requirement for the application. The first run of STOREQ without any ordering fixed shows that all dependencies that do not cross a t-line in Fig. 4.38, has the k dimension placed innermost in its optimal ordering.

For dependencies with t as an SD, that is all dependencies crossing a t-line in Fig. 4.38, the upper and lower bounds converge at the previous upper bound if t is fixed outermost. Due to the loop splitting caused by the t dimension, none of these dependencies is then alive simultaneously. Their individual sizes are hence determining the overall storage requirement. To reduce the storage requirement, the common iteration space must be rearranged so that the dependencies do not have t as an SD (do not cross a t-line). The removal of the t dimension from the set of SDs for dependencies corresponds to a loop merging. This may cause dependencies to be alive simultaneously, so that their combined sizes determine the global storage requirement.

The first objective is to reposition statement S.3 in the common iteration space so that the t dimension becomes an ND for dependency 3. This corresponds to a loop merging between the first and second loop nest in the original code. Dependency 3 has two negative dimensions, one in each DV, as shown in Table 4.16. In DV1 the k dimension is negative. This DV has two other SDs, t and y. It is hence possible to fulfill the requirement of having another SD outside a negative dimension, even if t is transformed into an ND. The requirement is fulfilled if y is placed outside k. For DV2, the t dimension is the only SD in addition to the negative y dimension. For a repositioning of S.3 to be legal, it must also result in a non-negative y dimension in dependency 3. The negative dimension and the length of its DVP in this dimension, $|DVP_y|$ in this case, determines in which direction and how far the ID must be moved. Since $|DVP_y| = 1$ for dependency 3, this requires a simple transformation so that the array elements of statement S.3 are produced at one iteration node further out along the y axis of the common iteration space compared to their original production. The transformation that is used is a form of *skewing*, [290], where the repositioning is done using if-clauses. The corresponding statement after the move is given in Fig. 4.39. S.4 and all statements depending on its elements are moved similarly to avoid additional negative y dimensions. Running STOREQ reveals that the lower bound of dependency 3 is now increased from 1 to 2 while the upper bound is reduced from 304964 to 956 since the DP of S.2 and its depending iteration nodes are now partially overlapping.

The next dependencies with t as SD are dependencies 6 and 7. As can be seen from Table 4.16, DV1 of dependency 6 has only one SD apart from t, and this dimension, k, is negative. The k dimen-

if(t==0& x>=GB & x<=N-1-GB & y>=GB+1 & y<=M-1-GB+1 &

k<=SIZE_TOT-1)

S.3   gauss_xy_compute[x][y-1][k+1] = f3(gauss_xy_compute[x][y-1][k],

gauss_x_image[x][y-1+k-1]);

Figure 4.39.   Statement S.3 after transformation.

		Dep. 3	Dep. 7	Dep. 11	Combined
a)	Original code (FD = {1,X,X,X})	304964/	304964/	304964/	304964/
		304964	304964	304964	304964
b)	y positive (FD = {1,3,2,4})	956	1435	1435	3826
c)	x positive (FD = {1,2,3,4})	2	1915	1915	3832
	No neg. dim. (FD = {1,X,X,4})	2/	959/	959/	1920/
		956	1279	1279	3514
d)	No neg. dim. (FD = {1,3,2,4})	956	959	959	2874
e)	No neg. dim. (FD = {1,2,3,4})	2	1279	1279	2560

Table 4.17.   Estimated dependency sizes resulting from alternative transformations (N = 480, M = 640).

sion must hence be made non-negative if t is to be made an ND. The situation is more complicated for dependency 7. It has four DVs, all with different combinations of SDs, NDs and negative dimensions. For DV1, x, y, and k are all negative. It is hence not possible to perform the loop merging and transform t into an ND without at the same time repositioning statement S.5 in such a way that either all negative SDs are removed or so that at least one of the SDs are made positive. A number of transformations exist that fulfills the requirements of dependency 6 and 7, each with different effects on the size of this and other dependencies. The STOREQ tool will now be used to evaluate the global consequences of a number of these alternative transformations. The current version of STOREQ does not handle dependencies with multiple negative dimensions, as is the case for dependency 7. A manual inspection using the methodology presented in this chapter results in an optimal ordering among the SDs with k innermost, x second innermost, and y outermost. The y dimension should hence be made positive for dependency 7 while the k dimension is made non-negative for dependency 6. The ordering enforced by this loop merging and skewing has negative consequences for dependency 3 however, as seen in row b) of Table 4.17. After the transformation, the dependencies cover some of the same iteration nodes, and are hence simultaneously alive, regardless of the ordering chosen. Their combined size will therefore determine the overall storage requirement. The elements carried by dependency 6 is a subset of those carried by dependency 7, so only dependency 7 needs to be included in the combined size. Similar reasoning regarding the loop merging, optimal ordering, transformations, and simultaneous aliveness can be used for dependency 10 and 11.

Alternative transformations, where x is made positive and placed outermost or where all negative dimensions are removed, are now investigated since this allows an ordering that is better for dependency 3. Both statements S.5 and S.7 and all statements depending on their elements are repositioned as needed. In all cases a complete merging of the original loops has been performed. Table 4.17 presents the STOREQ estimation results for dependencies 3, 7, and 11 for these transformation alternatives. Each row shows the estimated storage requirement for alternative legal placements and execution orderings in this new common iteration space. The storage requirement of the original code without loop merging is shown for comparison in the first row. Row e) holds the globally best solution. It has a storage requirement over two orders of magnitude lower than the original solution, and substantially lower than the other alternatives.

If the same experiment is performed using a different image format, for instance the panoramic format of the *Advanced Photo System*, the conclusion turns out to be somewhat different. The previously best solution is now second to worst as shown in Table 4.18 and Fig. 4.40. The storage required for buffering full image lines along the y dimension for dependency 7 and 11 are now larger than that of buffering full image lines along the x dimension for dependency 3. These somewhat surprising

		Dep. 3	Dep. 7	Dep. 11	Combined
a)	Original code (FD = $\{1,X,X,X\}$)	658684/	658684/	658684/	658684/
		658684	658684	658684	658684
b)	y positive (FD = $\{1,3,2,4\}$)	956	1435	1435	3826
c)	x positive (FD = $\{1,2,3,4\}$)	2	4135	4135	8272
	No neg. dim. (FD = $\{1,X,X,4\}$)	2/	959/	959/	1920/
		956	2759	2759	6474
d)	No neg. dim. (FD = $\{1,3,2,4\}$)	956	959	959	2874
e)	No neg. dim. (FD = $\{1,2,3,4\}$)	2	2759	2759	5520

Table 4.18.   Estimated dependency sizes resulting from alternative transformations for another image size (N = 480, M = 1380).

results demonstrate how important the storage requirement estimation tool is for the optimization of the memory usage.

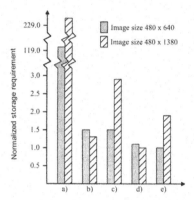

Figure 4.40.   Combined storage requirement for alternative cavity detection implementations. Normalized to best solution for each image size. Transformations as in Table 4.18.

4.6 SUMMARY

In this chapter we have presented a storage requirement estimation methodology to be used during the early system design steps. The execution ordering is at this point typically partially fixed, and this is taken into account to produce upper and lower bounds on the storage requirement of the final implementation. The methodology is divided into four steps. In the first step, a data-flow graph is generated that reflects the data dependencies in the application code. The second step places the polyhedral descriptions of the array accesses and their dependencies in a so-called common iteration space. The third step estimates the upper and lower bounds on the storage requirement of individual data dependencies in the code, taking into account the available execution ordering. As the execution ordering is gradually fixed, the upper and lower bounds on the data dependencies converge. This is a very useful and unique property of the methodology. Finally, simultaneously alive data dependencies are detected. Their maximal combined size at any time during execution equals the total storage requirement of the application. The feasibility and usefulness of the methodology are substantiated using several representative application demonstrators.

5 AUTOMATED DATA REUSE EXPLORATION TECHNIQUES

Efficient exploitation of temporal locality in the memory accesses on array signals can have a very large impact on the power consumption in embedded data dominated applications. The effective use of an optimised custom memory hierarchy or a customized software controlled mapping on a predefined hierarchy, is crucial for this. Only recently effective systematic techniques to deal with this specific design step have begun to appear. They were still limited in their exploration scope.

In this chapter we construct the design space by introducing three parameters which determine how and when copies are made between different levels in a hierarchy, and determine their impact on the total memory size, storage related power consumption and code complexity. Heuristics are then established for an efficient exploration, such that cost-effective solutions for the Memory size/Power trade-off can be achieved. The effectiveness of the techniques is demonstrated for several real-life image processing algorithms.

5.1 CONTEXT AND MOTIVATION

Memory hierarchy design has been introduced long ago to improve the data access (bandwidth) to match the increasing performance of the CPU [226][433][424]. The by now well-known idea of using memory hierarchy to minimise the power consumption, is based on the fact that memory power consumption depends primarily on the access frequency and the size of the memory [559, 561]. Performance and power savings can be obtained by accessing heavily used data from smaller memories instead of from large background memories. This can be controlled by hardware as in a cache controller but then only opportunities for reuse are exploited that are based on *local* access locality. Instead, also a compile-time analysis and code optimisation stage can be introduced to arrive at a *global* data reuse exploitation across the entire program where explicit data copies are introduced in the source code [154, 563]. Such a power-oriented optimisation requires architectural transformations that consist of adding layers of memories of decreasing size to which frequently used data will be copied, from which the data can then be accessed. This is shown in Fig. 5.1 for a simple example for all read operations to a given array A. The dots represent how the memory accesses to the array elements are ordered relatively to each other in time. Note that loops have been represented here as groups of dots, but the formal approach is completely parameterized and its complexity is nearby independent of the loop sizes. It can be seen that, in this example, most values are read multiple times. Over a very large time-frame all data values of the array are read

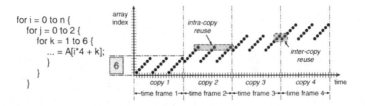

Figure 5.1.   Exploiting data reuse local in time to save power.

from memory, and therefore they have to be stored in a large background memory. However, when we look at smaller time-frames (indicated by the vertical dashed lines), we see that only part of the data is needed in each time-frame, so it would fit in a smaller, less power consuming memory. If there is sufficient reuse of the data in that time-frame, it can be advantageous to copy the data that is used frequently in this time-frame to a smaller memory, such that from the second usage on, a data element can be read from the smaller memory instead of the larger memory.

In a hardware controlled cache all data would be copied the first time into the cache and possibly overwrites existing data, based on a replacement policy which only uses knowledge about *previous* accesses. In contrast, in our approach it is checked at compile-time that the new data that will be copied has sufficient *future* reuse potential to improve the overall power cost, compared to leaving existing data or copying other data. This analysis is performed not for entire arrays only but also for uniformly accessed sub-arrays of data.

So, memory hierarchy optimisation has to introduce copies of data from larger to smaller memories in the Data Flow Graph (DFG). This means that apart from the already mentioned global trade-off also a local trade-off exists here: on the one hand, power consumption is decreased because data is now read mostly from smaller memories, while on the other hand, power consumption is increased because extra memory transfers are introduced. Moreover, adding another layer of hierarchy can also have a negative effect on the memory size and interconnect cost, and as a consequence also on the power. It is the task of the data reuse decision step to find the best solution for this overall trade-off [93]. This is required for a custom hierarchy, but also for efficiently using a predefined memory hierarchy with software cache control, where we have the design freedom to copy data at several moments with different copy sizes [154]. In the latter case, several of the virtual layers in the global copy-candidate chain of Fig. 5.2 (see further) can be collapsed to match the available memory layers.

5.2 RELATED WORK

The main related work to this problem lies in the parallel compiler area, especially related to the cache hierarchy itself. Many papers have focused on introducing more access locality by means of loop transformations (see e.g. [43][361][291][192]). That literature, as is chapter 3, is however complementary to the work presented here: it forms only an enabling step. The techniques proposed there on cache exploitation are not resulting yet in any formalisable method to guide the memory organisation issues that can be exploited at compile-time.

In most work on parallel MIMD processors, the emphasis in terms of storage hierarchy has been on hardware mechanisms based on cache coherence protocols (see e.g. [329] and its references). Some approaches address the hierarchical data organisation in processors for programs with loop nests, e.g. a quantitative approach based on approximate life-time window calculations to determine register allocation and cache usage [192]. The main focus is on CPU performance improvement though and not directly on memory related energy and memory size cost.

Also in an embedded system synthesis context, applying transformations to improve the cache usage has been addressed [280][405][260]. None of these approaches determine the best memory hierarchy organisation for a given (set of) applications and they do not directly address the power cost. An analysis of memory hierarchy choices based on statistical information to reach a given

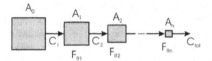

Figure 5.2.   Copy-candidate chain with memory sizes A_j, data reuse factors F_{Rj} and number of writes C_j.

throughput expectation has been discussed [245]. More recently, dynamic management of a scratch-pad memory for temporal and spatial reuse has been addressed [257].

In the hardware realization context, much less work has been performed, mainly oriented to distributed memory allocation and assignment [328][444][35] [479][51][408].

The impact of the data reuse exploration step turns out to be very large in our overall methodology for power reduction. This has been demonstrated for a H.263 video decoder [387], a motion estimation application [154] and a 3D image reconstruction algorithm [511].

A basic methodology for data reuse exploration in an embedded context [154] has been developed as part of the DTSE script for data transfer and storage exploration. A variant of this is discussed in [485]. The techniques described there were further extended in [512] to also handle applications with non-homogeneous access patterns. Examples of real-life image processing algorithms with holes in the access pattern are the "Binary Tree Predictive Coder" and "Cavity Detection" discussed further on. In this chapter we will first introduce a simple example to illustrate the problem, and deduce the required parameters to describe a complete search space. The relationship between these parameters and the relevant cost variables is explored, which helps us to establish heuristics to steer the search for a good solution.

The rest of this chapter is organized as follows. In section 5.3 we will give an overview of the basic concepts and cost functions to evaluate a possible hierarchy. Section 5.4 will give a simple example of an application with a hole in the access pattern. We will determine the search space parameters to deal with this problem. How to steer the exploration of the huge resulting search space is explained in section 5.5. In section 5.6 we will give results on the application of these techniques on four real-life applications. Section 5.7 concludes the chapter.

5.3 BASIC CONCEPTS

5.3.1 Data reuse factor

The usefulness of a memory hierarchy is strongly related to the signal reusability, because this is what determines the ratio between the number of read operations from a copy of a signal in a smaller memory, and the number of read operations from the signal in the larger memory on the higher hierarchical level.

In Fig. 5.2 a generic copy-candidate chain is shown where for each level j, we can determine a memory size A_j, a number of writes to the level C_j (equal to the number of reads from level (j-1)) and a data reuse factor F_{Rj} defined as

$$F_{Rj} = \frac{C_{tot}}{C_j}. \tag{5.1}$$

C_{tot} is the total number of reads from the signal in the lowest level in the hierarchy. This is a constant independent of the introduced copy candidate chain. The number of writes C_j is a constant for level j, independent from the presence of other levels in the hierarchy. As a consequence we can determine F_{Rj} as the fraction of the number of reads from level j to the number of writes to level j, as it would be the only sub-level in the hierarchy. As we will see further on, this definition enables us to define the power cost function in such a way that the cost contribution of each introduced sub-level is clearly visible. Since data reuse factors are independent of the presence of other copy-candidates, they do not have to be recomputed for each different hierarchy.

A reuse factor larger than 1 is the result of **intra**-*copy reuse* and **inter**-*copy reuse. Intra-Copy* reuse means that each data element is read several times from memory during one time-frame, visible because one (or more) iterators of the surrounding loop nest is missing in the index expression of the accessed array (e.g. iterator j in Fig. 5.1). *Inter-Copy* reuse means that advantage is taken from the fact that part of the data needed in the next time-frame could already be available in the memory from the previous time frame, and therefore does not have to be copied again.

If F_{Rj} equals 1 or is too low (dependent on the power models of the memories in the different layers), this sub-level is useless and would even lead to an increase of memory size and power, because the number of read operations from level (j-1) would remain unchanged while the data also has to be stored and read from level j. So these cases are pruned from the search space.

5.3.2 Assumptions

1. For the specific step addressed in this chapter, we assume the code has been pre-processed to single assignment code, where every array value can only be written once but read several times. A pre-processing step in our overall approach allows to achieve this systematically [93]. This makes the further analysis much easier.

2. We assume the code has been transformed to optimise the data locality during a previous data flow and loop transformation step . This is compatible with the position of the data reuse decision step in the complete ATOMIUM memory management methodology [93] . A certain freedom in loop nest ordering is still available after this step. During the data reuse step we estimate the memory hierarchy cost for each of the signals and each loop nest ordering separately. A global decision optimizing the total memory hierarchy including all signals, will then be taken in a subsequent *global hierarchy layer assignment* step [93].

3. Copy-candidates are selected based on the reuse possible inside one iteration of a certain loop in the complete loop nest, so one copy-candidate can be defined for each of the loop nest levels. Together these copy-candidates form *the copy-candidate chain* for the considered read instruction. An optimal subset of this copy-candidate chain can be selected, based on the possible gain acquired by introducing an intermediate copy-candidate, trading off the lower cost (speed and power) of accesses to a smaller memory and the higher cost of the extra memory size.

The basic methodology [154] constructs a limited search space by fixing the time-frame boundaries on the loop boundaries of loop nests. So, one possible copy-candidate per loop level is proposed. Intra-copy reuse and inter-copy reuse between subsequent time-frames is fully exploited.

5.3.3 Cost functions

Let $P_j(N_{bits}, N_{words}, F_{access})$ be the power function for read and write operations to a level j in a copy-candidate chain, which depends on the estimated size N_{words} and bit-width N_{bits} of the final memory, as well as on the real access frequency F_{access} (this is **not** the clock frequency) corresponding to the array signals, which is obtained by multiplying the number of memory accesses per frame for a given signal with the frame rate F_{frame} of the application. The total cost function for a copy-candidate chain with n sub-levels is defined as:

$$F_c = \alpha \sum_{j=0}^{n} P_j(N_{bits}, N_{words}, F_{access}) + \beta \sum_{j=1}^{n} A_j(N_{bits}, N_{words}) \tag{5.2}$$

Here α and β are weighing factors for Memory size/Power trade-off.

The power cost function $P_j(N_{bits}, N_{words}, F_{access})$ for one level j can be written as:

$$P_j = \frac{\#reads}{frame} * P_j^r + \frac{\#writes}{frame} * P_j^w \tag{5.3}$$

```
for (col=0; col<(MAXCOL-2); col++)
  for (row=0; row<(MAXROW-2); row++)
    {
      outpix = 0.0;
      for (i=0; i<3; i++)
        for (j=0; j<3; j++)
          if((i!=1) || (j!=1))   ⇐
              outpix += mask[i][j] * inim[row+i][col+j];
      outim[row][col] = (int)(outpix/2.0) + 127;
    }
```

Figure 5.3.   Example C code

where $P_j^{r(w)}$ is defined as

$$P_j^{r(w)} = F_{frame} * \frac{Energy}{1 \ read(write)} \tag{5.4}$$

Based on the previous definitions we can deduce the following expression for the total storage power cost for a complete copy candidate chain of n sub-levels:

$$
\begin{aligned}
\sum_{j=0}^{n} P_j() &= C_1(P_0^r + P_1^w) + C_2(P_1^r + P_2^w) + \cdots + C_{tot}(P_n^r) \\
&= C_{tot}[\frac{C_1}{C_{tot}}(P_0^r + P_1^w) + \frac{C_2}{C_{tot}}(P_1^r + P_2^w) + \cdots + (P_n^r)] \\
&= C_{tot}[\sum_{j=1}^{n}(\frac{1}{F_{Rj}}(P_{j-1}^r + P_j^w)) + P_n^r]
\end{aligned}
\tag{5.5}
$$

From equation (5.5) we learn that smaller memory sizes A_j, which reduce $P_j^{r(w)}$, and higher data reuse factors F_{Rj} reduce the total power cost of a copy candidate chain.

5.4   A HOLE IN THE PICTURE

In this section we will give a simple example of an application where assumptions made in the basic methodology on the homogeneous access properties of a copy-candidate lead to non-optimal results for the power and memory size cost. We will identify the parameters which allow us to explore a more complete search space.

5.4.1   A first simple example

In Fig. 5.3 an example in C code is given, where a square mask is moved over an image in vertical and horizontal direction. A pixel in the output image is based on the *surrounding* pixels of the input image and not the reference pixel itself, i.e. there is a *hole* in the access pattern of the two inner loops, as indicated by the arrow in the code. A data access graph of the input image can be constructed showing the accessed index values as a function of relative time. The graph for three subsequent mask iterations is illustrated in Fig. 5.4. We define a data reuse dependency (DRD) as the time interval between two accesses to the same index, represented in the graph by grey horizontal lines started and ended by a bullet representing the accesses.

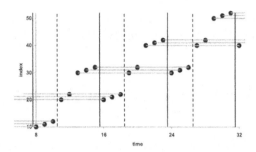

Figure 5.4.   Data access graph for 3 subsequent mask iterations, showing the influence of the time-frame on memory size and data reuse factors

In the basic methodology we fix the time-frame boundaries defining the copy-candidates on the iteration boundaries [154], shown in the graph as full vertical lines. One copy-candidate fits the data accessed during this time-frame, in this case 8 different indices. Data, not present in the copy-candidate during the previous time-frame, has to be copied from a higher level. Data, present in the previous time-frame but not accessed in the current time-frame, is discarded. In this way, the reference pixel (the *hole*) in the current time-frame, which was accessed in the previous time-frame, is discarded. However this index is accessed again in the *next* time-frame and has to be copied again from a higher level. By keeping this index in the copy-candidate over *multiple* time-frames until the next access, we can reduce the global number of writes to the copy-candidate, which results in a higher data reuse factor, but at the same time increases the memory size to 9 since more data is kept in the copy-candidate. However, when we move the time-frame with an offset of 3, shown in Fig. 5.4 by dashed vertical lines, the number of indices kept in the copy-candidate (or the memory size) is only 6, while retaining the higher data reuse factor. In general, we also have the freedom to change the size of the time-frame, i.e. the number of reads during a time-frame. In this specific example, no more gain for memory size and power cost can be achieved though.

By varying the following three parameters, a complete search space is explored:

- The time-frame size $T_{timeframe}$

- The time-frame *offset*

- The included data reuse dependencies (*DRD*)

The last parameter defines whether and during which time intervals a certain index is kept in the copy-candidate over multiple time-frames, trading off higher data reuse for higher memory size. The effect will be illustrated on the application but first we have to define the equations for reuse factors and memory size, to take into account the effect of the included data reuse dependencies.

5.4.2   Definition of reuse factors and memory size

A graphical representation of the following notions is given in Fig. 5.5. Let D_i be defined as the set of different indices accessed during $timeframe_i$ (denoted by squares), and A_i as the set of different indices actually present in the copy-candidate during $timeframe_i$. Then A_i is equal to the union of D_i and the set of indices kept in the copy-candidate by including certain data reuse dependencies in $timeframe_i$, i.e. data that will be reused in a later time-frame but not in $timeframe_i$ (denoted by triangles). The data accessed during $timeframe_i$, which was not present in the copy-candidate during the previous $timeframe_{i-1}$, is copied from a higher level, these elements have a vertical arrow assigned in Fig. 5.5. The number of writes to the copy-candidate at the start of $timeframe_i$ is thus equal to $\#(D_i \setminus A_{i-1})$. Note that in the basic methodology the set A_i would always be equal to the set D_i.

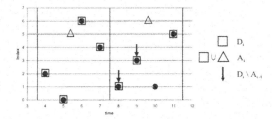

Figure 5.5.   Definition of D_i, A_i, $\#(D_i \setminus A_{i-1})$: graphical representation in a data access graph

$$F_{R_i} \quad = \quad \frac{T_{timeframe}}{\#(D_i \setminus A_{i-1})}$$

$$= \quad NIaC_i * NIeC_i$$

$$NIaC_i \quad = \quad \frac{T_{timeframe}}{\#(D_i)}$$

$$NIeC_i \quad = \quad \frac{\#(D_i)}{\#(D_i \setminus A_{i-1})}$$

Figure 5.6.   Definition of data-reuse factors: per time-frame

$$F_R \quad = \quad \frac{\sum_i T_{timeframe}}{\sum_i \#(D_i \setminus A_{i-1})} = \frac{C_{tot}}{C}$$

$$= \quad NIaC * NIeC$$

$$NIaC \quad = \quad \frac{\sum_i T_{timeframe}}{\sum_i \#(D_i)} = \frac{C_{tot}}{\sum_i \#(D_i)}$$

$$NIeC \quad = \quad \frac{\sum_i \#(D_i)}{\sum_i \#(D_i \setminus A_{i-1})}$$

$$A \quad = \quad Max_i(A_i)$$

Figure 5.7.   Definition of data reuse factors: per copy-candidate

A set of reuse factor definitions per time-frame is given in Fig. 5.6. The *data reuse factor* F_{R_i} is equal to the fraction of the number of reads from the lowest level during this time-frame to the number of writes to the copy-candidate from a higher level at the start of this time-frame. The effect of reuse inside the time-frame, during which the same index can be accessed multiple times, is expressed by the **intra**-*copy reuse factor* $NIaC_i$ (see section 5.3.1). The **inter**-*copy reuse factor* $NIeC_i$ illustrates the reuse of data already available in the copy-candidates during the previous time-frame and accessed during this time-frame. The overall data reuse factor F_{R_i} is again equal to the product of the intra-copy and inter-copy reuse factor. The global reuse factors over all time-frames and the needed memory size of a certain copy-candidate are given in Fig. 5.7. C_{tot} is the total number of reads from the lowest level in the copy-candidate chain, C is the total number of writes to this copy-candidate.

	(A) Original	(B) DRD	(C) DRD, Offset=3
$timeframe_i$	8	8	8
$\#(D_i)$	8	8	6
$\#(D_i \setminus A_{i-1})$	4	3	3
$NIaC_i$	$\frac{8}{8} = 1$	$\frac{8}{8} = 1$	$\frac{8}{6}$
$NIeC_i$	$\frac{8}{4} = 2$	$\frac{8}{3}$	$\frac{6}{3} = 2$
F_{R_i}	$\frac{8}{4} = 2$	$\frac{8}{3}$	$\frac{8}{3}$
$\#(A_i)$	8	9	6

Table 5.1.   Data reuse factors and memory size values for basic methodology (a), including data reuse dependencies (b) and offsetting the time-frame (c)

5.4.3   Application to simple example

We have applied these definitions of Fig. 5.6 for reuse factors and memory size on the example given in section 5.4.1, which is derived from a real-life image processing application. The results are given in Table 5.1. The first column shows the outcome for the basic methodology. The second column shows a 33,3% increase of the data reuse factor F_{R_i} when including data reuse dependencies, but a 12,5% increase of the memory size. Finally, the third column shows both an 33,3% increase of the data reuse factor, as well as a 25,0% decrease of the memory size, when including data reuse dependencies and offsetting the time-frame.

5.4.4   Multiple levels of reuse

In general a copy-candidate chain consists of copy-candidates of decreasing memory size, time-frame and data reuse factor. Constraints exist on which copy-candidates can be placed in a hierarchy. The time-frame boundaries of a higher level should coincide with the smaller time-frame boundaries of the lower hierarchy. This is to ensure that at the start of a smaller time-frame, all data accessed during this time-frame which should be loaded from the higher level is also available at that time. Thus a start of a higher level time-frame cannot be located in between time-frame boundaries of the lower level.

5.5   STEERING THE GLOBAL EXPLORATION OF THE SEARCH SPACE

As explained in section 5.3, the basic methodology defines the search space as all possible subsets of a copy-candidate chain with *fixed* copy-candidates determined by the loop boundaries. By adding the freedom of re-sizing and offsetting time-frames and including data reuse dependencies for *each* copy-candidate, we deal now with a search space explosion. Heuristics have to be found which steer the search for optimal solutions.

5.5.1   Extension of Belady's MIN Algorithm to larger time-frames

Belady's MIN Algorithm [49] is a replacement algorithm for local memory which optimally exploits temporal locality, and assumes knowledge of the future. Given an address trace, and a fixed local memory size A_{fixed}, at each time instance the algorithm selects which references will be kept in local memory and which not in order to minimize the total number of memory fetches.

The algorithm can be explained by using the definitions of Fig. 5.6. Consider a time-frame of size 1. Then $\#(D_i)$ is always equal to 1, and $\#(A_i)$ depends on which DRD's are kept in the local memory. When the value for $\#(A_i)$ exceeds A_{fixed}, certain DRD's present in $timeframe_i$ are cut to meet the memory size constraint. This means that the data corresponding to a cut DRD is not kept in the local memory during the time interval between two accesses. Belady's algorithm now states that those DRD's are cut which second access is furthest in the future.

For a time-frame size bigger than 1, this algorithm can still be applied. Since it should always be possible to fit the data accessed during one time-frame ($=D_i$) the minimum memory size possible for A_{fixed} is $Max_i(\#(D_i))$, where the minimum memory size A_{fixed} for Belady's algorithm was 1. The same selection procedure for cutting DRD's to meet the memory size constraints can be applied. Since Belady's algorithm results in the minimum number of writes to the copy-candidate or the maximum data reuse factor F_R for a certain fixed memory size, larger time-frames will always lead to equal or smaller data reuse factors for the same fixed memory size, or if the same data reuse factor is required, the fixed memory size should possibly increase.

5.5.2   DRD, $T_{timeframe}$, offset: influence on relevant cost variables

We will now investigate how the three parameters - time-frame size $T_{timeframe}$, time-frame **offset**, included data reuse dependencies **DRD** - influence the relevant cost variables:

- $F_R = f(DRD)$
  For a fixed set of data reuse dependencies included, the data reuse factor is fixed for all possible time-frame sizes and offsets, since changing the time-frame properties only provoke an exchange of intra-copy reuse for inter-copy reuse or vice versa. The product of the corresponding reuse factors $NIaC$ and $NIeC$ is a constant for constant DRD. More DRD results in a higher F_R. When all possible data reuse dependencies are included, which means that each data element is kept in the copy-candidate from the first to the last read, then a maximal value for the data reuse factor is reached where $F_{Rmax} = \frac{C_{tot}}{signal\_size}$[1]. Then each accessed element of the signal is written to the sub-level only once.

- $A = f(DRD, T_{timeframe}, offset)$
  A larger time-frame results in a higher $\#(D_i)$, more included data reuse dependencies result in a higher $\#(A_i)$, both leading to a higher memory size A. Changing the offset, changes both $\#(D_i)$ and $\#(A_i)$, depending on the access pattern. For a fixed data reuse factor F_R, the minimum needed memory size is found with Belady's algorithm ($T_{timeframe} = 1$).

- copy behaviour cost $= f(T_{timeframe}, offset)$
  The time-frame size and offset define when copies are made to copy-candidates during the execution of the algorithm. As discussed above, the timing for the optimal memory size is a time-frame size equal to 1. However, the implementation of this timing tends to lead to conditional statements in the inner loop kernel, and inhibits software pipelining for architectures with parallel functional units. Larger time-frame sizes corresponding to the iteration lengths lead to more regular copy-behaviour and are thus preferred for performance.

5.5.3   Heuristics to speed up the search

Enumerating and computing the cost of all possible combinations of time-frame sizes, offsets, and DRD for each of the copy-candidates, is not feasible for real-life applications. Therefore a reduced search space will be explored, taking into account the constraints given by the implementation platform.

We have seen in section 5.5.1 that, for a fixed memory size, the solution provided by Belady's algorithm with $T_{timeframe} = 1$ gives an equal or better power-memory size trade-off than a solution with the time-frame size larger than 1. However, more regular time-frames corresponding to the iteration lengths are generally better for the copy behaviour cost.

We propose to search for a good solution in the following way. Firstly, find the copy-candidate with optimal memory size A_{Max} for maximum data reuse F_{RMax} with Belady's algorithm. Then also with Belady's algorithm, for values of $A < A_{Max}$ one can find a (smaller) F_R. This results in a data reuse factor curve as a function of the copy candidate size, of which an example is shown

---

[1] *signal_size* comprises only the signal elements accessed during the program.

in Fig. 5.8(a). Each subset of the corresponding copy-candidates can be combined to a possible hierarchy. Time-frame constraints for multiple levels as discussed in section 5.4.4 do not exist here, since the time-frame size is equal to 1 for all levels.

By considering all possible hierarchies combining points on the data reuse factor curve, one obtains a power - total memory size trade-off curve (which is called a Pareto curve in mathematics), for an example see Fig. 5.8(b). A good solution should be chosen on this Pareto curve because all points above it are sub-optimal and below only unfeasible points exist. The choice is based on the weighing factors in the total cost function (equation (5.2)), possibly taking into account a fixed hierarchy constraint when the implementation platform is already known.

The cost of power and memory size are thus estimated by assuming an optimal replacement strategy. Upper and lower bounds on required storage sizes when a more regular program is assumed can be produced by a *high-level memory size estimation* tool (see chapter 4). Indeed, the exact timing of the copies will be optimised in a subsequent *storage cycle budget distribution (SCBD)* step as described in chapter 6, where the same data reuse factor is kept, but the memory size possibly increases, trading off the power-memory size cost with performance.

5.6 EXPERIMENTAL RESULTS

To verify and demonstrate the usefulness of the techniques described in this chapter, we developed a simulation tool to explore the search space for accesses to one signal in nested loops. First, an array access trace is generated for the instrumented signal in the source code. Based on this trace a data reuse factor curve is generated when optimal replacement (Belady) is assumed, using the definitions in Fig. 5.7. Then, by enumerating all possible combinations of points on the data reuse factor curve, a power-memory size Pareto curve is constructed, based on the power cost function defined in section 5.3.3. Also, for a chosen copy-candidate with a fixed data reuse factor, it is possible to generate a graph for the increase in memory size as a function of the timing of the transfers, by applying the extended Belady's algorithm described in section 5.5.1. The complexity of this tool depends on the length of the access trace, and is consequently not yet applicable for applications with large loop nests and signal sizes.

We show here the results for four real-life applications. We are using a proprietary industrial power model for on-chip memory, therefore only relative values for the power cost are given, which are normalised to the cost when all accesses for this signal are external memory accesses.

5.6.1 Binary Tree Predictive Coder

The Binary Tree Predictive Coder (BTPC) [449] is an algorithm for lossless or lossy image compression, effective for both photos and graphics.

We have analyzed the accesses to an input image (256x256 bytes) for lossless encoding in a 3-level nested loop, the results are shown in Fig. 5.8. In the data reuse factor curve we see 3 discontinuities, corresponding to the maximum reuse factors obtained in each of the 3 nested loops. For example in the inner loop a reuse factor $F_R = 1,23$ is reached for a memory size $A_2 = 8$. When we apply the *basic* methodology in the inner loop, we only reach a reuse factor of $F_R = 1,07$ for a memory size $A_{p2} = 12$. In the power-memory size Pareto curve we see that a hierarchy with a sub-level (A_{p2}) is not useful since the power consumption is higher than when no sub-level is used at all (P=1).

In a first step one can optimise the timing (time-frame/offset) of the copies, and the memory size can be reduced to A=3 (first arrow), such that the solution lies closer to the optimal data reuse factor curve. In a second step one can reuse data over multiple iterations by including more DRD to come to solution (A_2) where about 8% power gain is obtained (second arrow). Moreover this power gain is obtained for a memory size of 8, 33,3% smaller than the copy-candidate proposed by the basic methodology.

Analogously, the copy-candidate proposed by the basic methodology for the second loop nest (A_{p1}) does not yield any power gain. Timing optimisations reduce the memory size by 66,8% and a power gain of about 16% can be obtained. In this case there is no further gain from reuse over mul-

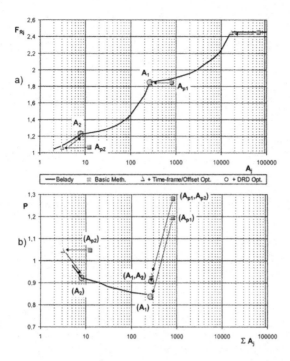

Figure 5.8.   Data reuse factor curve (a) and Power-Memory size Pareto curve (b) for the BTPC on a 256x256 image

```
for (i1=0;i1<H/n; i1++)              /*Vert.  CB counter*/
  for (i2=0; i2<W/n; i2++) {         /*Horz.  CB counter*/
    Δₒₚₜ[i1][i2] = +∞;
    for (i3=-m; i3<m; i3++)          /*Vert. in RW*/
      for (i4=-m; i4<m; i4++) {      /*Horz. in RW*/
        Δ = 0;
        for (i5=0; i5<n; i5++)       /*Horz. in CB*/
          for (i6=0; i6<n; i6++) { /*Vert. in CB*/
            Δ += ABS(New[i1*n+i5][i2*n+i6]
                   - Old[i1*n+i3+i5][i2*n+i4+i6]);
          }
        Δₒₚₜ[i1][i2] = MIN(Δ, Δₒₚₜ[i1][i2]);
      }
  }
```

Figure 5.9.   Motion estimation kernel used in data reuse experiment.

tiple iterations. A hierarchy including both copy-candidates gives a power consumption which lies above the power-memory size Pareto curve and is consequently not interesting for implementation.

Since there is little data reuse in this application, the power gain is still modest. As illustrated by the test-vehicles in the following sections, a considerable reduction in memory power consumption is obtained when higher data reuse factors are present.

5.6.2   MPEG4 Motion estimation

We show here the results for the *Old* frame in a "full-search full-pixel" motion estimation kernel [281] used in moving image compression algorithms. It allows to estimate the motion vector of

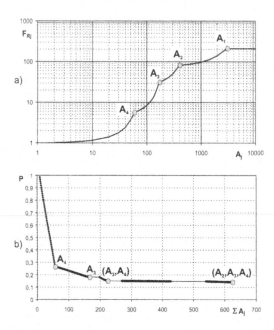

Figure 5.10.   MPEG4 Motion estimation: (a) Data reuse factor for array Old[][] as a function of the copy-candidate size. (b) Power - memory size Pareto curve for array Old[][].

small blocks of successive image frames. The principle was shown already in Fig. 4.31 and the algorithm is shown in Fig. 5.9. The simulations were done for the following parameter values: H=144, W=176, n=m=8.

In Fig. 5.10a the data reuse factor is shown as a function of the copy-candidate size. The maximum (average) reuse factor of 209,5 is obtained at a size of 2745 (A_1) which is about 16 lines of the *Old* frame. This is the maximum reuse for the iterations in the outer loop. We also see several discontinuities in the curve for smaller copy-candidate sizes ($A_2 - A_4$). These are the sizes where maximum reuse is obtained for a subset of inner loops in the total loop nest. Several combinations of levels $A_2 - A_4$ lie on the Pareto curve. Memory power consumption reductions of about 75 to 85% are reached.

5.6.3 Cavity detection

Cavity detection is an image processing algorithm for the segmentation of medical images [150].

Two dominant signals with a size equal to the size of the input image (320x320 bytes) exhibit a similar access pattern as given in the example in section 5.4.1. With a maximum data reuse factor of 8, memory power consumption reductions of about 55 to 60% are reached when introducing respectively a 1 or 2-level memory hierarchy (see Fig. 5.11).

For a fixed data reuse factor, the required memory size changes with the chosen time-frame and offset. To illustrate this, we show the contour graph of the memory size as a function of the time-frame size and offset for the copy-candidate A_2 in Fig. 5.12. For a time-frame=1 and offset=0 (Belady), the memory size becomes A=6. A more regular implementation with a time-frame=8 equal to the iteration size and an offset=0, leads to a memory size A=9 (see also Table 5.1). However, for offsets 3, 4 and 5, a local minimum is reached and the increase in memory size can be eliminated: A=6. In general, for a regular access pattern, the memory size contours are repeated for larger time-frame sizes and offsets with a period equal to the iteration size and with a gradually increasing memory size.

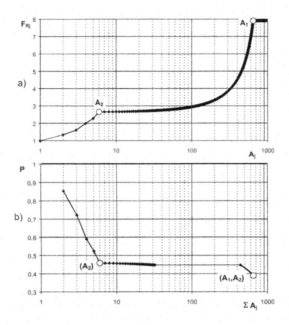

Figure 5.11.   Data reuse factor curve (a) and Power-Memory size Pareto curve (b) for Cavity detection on a 320x320 image

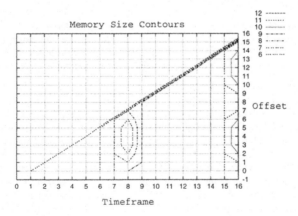

Figure 5.12.   Cavity detection: Memory size contours for constant data reuse as a function of the time-frame and offset

5.6.4 SUSAN principle

The SUSAN principle [484] is a basis for algorithms to perform edge detection, corner detection and structure-preserving image noise reduction. The image is accessed by moving a reference pixel in the horizontal and vertical direction, where each time the difference is computed with the surrounding pixels lying on a circular mask (see Fig. 5.13a). The original unfolded pointer-based loop body first has been pre-processed to a series of loops with different accesses to an array *image*. A memory power consumption reduction of about 83% is reached for a 1-level hierarchy with a memory size of

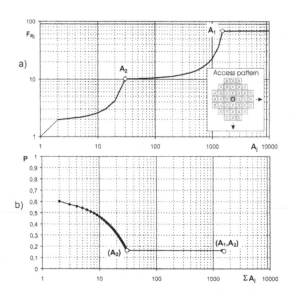

Figure 5.13.   Data reuse factor curve (a) and Power-Memory size Pareto curve (b) for the SUSAN principle on a 256x256 image

30 (A_2) (see Fig. 5.13b).

As a summary, the results for these four applications clearly demonstrate that constructing a search space based on the introduction of timing and DRD parametric explorations, is essential to find cost-effective memory hierarchy solutions. They also show that a trade-off exists between on-chip memory size and power. Clearly, since a larger data reuse factor results in less accesses to slower higher memory levels, this also leads to a higher memory access performance.

5.7   SUMMARY

In this chapter, we have discussed techniques for data reuse exploration to find cost-effective memory hierarchy solutions for data-dominated applications. We have identified the parameters which define the huge search space and indicated their relationship with the relevant cost variables. The power and memory size costs have to be traded off with the additional performance cost of more complex copy behaviour. A strategy to steer the search for a good solution is based on these relationships and the weight assigned to each of the different costs. The effectiveness of the techniques has been demonstrated for several real-life applications. They have been implemented in a prototype tool as a feasibility proof of the automation.

6 STORAGE CYCLE BUDGET DISTRIBUTION AND ACCESS ORDERING

In many cases, a fully customized (on-chip) memory architecture can give superior memory band-width and power characteristics over traditional hierarchical memory architecture including data caches. This is especially so when the application is very well analyzable at compile-time. In an embedded context this is typically quite well achievable because the application (set) to be mapped is usually fully fixed and the on-chip memory organisation can be at least partly tuned towards this application (set).

Although such a custom memory organization has the potential to significantly reduce the system cost, achieving these cost reductions is not trivial, especially manually. Designing a custom memory architecture means deciding how many memories to use and of which type (single-port, dual-port, etc). In addition, the memory accesses have to be ordered in time, so that the real-time constraints (the cycle budget) are met. Finally, each array must be assigned to a memory, so that arrays can be accessed in parallel as required to meet the real-time constraints while optimizing for low power [93]. This is a challenging task in modern low power processors, such as the TI C55x series [501], which do have a very distributed memory hierarchy. Even when the basic memory hierarchy is fixed, e.g. based on two cache levels and an (S)DRAM memory, the modern low-power memories [243] that are being introduced allow much customization, both in terms of bank assignment (e.g. SDRAMs) and even sizes (see e.g. [339]) or ports (see data sheets of modern SDRAMs). So the steps discussed in the next subsections also have a large impact on memory bandwidth and power in a fully predefined memory architecture context. The least amount of freedom that is available during the mapping includes the access order over the ports and busses and the assignment of the individual arrays (and basic groups in general) to the memory partitions (including banks).

6.1 SUMMARY OF RESEARCH ON CUSTOMIZED MEMORY ORGANISATIONS

Most of the memory organisation work focuses on scalar storage (see e.g. [510, 301, 33, 207, 7, 454, 488, 464]). As the major goal is typically the minimization of the number of registers for storing scalars, the scheduling-driven strategy that is applied in these techniques is well-fitted to solve a *register allocation* problem, but not geared towards a *background memory allocation and assignment* problem.

In the compiler literature, the register allocation problem has also lead to many publications. One of the first techniques is reported in [99], where colouring of the interference (or conflict) graph is applied. Many other variants of this colouring have been proposed. Usually also spilling from foreground to background memory is incorporated. The fact that scheduling and register allocation are heavily interdependent has lead to a number of compiler approaches where register allocation is done in 2 phases: once before and once after scheduling [506], or where scheduling is first done for minimizing the registers and then again after register allocation for the other cost functions [205]. Sometimes, scheduling and register allocation are fully combined [62]. Also an approach as [107, 108] also looks at array variables but then only in the context of register pressure and hiding memory latency during scheduling. Practically all these techniques are restricted however to the scope of basic blocks. Only recently, a few more general approaches (like [429]) have been proposed which offer a more general framework and heuristics to deal also with across basic-block analysis and inter-procedural considerations.

Several approaches for specific tasks in custom memory management oriented to non-scalar signals have been published. The MeSA [444] approach has been introduced at U.C. Irvine. Here, the emphasis is on memory allocation where the cost is based both on a layout model and on the expected performance, but where the possibility of M-D signals to share common storage locations when their life times are disjoint is neglected. Also some manual methodologies oriented to image processing applications have been investigated at the Univ. of Milano [25]. For memory packing data are assigned to physical memories with a cost function while respecting performance constraints [262]. A more general scheme has been introduced in [249]. Strategies to mapping arrays in an algorithm on horizontally and vertically partitioned dedicated memory organisations have been proposed at CMU [463]. Other examples are the memory selection technique in GAUT [472] the memory allocation for pipelined processors in [32], and the synthesis of intermediate memories between systolic processor arrays [464]. A recursive strategy of memory splitting (resulting in a distributed assignment) starting from a single port solution for a sequential program has been proposed in [51]. Interleaved memory organisations for custom memories are addressed in [104]. Memory bank assignment in DRAMs is addressed in [415]. Also the modification of loops to better match SDRAM modes has started to be addressed very recently [257]. Exploiting quadword access for interleaved memory bank access is supported in [194].

A nice overview of the state-of-the-art related to the low power cost aspect can be found in section 4.1.2 of [52]. Also [404] provides a good overview of the available work in memory management related issues.

However, all these approaches (both in the custom hardware and compiler domains) still consider the data to be stored as if they were "scalar units" because partly overlapping life-times and or exploiting the shape of the index space addressed by M-D signals is not done yet.

Also in the parallel compiler context, similar work has been performed (see chapter 2, in the software oriented memory management section) with the same focus and restrictions.

In the less explored domain of true background memory allocation and assignment for hardware systems, only few techniques are available. The pioneering techniques start from a given schedule [328], or perform first a bandwidth estimation step [35] which is a kind of crude ordering that does not really optimize the conflict graph either. These techniques also have to operate on groups of signals instead of on scalars to keep the complexity acceptable, e.g., the *stream model* of Phideo [328] or the *basic sets* in the ATOMIUM environment [35, 39, 38]. But at the same time, they are not operating any more on a "scalar unit model" for the arrays. The techniques of this chapter reuse these basic principles, but extend it further (see below).

In the scheduling domain, the techniques for optimizing the number of resources given the cycle budget are of interest to us. Also here most of the early techniques operate on the scalar-level (see e.g. [426] and its references). Many of these scalar techniques try to reduce the memory related cost by estimating the required number of registers for a given schedule but that does not scale for large array data. The few exceptions that have been published are the Phideo stream scheduler [530, 528], the Notre-Dame rotation scheduler [420] and the percolation scheduler of U.C.Irvine [212]. They schedule both accesses and operations in a compiler-like context, but with more emphasis on the cost and performance impact of the array accesses and including a more accurate model of their timing. Only few of them try to reduce the required memory bandwidth, which they do by minimizing the *number* of simultaneous data accesses [528]. They do not take into account *which* data is being accessed simultaneously. Also no real effort is spent to optimize the data access conflict graphs such that subsequent register/memory allocation tasks can do a better job.

The main difference between the Storage Cycle Budget Distribution (SCBD) approach at IMEC and all the related work discussed here is that it tries to minimize the required memory bandwidth in advance by optimizing the access conflict graph for groups of scalars across the different loop nests within a given cycle budget. This is achieved by putting ordering constraints on the hierarchical flow-graph, taking into account *which* data accesses are performed in parallel (i.e., will show up as a conflict in the access conflict graph). These ordering constraints are then passed to the (more) conventional compiler that is used as "back-end".

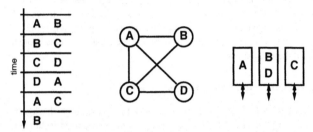

Figure 6.1.   Conflict graph corresponding to ordered memory accesses and a possible memory organization obeying the constraints

6.2   MEMORY BANDWIDTH AND MEMORY ORGANISATION PRINCIPLES

6.2.1   *Memory access ordering and conflict graphs*

The memory bandwidth requirement can be modeled as a conflict graph derived from the ordering of the memory accesses, as shown in Fig. 6.1. Its nodes represent the application's arrays, while its edges indicate that two arrays are accessed in parallel. The graph models the constraints on the memory architecture. An example is given in Fig. 6.1 where array A cannot be stored in the same single-port memory as any of the other arrays: it is in conflict with them. However, there is no edge between B and D, so these two arrays can be stored in the same memory. In order to support multi-port memories, this model must be extended with hyper-edges [93, 566].

From the conflict graph, it is possible to estimate the cost of the memory bandwidth required by the specified ordering of the memory accesses. Three main components contribute to the cost.

1. When two accesses to the same array occur in parallel, it means there is a self-conflict, a loop in the conflict graph. Such a self-conflict forces a two-ported memory in the memory architecture. Two- and multi-ported memories are much more costly, therefore self-conflicts carry a very heavy weight in the cost function.

2. In the absence of self-conflicts, the conflict graph's chromatic number indicates the minimal number of single-port memories required to provide the necessary bandwidth. For example, in Fig. 6.1, the chromatic number of the graph is three: e.g., A, B, and C need three different colours to colour the graph. That means also a minimum of three single-port memories is required, even though three accesses never occur in parallel. More memories make the design potentially more costly, therefore the conflict graph's chromatic number is the second component of the cost.

3. Finally, the more conflicts exist in the conflict graph, the more costly the graph is. Indeed, every conflict takes away some freedom for the assignment of arrays to memories. In addition, not every conflict carries the same weight. For instance, a conflict between a very small, very heavily accessed array and a very large, lightly accessed array is not so costly, because it is anyway a good idea for energy efficiency to split these two up over different memories. A conflict between two arrays of similar size, however, is fairly costly because then the two cannot be stored in the same memory, and potentially have to be combined with other arrays which do not match very well.

This abstract cost model makes it possible to evaluate and explore the ordering of memory accesses, without going to a detailed memory architecture yet. Indeed, the latter is in itself a rather complex task, so combining the two is not feasible. The conflict graph is heavily influenced by the access ordering and the tool support is crucial to achieve this tedious and error-prone task. Efficient techniques for the SCBD step based on the conflict graph approach have been published since 96 by IMEC. Here we determine both the bandwidth requirements [562, 566] and the balancing of the available cycle budget over the different memory accesses in the algorithmic specification [566, 66, 64]. Mature tool support for this has become available since then. Very fast algorithms to support the basic access ordering in loop bodies have been developed with different variants [394, 398, 400] including more adaptive [395] and space-time extensions [396] of the underlying access ordering technique. The combination of these tools allows us to derive real Pareto curves of the background memory related cost (e.g. power) versus the cycle budget [63]. This systematic exploration step is not present in conventional compilers. In the next sections, more information will be provided about these techniques.

6.2.2 Memory/Bank Allocation and Assignment

In data-transfer intensive applications, a very high peak bandwidth to the background memories is needed. Spreading out the data transfers over time, as will be discussed in subsection 6.3.4, helps to alleviate this problem, but the required bandwidth can still be high. The high-speed memory required to meet this bandwidth is a very important energy consumer, and takes up much of the chip or board area (implying a higher manufacturing or assembly cost). In addition, purchasing a high-performance memory component is several times more expensive than a slower memory, designed in older technology. However, the memory organization cost can be heavily reduced if there is a possibility to customize the memory architecture itself.

Instead of having a single memory operating at high speed, it is better in terms of e.g. energy consumption to have two or more memories which can be accessed in parallel. Also for embedded instruction-set processors, a partly customized memory architecture is possible: the processor core can be combined with one or more SRAMs or embedded DRAMs on the chip, and different configurations of the off-chip memories can also be explored. For partial predefined memory hierarchy, where the connections and the amount of parallelism is fixed, it is still important to determine the optimal sizes of each individual memory.

Even when the memory hierarchy is fully fixed, the array to memory/bank assignment can bring large gains is power. Typically, low power processors platforms do have a (very) distributed memory subsystem containing many small memories [501] which have to be used effectively. Although an instruction-set processor usually has only a single bus to connect with the large (off-chip) memories, the clock frequency of the processor and the bus is often much higher than that of the memories, thus the accesses to different memories and/or banks can be interleaved over the bus.

An example of the impact of a custom memory organization is shown in table 6.1, for the BTPC application discussed in more detail in subsection 6.4.2. The application's many memory accesses put a heavy load on the memory system. Therefore, the 'default' memory configuration contains a high-speed off-chip DRAM as bulk memory, and a dual-port SRAM for the arrays which are small enough to fit there. For this configuration, we have estimated the area of the on-chip SRAM and the energy consumption of both, based on the vendor's data-sheets and how many times each array is accessed. Then, we have also designed a custom memory organization for this application. In contrast to the default memory architecture, the custom memory organization can meet the real-time constraint with a much lower energy consumption and area cost. Slower, smaller and more energy-efficient memories can be used. However, these results are only attainable by careful systematic exploration of the alternatives, as the following subsection shows.

In a custom memory organization, the designer can choose memory parameters such as the number, size, and the number of ports. Even in a partly predefined (embedded) memory organisation some freedom is usually left for the on-chip organisation (e.g. the addition of one or more scratch pad memories). Moreover, the bank organisation inside a (S)DRAM is usually fixed in terms of the maximal configuration but freedom can be available on the number of banks that is powered

Memory architecture	on-chip area	on-chip energy	off-chip energy
Default memory architecture 1 off-chip memory, 1 dual-port on-chip memory	200 mm^2	334 mJ	380 mJ
Custom memory organization 3 off-chip memories, 4 single-port on-chip memories	111 mm^2	87 mJ	128 mJ

Table 6.1.   Impact of a custom memory organization for the BTPC application

up. This again leads to opportunities to customize the memory bandwidth and certainly also the signal-to-bank assignment becomes a crucial design decision in that case.

These decisions, which should take into account the constraints derived in the previous section, are the focus of the memory allocation and assignment (MAA) step. The problem can be subdivided into two sub problems. First, memories must be allocated: a number of memories is chosen from the available memory types and the different port configurations, possibly different types are mixed with each other, and some memories may be multi-ported. The dimensions of the memories are however only determined in the second stage. When arrays are assigned to memories, their sizes can be added up and the maximal bit-width can be taken to determine the required size and bit-width of the memory. With this decision, the memory organization is fully determined.

Both the memory/bank allocation and for the signal-to-memory/bank assignment have to take into account the constraints generated to meet the real-time requirements. For the memory allocation, this means that a certain minimum number of memories is needed. This minimum is derived from the conflict graph's chromatic number. For the signal-to-memory assignment, conflicting arrays cannot be assigned to the same memory. When there is a conflict between two arrays which in the optimal assignment were stored in the same memory, a less efficient assignment must be accepted. Thus, more conflicts make the memory organization potentially more expensive.

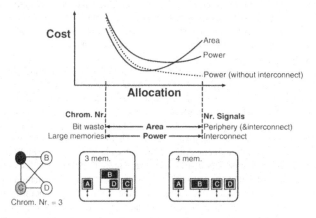

Figure 6.2.   Tradeoff memories versus cost during allocation and assignment.

Allocating more or less memories has an effect on the chip area and the energy consumption of the memory architecture (see Fig. 6.2). Large memories consume more energy per access than small memories, due to the longer word- and bit-lines. Therefore, the energy consumed by a single large memory containing all the data is much larger than when the data is distributed over several smaller memories. Also the area of the one-memory solution is often higher when different arrays have different bit-widths. For example, when a 6-bit and an 8-bit array are stored in the same memory, two bits are unused for every 6-bit word. By storing the arrays in different memories, one 6 bits wide and the other 8 bits wide, this overhead can be avoided.

The other end of the spectrum is to store all the arrays in different memories. This leads to relatively high energy consumption also, because the external global interconnection lines connecting all these (small) memories with each other and with the data-paths become an important contributor. Likewise, the area occupied by the memory system goes up, due to the interconnections and due to the fixed address decoding and other overhead per memory.

Obviously, the interesting memory allocations lie somewhere between the two extremes. The area and the energy function reach a minimum somewhere, but at different points. The useful exploration region to trade off area with energy consumption lies in between the two minima.

The cost of the memory organization depends not only on the allocation of memories, but also on the assignment of arrays to the memories (the discussion above assumes an optimal assignment). When several memories are available, many ways exist to assign the arrays to them. In addition to the conflict cost mentioned earlier, the optimal assignment of arrays to memories depends on the memories used. For example, the energy consumption of some memories is very sensitive to their size, while for others it is not. In the former case, it may be advantageous to accept some wasted bits in order to keep the heavily accessed memories very small, while in the latter case the reverse may be true. To find a (near-)optimal signal-to-memory assignment, a large number of possibilities have to be considered. Therefore, a tool to explore the memory organization possibilities based on a good memory library is indispensable for this step. A mature software tool to support this MAA step is available at IMEC now, based on earlier research [483, 524]. Also extensions to the network component (e.g. ATM) application domain [483] and to mapping on embedded processor cores [388] have been developed. See also [479] for alternative techniques based on the conflict graph produced by a SCBD approach.

A demo version of the SCBD-MAA tools can be downloaded from: HTTP://WWW.IMEC.BE/ACROPOLIS/ There, also more information is available.

6.3 HOW TO MEET REAL-TIME BANDWIDTH CONSTRAINTS

In data transfer intensive applications, a costly and difficult to solve issue is to get the data to the processor quickly enough. A certain amount of parallelism in the data transfers is usually required to meet the application's real-time constraints. Parallel data transfers, however, can be very costly. Therefore, the tradeoff between data transfer bandwidth and data transfer cost should be carefully explored. This section describes the potential tradeoffs involved, and also introduces a new way to systematically tradeoff the data transfer cost with other components (illustrated for total system energy consumption).

In our application domain, an overall target storage cycle budget is typically imposed, corresponding to the overall throughput. In addition, other real-time constraints can be present which restrict the ordering freedom. In data transfer and storage intensive applications, the memory accesses are often the limiting factor to the execution speed, both in custom "hardware" and instruction-set processors ("software"). Data processing can easily be sped up through pipelining and other forms of parallelism. Increasing the memory bandwidth, on the other hand, is much more expensive and requires the introduction of different hierarchical layers, typically involving also multi-port memories. These memories cause a large penalty in area and energy though. Because memory accesses are so important, it is even possible to make an initial system level performance evaluation based solely on the memory accesses to complex data types [93]. Data processing is then temporarily ignored except for the fact that it introduces dependencies between memory accesses. This section focuses on the tradeoff between cycle budget distribution over different system components and gains in the total system energy consumption. Before going into that, a number of other issues need to be introduced though.

Defining such a memory system based on a high-level specification is far from trivial when taking the real time constraints into account. High-level tools will have to support the definition of the memory system (see section 6.2.2). An accurate data transfer scheduling is needed to come up with a suitable memory architecture within the given (timing) constraints. This as such is an impossible task to do by hand, especially when taking sophisticated cost models into account.

The cost models can be area or power based. Using both the models will provide a clear tradeoff in the power, area and performance design space. In this section, we will focus on the (cost) consideration and optimisation freedom of all the three axes (Subsection 6.3.1 to 6.3.4). Moreover, the effective usage of the tradeoff is shown in subsection 6.4. The techniques, which allow to come up with Pareto curves visualizing the useful tradeoff space, will be explained in subsection 6.5. We have implemented a tool based on these techniques making it possible to make the tradeoff on real life examples used as illustration in this chapter. As far as we know, no other systematic (automatable) approach is available in literature to solve this important design problem.

6.3.1 The cost of data transfer bandwidth

High performance applications do need a high bandwidth to their data. This high memory bandwidth requirement will typically involve a high system cost. A clear insight in these costs is necessary to optimize the memory subsystem. The cost will nearly always boil down to energy and area/size. These are the relevant costs for the end user and manufacturer. A high memory bandwidth can be obtained in several ways. Basically, two classes of solutions exist to obtain a high physical memory bandwidth: high access speed over a single port (see subsection 6.3.2) and high parallelism. (see subsection 6.3.3) Both come with large costs in energy and area. Increasing the access speed alone will not solve the problem properly, the application has to be adapted to use the provided bandwidth more effective.

A novel approach, discussed in subsection 6.3.4, is to lower the application demand in terms of peak bandwidth and is balancing the load. Note that not the maximal power consumption is relevant in this context, but the actual energy consumed to execute a given application in a specific time period (cycle budget). Lowering the peak bandwidth does not have to affect the overall application performance or functionality. Note also that the larger the available cycle budget, the more freedom exists to optimize the memory access ordering. Hence, the cycle budget (or performance) can be traded off for energy reduction. Tool support is then needed to fully exploit this.

6.3.2 Cost model for high-speed memory architectures

Technology-wise, the basic access speed of memory planes increases slowly (less than 10% per year [434]). In contrast, the data-path speed is doubling every 18 months (Moore's law). As a result, the performance gap between memory and data-path increases rapidly. Advanced memory implementations are used to keep up with the data-path speed, but in many cases these can be quite energy consuming.

The new memory architectures are developed to provide a higher basic access speed. To have a better understanding, we first briefly explain a model reflecting the virtual operation of a modern SDRAM[1] (see Fig. 6.3). We abstract away the physical implementation details which we cannot influence, e.g. the bit width, the circuits used, the floor-plan and the memory plane organisation. The SDRAM consists out of several banks which can be accessed in an interleaved fashion to sustain a high throughput (pipelining principle). Every bank typically has several planes which can be activated separately. The "virtual" plane is defined as the total storage matrix entity which is activated (and consumes energy) when a particular memory access is performed. Within the planes there is a page, which is copied to a local latch register for fast access.

It is also beneficial to have successive accesses located in a single page. Typically a factor 2 to 4 in energy can be saved this way. In an initial phase of a memory access, the row is selected. This row (=page) is sensed[2] and partly or entirely stored in a local register. Next, the proper column is transferred to the output buffer. When the next access is within the same page, the row selection and sensing can be skipped. A high page hit rate is obtained when the next three conditions are fulfilled. First, the accesses need to be localized (local within every array). Second, the data layout should match the page mapping. And third, the memory access order within different arrays should access

---

[1]RAMBUS and other alternative architectures work with the same basic concept.
[2]Note that the sensing is an energy-consuming phase of an access.

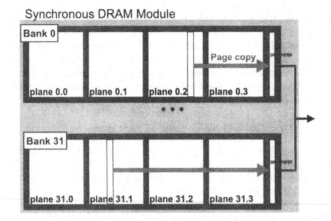

Figure 6.3.   High-level synchronous DRAM organisation.

these page members together. Obviously, array access ordering and data layout freedom form the key factors.

Large memories consist out of multiple planes which can be powered up separately. Multiple planes are used to avoid too long bit lines. By lowering switching between different planes the energy is saved further. This should again be enabled by optimizing the access ordering and the data placement in the memories.

6.3.3 Costs in highly parallel memory architectures

The memory bandwidth can be enlarged by increasing the number of bits being transferred per memory cycle. Different signals can be transferred from different memories. Moreover, multi-port memories can provide multiple data elements from the same memory. Alternatively, multiple data elements can be grouped together to form a new data element which are always accessed together. The main problem here is creating the parallelism and efficient usage of parallel data.

Increasing the number of memories has a positive influence on both bandwidth and memory energy consumption. But the interconnect complexity increases, causing higher costs. Obviously, there must be a tradeoff. The definition of the memory architecture and assignment of signals to these memories has a large impact on the available bandwidth.

Multi-ported memories increase the bandwidth roughly linearly with the number of ports. A complete double wiring accomplishes the improvement. Unfortunately, this will also heavily increase the area of the memory matrix. Due to this increase, the average wire length increases as well. Therefore the energy consumption and the cost in general will go up rapidly when using multi-port memories. To this end, multi-port memories have to be avoided as much as possible in energy-efficient solutions.

By packing different data elements together, a larger bandwidth can be obtained[524, 166]. Consequently, the data bus widens and more data is transferred. These elements can belong to the same signal (packing) or they can belong to different signals (merging). In addition to the wider bus, such a coupling of two elements can cause many redundant transfers when only one of the elements is needed. Needless to say, this option should be considered very carefully.

6.3.4 Balance memory bandwidth

The previous two subsections have focused on how to use the memories efficiently and what can be done when we change the memory (architecture). In addition much can gained by changing the application and balancing the bandwidth load over the available time.

The required bandwidth is as large as the maximum bandwidth needed by the application (see Fig. 6.4). When a very high demand is present in a certain part of the code (e.g. a certain loop nest), the entire application suffers. By flattening this peak, a reduced overall bandwidth requirement is obtained (see lower part of Fig. 6.4).

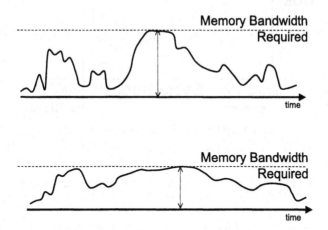

Figure 6.4.   Balancing the bandwidth lowers the cost.

The rebalancing of the bandwidth load is again performed by reordering the memory accesses (also across loop scopes). Moreover, the overall cycle distribution over all the loops has a large impact on the peak bandwidth. Tool support is indispensable for this difficult task. An accurate memory access ordering is needed for defining the needed memory architecture.

6.4 SYSTEM-WIDE ENERGY COST VERSUS CYCLE BUDGET TRADEOFF

The previous subsection has shown the most crucial factors influencing the data transfer and storage ("memory") related cost. It clearly increases when a higher memory bandwidth is needed. In most cases, designers of real-time systems will assume that the time for executing a given (concurrently executed) task is predefined by "some" initial system decision step. Usually, this is only seemingly so though. When several communicating tasks or threads are co-operating concurrently, then tradeoffs can be made between the cycle budget and timing constraints assigned to each of them, which makes the overall problem much more complex (see Subsection 6.4.3.2). In this case, making an optimal decision manually is unfeasible even by experienced designers. Moreover, also in data-dominated applications the memory subsystem is not the only cost for the complete system. In particular after our DTSE optimisation steps, the data-path is typically not negligible any more, especially for instruction-set processor realisations (see Subsection 6.4.3.1).

This subsection explains how to use the available tradeoffs between cycle budget for a given task, and the memory cost related to it. This leads to the use of Pareto curves. Moreover, several detailed real-life demonstrators are worked out.

6.4.1 Basic Trade-off Principles

The data transfer and storage related cycle budget ("memory cycle budget") is strongly coupled to the memory system cost (both energy and memory size/area are important here, as mentioned earlier). A Pareto curve is a powerful instrument to be able to make the right tradeoffs. The Pareto curve only represents the potentially interesting points in a search space with multiple axes and excludes all the solutions which have an equally good or worse solution for all the axes.

The memory cost increases when lowering the cycle budget (see Fig. 6.5). When lowering the cycle budget, a more energy consuming and more complex (larger) memory architecture is needed to deliver the required bandwidth. Note that the lower cycle budget is not obtained by reducing the

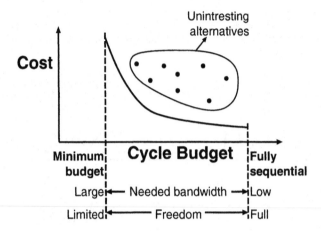

Figure 6.5.   Pareto curve for trading of memory cycle budget vs. cost

number of memory accesses, as in the case of the algorithmic changes and data-flow transformations (discussed in subsection 1.7.3). These higher-level code transformation related optimisations can be performed platform and cycle budget independent, prior to the tradeoff which is the focus here. Therefore, during the currently discussed step in the system design trajectory, the amount of data transferred to the data-path remains equal over the complete cycle budget range. Nevertheless, some control flow transformations and data reuse optimisations can be beneficial for energy consumption and not for the cycle budget (or vice versa). These type of optimisations still have to be considered in the tradeoff.

The interesting range of such a Pareto graph on the cycle budget axis, should be defined from the critical path up to the fully sequential memory access path. In the fully sequential case all the memory accesses can be transferred over a single port (lowest bandwidth). However, the number of memories is fully free and the memory architecture can be freely defined. Therefore, the energy is the lowest. In the other extreme, the critical path, only a limited memory access ordering is valid. Many memory accesses are performed in parallel. Consequently, the constraints on the memory system are large and a limited freedom is available. As a result, the energy consumption goes up. Due to the more restricted cycle budget, the power itself increases even more of course, but the latter involves a more traditional tradeoff which is not our focus here.

6.4.2 Binary Tree Predictive Coder Demonstrator

This energy-performance tradeoff can be made on real-life applications thanks to the use of the powerful SCBD-MAA tools that have been discussed earlier in this section. The use in this new Pareto curve context is now illustrated on a Binary Tree Predictive Coder (BTPC), used in offset image compression. Another demonstrator based on a Digital Audio Broadcast (DAB) receiver (see [69, 66]) is discussed in the demonstrator chapter 8. Both demonstrators, consist of several pages of *real* code containing multiple loops and arrays (in the range of 10 to 20) and a large number of memory access instructions (100 to 200).

Algorithm overview.   Binary Tree Predictive Coding (BTPC) [449] is a loss less or lossy image compression algorithm based on multi-resolution. The image is successively split into a high-resolution image and a low-resolution quarter image, where the low-resolution image is split up further. The pixels in the high-resolution image are predicted based on patterns in the neighbouring pixels. The remaining error is then expected to achieve high compression ratios with an adaptive Huffman coder. Six different Huffman coders are used, depending on the neighbourhood pattern. For lossy compression, the predictors are quantized before the Huffman coding. Fig. 6.6 shows the structure of the considered algorithm. Only the most important parts, and the loop structure around them, are shown. The loop structure is manifest, but inside the loop bodies, many data-dependent conditionals are present. This shows that our approach can handle such heavily data-dependent code.

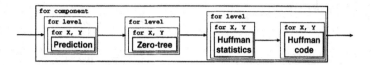

Figure 6.6.   Most important blocks and loop structure of the BTPC application

For a hardware implementation, some of the parameters of the original (software) specification have to be fixed. For example, the maximum input image size is set at 1024 × 1024 pixels. Fixing these parameters is necessary to be able to determine the sizes or the bit-widths of some of the arrays. For example, the size of the input image determines the size of three arrays containing image data (in total 18 Mbit), and it determines the bit-width of the arrays gathering the Huffman statistics (in total 80 kbit, distributed over 16 signals).

General Pareto curves discussion.   The platform-independent code transformation steps [93] that have been discussed in chapter 3 and 5, have been applied manually on this example and the platform-dependent steps (using tools) give accurate feedback about performance and cost [524]. Fig. 6.7 shows the relation between speed and power for the original, optimized and intermediate transformed specifications. The off-chip signals are stored in separate memory components. Four memories are allocated for the on-chip signals. Every step leads to a significant performance improvement without increasing the system cost. For every description, the cycle budget can be traded for system cost.

The performance versus cost Pareto curve is a useful instrument to visualize the relevant tradeoff. From the curve in Fig. 6.7 it can be concluded that the original implementation needs 28 million cycles when executed fully sequentially. In principle, all the signals could be stored in one single memory having one port. However, more memories will decrease the power consumption. The original description can be sped up to 18 million cycles by customizing the memory architecture. The power overhead of this change is negligible due to an optimized ordering.

Further cycle reduction forces a dual-ported off chip memory, introducing large costs (power, area, pin count). We cannot accurately estimate this part of the impact since no appropriate model is available from vendors. A reasonable assumption is a power overhead of 50% for a dual-ported memory. Using this assumption we can estimate the power increase for lower budgets. Adding this dual-ported off-chip memory dramatically increases the global power (see dashed part in Fig. 6.7). We didn't redo this (useless) exercise for the other applied transformations.

Detailed Pareto feedback when transforming BTPC application.   By global loop merging, the access count is reduced in the zero-tree generation. This has a positive effect on both cycle budget and power consumption. Moreover, this is an enabling transformation for the SCBD step. The more memory accesses in a basic block, the more ordering freedom available.

The data reuse step [154] can be applied on two levels: firstly register level data reuse, secondly line buffer based. Here, the register level reuse decreases the number of accesses little. Not much data is reused on the average actually. The main benefit is the reduction of the critical path in the worst case conditional branch (which is not executed very often) but has a large memory impact. The large predicting condition trees cause a large cycle budget, and the worst case path had to be taken into account. By introducing the register reuse, all data dependent background memory accesses can be removed, leading to a more manifest behaviour. The line buffer data reuse is not introduced in this step because it has a too negative impact on power.

Basic group matching [524, 166, 164] has a big positive impact on the off-chip power. The signals *pyr* and *ridge* are nearly always accessed together (same time, same address). These two main signals are merged together, decreasing the off-chip memory count from three to two. Also

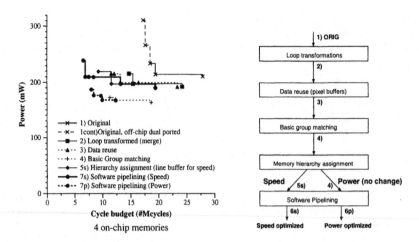

Figure 6.7.   Power versus performance trade-offs (Pareto curve) for different transformed BTPC code.

some on-chip signals are merged. Though this has a large possible power impact, the minimal cycle budget is nearly unchanged.

In the hierarchy assignment the data reuse copies are grouped into a final memory hierarchy. The on-chip memories in this hierarchy can have multiple ports at a relatively low cost. For these performance reasons, we have added the line-based data-reuse level. Here, we were able to introduce multiple line based signals for which we could prevent dual-ported memories. This transformation is not applied when the main emphasis is power. The power and performance exploration path do separated here.

The power oriented path (without the memory hierarchy assignment) is given with dotted line in the graph. It does not reach the same cycle budget but the power remains low (the last 18% performance gain costs 28% in power). Of course it is a matter of constraints and tradeoffs which implementation is chosen finally.

In a last step, the code has been software pipelined. Data dependencies initially exclude optimal off-chip memory bandwidth usage. A certain time-slot can be blocked to be used efficiently. For instance when a loop body needs data from an on-chip memory before writing to an off-chip memory then the first cycle of the loop body cannot be used for the off-chip access. By moving the memory accesses to a next or previous loop iteration, the dependency can be broken.

The trade-off between the power and performance optimisation directions is shown more clearly in Fig. 6.8. The overall Pareto curve (the fat dotted one) obviously combines both differently optimized "sub-Pareto" results. For the high cycle budget and low performance situations, the low energy implementation is better. Only for a very high performance, the speed optimized implementation should be used.

In total, a minimal cycle budget of 6.5 million cycles is achieved. The manual transformations have lead to a factor 4 performance improvement using the same power budget. This is the fastest achievable when having a single-ported off-chip memory (every memory cycle contains an off-chip access to the *in* signal). We will now also show how these curves can be effectively used to significantly improve the overall system design of data-dominated applications.

6.4.3 Exploiting the use of the Pareto curves

The previous subsections have shown which factors do influence the memory cost. The memory costs increases when a higher memory bandwidth is needed. However, the memory subsystem is not the only cost for the complete system. Data-path is not neglectable always. Moreover, communicating tasks or threads make the problem more complex.

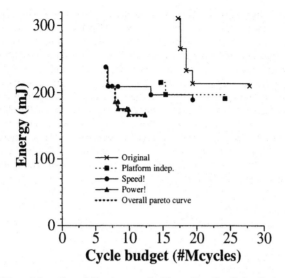

Figure 6.8.   Overall Pareto curve for Binary Tree Predictive Coder.

In this subsection, we will not go into detail on how to obtain the information for the curve itself yet (see section 6.2.2, 6.5 and [2, 66]), but we will show how to use them effectively in a low-power design context. Moreover, detailed real life examples are worked out.

6.4.3.1   Trading off cycles between memory organisation and data-path.   In case of a single thread and the memory subsystem being energy dominant, the tradeoff can be based solely on the memory organisation Pareto curve. The given cycle budget defines the best memory architecture and its corresponding energy and area cost. Due to the use of latency hiding and a "system-level pipeline" between the computation and data communication, most of the cycles where memory access takes place are also useful for the data-path. Still, to reduce the cost of the data-path realisation, it is important to leave some of the total cycle budget purely available for computation cycles (see further).

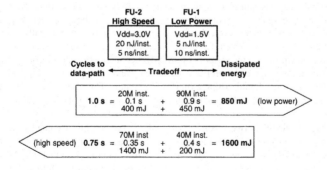

Figure 6.9.   Datapath example of a performance vs. energy tradeoff.

We will now illustrate how the assignment of spare cycles to the data-path can have a significant impact on its energy consumption. Assigning too little time to the functional units indeed introduces an energy waste in the functional units when implemented in a low-power set-up. For instance, in several previous papers (see e.g. [101] and its refs) it has been motivated that functional units should operate at several voltages. Especially in recent papers, the number of different Vdd's has been restricted to a few specific values only because of practical reasons. Assume now that we either have a single unit that can operate part of the time at a higher Vdd and partly at a lower Vdd, or similarly two parallel units with different Vdd's over which all the operations can be distributed (see Fig. 6.9). This duplication will require some extra area but on the entire chip this will involve a (very) small overhead and for energy reasons this can be motivated. The high Vdd unit will clearly allow a higher speed. In the simple example of Fig. 6.9, 110 million instructions need to be executed for the entire algorithm. Most instructions can be executed in the low energy functional unit in a very "loose" data-path schedule (in total 1.0s results in consuming 850mJ). The same functionality executed 25% faster, consumes twice as much energy due to the requirement to execute more on the high speed functional unit (in total 0.75s results in 1600mJ). The assignment of cycles to memory accesses and to the data-path is important for the overall energy consumption. Obviously, the contributions of both functions, the functional units and memory subsystem, are summed. This motivates a tradeoff between the memory system and the data-path itself.

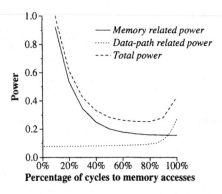

Figure 6.10.   Trading off cycles for memory accesses versus cycles for data-path operations that due to dependency and ordering constraints cannot be executed in parallel with any data transfer.

The reality is still more complex though. A certain percentage of the overall time can be spent to memory accesses and the remaining part to computational issues, taking into account the system pipelining and the (large) overlap between memory accesses and computation. However, the more time of the overall budget is committed to the memory accesses, the more restrictions are imposed on the data-path schedule which has less freedom for the ordering and the assignment to the functional units. As a result (see dotted curve), the Pareto curve for the data-path usually exhibits a cost overhead which is small and constant for small to medium memory cycle budgets, and which only starts to increase sharply when the memory cycle budget comes close to the globally available cycle budget (see Fig. 6.10). Due to the property that Pareto curve are monotonous in their behaviour, the sum of these 2 curves should exhibit a (unique) minimum. This also advocates for an accurate energy versus computation cycle budget modeling for the complex data-path functions.

6.4.3.2   Energy tradeoffs between concurrent tasks.   The performance-energy function can be used even better to tradeoff cycles assigned to different tasks in a complex system. Our Pareto curves should indeed be generated per concurrently executed task. It is then clearly visible that assigning too few cycles to a single task causes the cost of the entire system to go up. The Pareto curves also show the minimal (realistic) cycle budget per task. The cycle and energy estimates can clearly help the designer to assign concurrent tasks to processors and to distribute the cycles within the processors over the different tasks (see Fig. 6.11). Minimizing the overall energy within a heterogeneous multi-processor is then possible by applying Pareto function minimization on all the energy-cycle functions together. Even when multiple tasks have to be assigned to a single processor under control of an operating system, interesting trade-offs are feasible (a detailed example is givin in [66]).

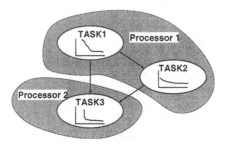

Figure 6.11.   Tradeoff cycles assigned tasks.

6.4.4 Demonstrator: Synchro core of DAB

In the near future mobile radios with Digital Audio Broadcast (DAB) reception will be produced. A DAB broadcaster is able to provide either 6 high quality radio programs with associated data, or for instance one MPEG 1 video signal. DAB provides any combination of services up to a total of about 1.8Mb/s. DAB uses Orthogonal Frequency Division Multiplex (OFDM) modulation which is resistant against multi path interference and frequency-selective fading.

Here, we will focus on the synchronization processor only [69]. The synchronization is performed in two steps. In the first step a correlation to a known reference signal determines the frequency error. Next, an IFFT is performed obtaining the channel impulse response.

Fig. 6.12 shows the power curve representing the useful tradeoff between cycle budget and power cost, from the fully sequential case down to the case close to the critical path. An equal graph can be made for the area cost. Note the discontinuities in the cost function when a dual ported memory is added or changes in the array to memory assignment.

The rightmost point of the graph is the fully sequential case. All arrays can be assigned to a single memory because no conflicts are available. However, for power and area considerations four single-port memories are allocated. In case of multiple memories, the signals are stored in smaller memories and less bit-waste due to bit-width differences in the signals. The cycle budget can be decreased down to 85 kCycles without increasing the power nor the area. Note that the SCBD step generates more points in the graph but the MAA step could sustain an equally (good) memory hierarchy so the extra points are not useful in the Pareto curve. In order to reduce the budget below 85 kCycles, dual-ported memories are needed. Allocation of 4 dual-ported memories allows to decrease the cycle budget close to the critical path (55047 cycles).

The platform independent DTSE optimisation stages of chapter 3 and 5 have been manually applied on the DAB application. As a result, the optimized implementation needs less accesses and memories for the same function and cycle budget. The optimized implementation also consumes much less power. Moreover, it can sustain this low power by a much lower cycle budget. It can nearly reach the critical path (45378 cycles) without sacrificing memory power. To obtain the SCBD results, the tool execution time is in the order of several seconds. So it can easily be used to give feedback to the designer.

A tradeoff is involved in how much time is assigned to the synchronization process and to the rest of the application. We focus here on three tasks of the DAB decoder. First there is an FFT task, to be executed for the 76 symbol of a frame. Concurrent with this FFT is the synchronization; it works with the reference symbol (the 1st symbol) of every frame, and consists of a correlation task followed by an Impulse Response Analysis (IR-Analysis) task. An overall real-time constraint of 96ms, the frame length, applies for all three tasks together. Assuming a speed of 40MHz, a low-speed and low-energy mode, 3840K cycles are available in a frame. In addition, constraints exist due to the availability of the incoming data.

A naive design would allocate one separate processor for the FFT task and one for the synchronisation tasks. This would also require a separate memory organization and communication channels between the two processors. The top figure in Fig. 6.13 shows how the tasks are organized in time,

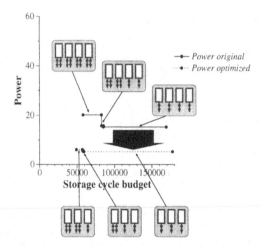

Figure 6.12.   Power cost and memory architecture given the cycle budget for original and optimized description.

and what the memory architecture is for this processor. The arrows depicted with a "1" in Fig. 6.14 are the corresponding points in the task Pareto curves.

With the correlation done in parallel to the FFT, the processing and memory resources are very much under utilized. A designer can then decide to map all three tasks onto one processor, and to make use of the "spare time" in the NULL symbol time to perform the correlation. However, the Pareto curve in Fig. 6.14) shows that this leads to a very costly solution (the correlator task cost increases). A very high bandwidth to the memories is required to complete the correlation in time, and a four-port memory must be allocated to provide this bandwidth. The complexity of such a memory architecture makes this option simply unfeasible.

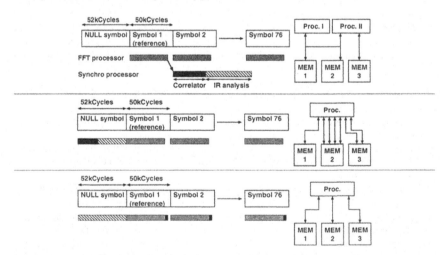

Figure 6.13.   The original task to processor mapping needs 2 processors (top); A naive more energy consuming combination of the two tasks to a single processor (middle); An improved task rescheduling overcomes the cost increase (bottom).

From the FFT pareto curve of Fig. 6.14 it is visible that there is some slack. This time can be used to do the correlation. Then only the IFFT and IR-analysis are performed during the NULL symbol time interval. A redefinition of the tasks corresponding to this decision, and the associated memory architecture, are shown at the bottom part of Fig. 6.13. Finally the points annotated with "3" are chosen in the pareto curves of Fig. 6.14 allowing a single processor with three memories.

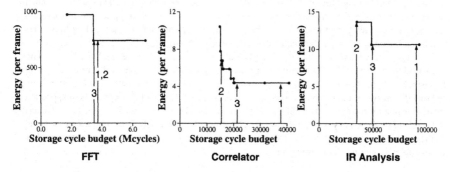

Figure 6.14.   Pareto curves of three different sub tasks of the DAB. The numbers at the arrows correspond to the implementation number of the previous figure.

Thus, by exploring the trade-off with the aid of the Pareto curves, the global system cost is effectively reduced (from two processors to one), while still meeting the real-time constraint and without an energy penalty. This exploration clearly shows the different solutions and design tradeoffs and is very useful for the system designer to allow a globally motivated distribution of the timing constraints over the system modules. This indirectly also has a significant impact on overall power.

6.5   STORAGE CYCLE BUDGET DISTRIBUTION TECHNIQUES

As already indicated, this step is needed to make sure that the real-time constraints are met at minimal cost. All the memory accesses in the application have to be made within the given real-time constraint. A certain amount of memory bandwidth is needed to meet this constraint. The goal of the SCBD step is to order the memory accesses within the available time, such that the resulting memory bandwidth is as cheap as possible. The small example in Fig. 6.15 illustrates the problem and will be used through the entire section.

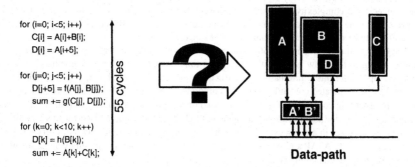

Figure 6.15.   Problem definition: map an application description to on a (predefined) memory architecture while still meeting the timing constraints.

To meet the real-time constraints, the data storage components of the system (i.e., the memory architecture) must be able to provide the required data to the data processing components fast enough. Also the data processing components must work fast enough. However, the two can be cleanly separated. Most of the transfers can take place while processing of other data is going on; there may be a small amount of processing or a small amount of transfers which cannot be overlapped, but between these there is a broad range of exploration, as indicated in subsection 6.4.3.1.

It is important to explore the ordering freedom of the data transfers before scheduling the data processing itself, and not the other way around. First of all, the cost of the data transfers is much higher than the cost of data processing, especially in data-transfer intensive applications. In addition, the constraints coming from first ordering the data transfers do not limit the scheduling freedom of the data processing too much, while the reverse is certainly not true.

To keep complexity reasonable, it is necessary at this stage to work at the level of arrays, and not to expand them into scalars. In data-transfer intensive applications, the size of the arrays can go up to several millions of words, making scalar approaches completely unfeasible. On the other hand, treating the arrays as a unit hardly limits the freedom; indeed, they will in any case be mapped into memories in a uniform way. The only disadvantage is that the analysis needs to be more advanced to be able to deal correctly with arrays [40, 93]. So array data flow analysis techniques are essential, as they were also for several other DTSE steps already. A detailed discussion on that literature falls outside the scope of this book, however.

6.5.1   Summary of the flat flow graph technique

The main goal of the approach is to find which arrays must be stored in different memories in order to meet the cycle budget (for details see [90, 566]). Ordering the memory accesses within a certain number of memory cycles determines the required memory bandwidth. The concept of memory cycles does not have to equal the (data-path) clock nor the maximum access frequency of the memory, it is only used to describe the relative ordering of memory accesses. Every memory access is assumed to take up an integer number of abstract cycles. A memory access can only take place in a cycle after any access on which it depends. The found ordering does not necessarily have to match exactly with the final schedule (complete scheduling, including data-path related issues). It is produced to make sure that it is possible to meet the real time constraints with the derived memory architecture.

If two memory accesses are ordered in the same memory cycle, they are in conflict and parallelism is required to perform both accesses. These simultaneous memory accesses can be assigned to two different memories, or, if it is the same array, to a dual port memory. Thus, the conflict constrains the signal to memory assignment and incurs a certain cost. A tradeoff has to be made between the performance gain of every conflict and the cost it incurs [66, 63]. Hence we call this substep Conflict Directed Ordering (CDO). On the one hand, every array in a separate memory is optimal for speed and seemingly also for power. But having many memories is very costly for area, interconnect and complexity and due to the routing overhead also for power in the end. Due the to presence of the real time constraints and the complex control and data dependencies, a difficult tradeoff has to be made. Therefore automatic tool support is crucial.

The CDO tool described in [90, 566] and our previous book [93] orders by "balancing the flow graph" a single (flat) flow graph within the given time constraints. The flow graph is extracted from a C input description. Internally, it constructively generates a (partial) memory access ordering steered by a sophisticated cost model which incorporated global tradeoffs and access conflicts over the entire algorithm code. The cost model is based on size, access frequency and other high level estimates (eg. possibilities for array size reduction [146]). The technique used for this is to order the memory accesses in a given cycle budget by iteratively reducing the intervals of every memory access (starting with ASAP and ALAP) [566]. The interval reductions are driven by the probability and cost of the potential conflicts between the accesses.

An Extended Conflict Graph (ECG) is generated which follows out of the memory access ordering. That provides the necessary constraints for the MAA step and the subsequent compiler phases. It contains the conflicting arrays which are accessed simultaneously. These arrays need to be stored in different memories. Note that many possible orderings (and also schedules) are compatible with a given ECG.

6.5.2 Dealing with hierarchical graphs

The technique mentioned above can only be directly used for flat flow graphs, i.e. when no loops and no function calls are present. Loops could be unrolled and function calls could be inlined, but this destroys the hardware or code reuse possibilities present in the original specification. Moreover, the complexity would explode. Therefore, ordering the memory accesses has to be done hierarchical.

We define a *block* as a code part which contains a flat flow graph. It can contain multiple basic blocks and condition scopes. Even the function hierarchy can be adapted to the needs of ordering freedom. In practice, the term loop body and block coincide; all loop bodies are defined as a separate block and the body of a nested loop is part of an other block.

Because the number of iterations to the blocks is mostly different, the problem is increased in two directions. First, the distribution of the global cycle count over the blocks need to be found (see subsection 6.5.3), within the timing constraints and while optimizing a cost function. Second, a single memory architecture must satisfy the constraints of all blocks, and therefore the global ECG cost must be minimized. Combining all the conflicts of the locally optimized blocks in the global conflict graph will lead to a poor result (see subsection 6.5.4). Reuse of the same conflict over different blocks is essential.

In our application domains, an overall throughput constraint (maximal or average) is put forward for the entire application. For instance, in a video application the timing constraint is 40ms to arrive at 25 frames per second. Sometimes, additional timing constraints are given. Here, we will only deal with one global timing constraint but extensions to support additional internal constraints are feasible on top of this.

6.5.3 Cycle distribution across blocks

The distribution of the cycles over the blocks is crucial. A wrong distribution will produce a too expensive memory architecture because the memory access ordering cannot be made nicely in some of the blocks while there are "cheap" cycles available in other blocks. Every single block affects the global cost of the memory subsystem. Therefore, if the cycle budget of one block is too tight (while there is space in other blocks) it will cause additional cost.

The global cycle budget can be distributed over different blocks in many different ways, as shown Fig. 6.16. At the left side of the figure a program is given containing 3 consecutive loops, to be ordered in 55 cycles. The table of Fig. 6.16 shows three different distributions and a good ordering matching the distribution. The resulting conflict graph and cheapest memory architecture is given in the last two rows. Obviously, the first solution (*loop-i* 3 cycles, *loop-j* 3 cycle and *loop-k* 2 cycles) is the cheapest solution. The very poor third distribution even forces a dual-ported memory (due to assignment of one cycle for *loop-i*).

The illustration here is based on an academic example to show the problem. But in fact, the problem in real-life applications is much more difficult. First, because more signals, accesses and blocks are involved, the number of different possible distributions increases. Second, the number of block iterations will be very different for each block. The impact on the global time elapse of one block is much bigger than another block. Hence, it is a very complex search space to explore.

6.5.4 Global optimum for ECG

A global optimisation over all blocks is needed to obtain the global optimal conflict graph. Ordering the memory accesses on a block per block basis will result in a poor global result. The local solution of one block will typically not match the (local) solutions found in other blocks. Together, the local optima will then sum up to an expensive global solution (see Fig. 6.17).

Solving the problem locally per block will lead to different local conflict graphs. The total application has one memory architecture and therefore one global conflict graph only. The memory architecture cannot change from one block to the next. Therefore, all the block (local) conflicts graphs should be added to one single global conflict graph. A typical conflict mismatch is shown in the right side Fig. 6.17. A fully connected global graph is the result, requiring four memories.

Ordering the memory accesses with a global view can potentially "reuse" conflicts over different blocks. When the different blocks use the same conflicts, as shown at the left hand side of

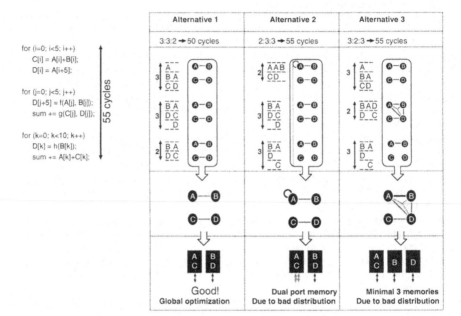

Figure 6.16.   The distribution over loops has a big impact on the memory architecture.

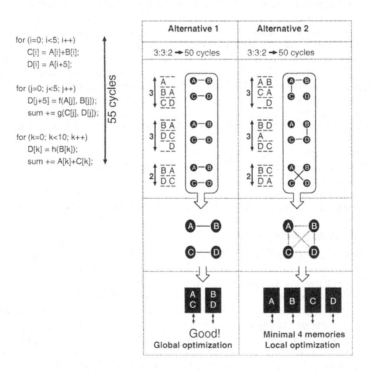

Figure 6.17.   Global optimum instead of local optima.

Fig. 6.17, the global conflict graph and the resulting memory architecture are much cheaper. Again, the academic example shows the essence of the problem. However, the real problem is much more complex.

6.5.5 Incremental SCBD

For solving the storage cycle budget distribution over multiple blocks we have placed the flat-graph solver (described in subsection 6.5.1) in a loop, iterating over all the blocks. As motivated in Subsection 6.5.4 as such, this leads to a poor global optimum. This is solved by suggesting reusable conflicts to the flat-graph solver. Moreover, additional constraints are put forward to steer the process.

The incremental SCBD reduces the global cycle budget every step until the target cycle budget is met. Initially, the memory access ordering is sequential. Therefore every block can be ordered without any conflicts. During the iteration over the blocks, the cycle budget is made smaller for every individual block. Gradually, more conflicts will have to be introduced. The SCBD approach decides which block(s) are reduced in local cycle budget and so which conflicts are added globally. Finally, after multiple steps of decreasing the budgets for the blocks, the global cycle budget is met and a global conflict graph is produced. This is less trivial than it looks, as shown next.

The proposed SCBD algorithm initializes with a sequential ordering (see Fig. 6.18). Every memory access has its own time slot. A block containing X memory accesses will have an initial cycle budget of X cycles (assuming one cycle per memory access). Due to this type of ordering, the global conflict graph will not contain any conflicts.

Initialisation:

```
for (all blocks)
   Use sequential ordering

The found ECG will contain NO conflicts
```

Iteration:

```
While (target cycle budget > cycle budget)  // step
   for (all blocks)
      reduce cycle budget of block
      return (some) ordering freedom
      execute flat-graph optimizer
   update one or multiple blocks (based on gain/cost)
```

Figure 6.18.   Basic algorithm for incremental SCBD.

In the successive steps of the algorithm the global budget will shrink. Every step, (at least) one of the blocks will reduce in length. The reduction of the block is at least one cycle. However, depending on the number of iterations of the concerning block, the impact on the global budget is much bigger. The global conflict cost change is calculated in the case of the block reduction. But the reduction is not approved yet. The block(s) having the biggest gain (based on a change in total cycle budget and/or change in conflict cost) is actually reduced in size. All the other ordering results in this step are discarded. The cycle budget reduction is continued until the target cycle budget is reached. Due to the block cycle budget reduction, additional conflicts arise. Note that this basic algorithm is greedy, since only a single path is explored to reach the target cycle budget.

The traversal of the cycle search space can be made less greedy however. At every step, multiple reduction possibilities exist. Instead of discarding non-selected block ordering information (as proposed in the previous paragraph), these can be selected and explored further. Different block reductions (paths) and finally the entire solution tree can be built. Many of the branches will be equivalent to the already found "greedy solutions". Due to this property, the exploration will not explode but still extra solutions can be found. Moreover, a lower bound can also cut off some of the potential branching paths. Note however, due to the heuristics in the flat graph solver, the solution may be different even though the distribution of cycles over the blocks is equal. In this way, the longer the tool will run, the more and (maybe) better solutions can be found.

The incremental nature is further explained in Fig. 6.19. The source code is shown left. The first loop contains five memory accesses and has five iterations. Therefore, in total 25 cycles are spent in *loop-i* when executed sequentially. Similarly, 25 cycles in *loop-j* and 40 cycles in *loop-k*. In total 90 cycles are spent for the entire algorithm. In this example, *loop-i* is selected in the first step and reduced from five to four cycles, causing a total cycle budget reduction from 90 to 85. In the next step, *loop-j* and *loop-k* are reduced, leading to a total cycle budget of 70 cycles. In the remaining two steps the cycle budget is reduced 50 cycles by adding one extra conflict. In the last step, the target budget is met and the algorithm ends.

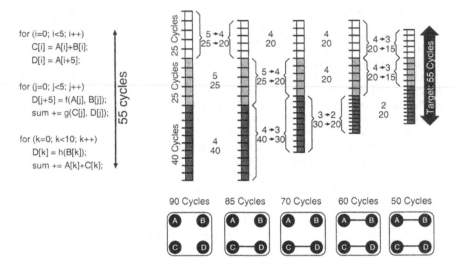

Figure 6.19.   Graphical representation of incremental SCBD.

To improve the global result and to avoid a long execution time and unstable behaviour, two new inputs have to be entered to the flat-graph scheduler. First of all, a list of reusable conflicts is specified. The internal cost function is adapted to (re)use conflicts which are already used by other blocks if possible. Second, the ordering freedom is limited. The returned ordering freedom to a block is based on the final ordering of the previous step[3].

The algorithm is sped up further by reusing ordering results which did not change. Since much ordering information is discarded in a step, this does not mean it is useless. By keeping track of which blocks have to be rescheduled, the tool execution time can be decreased drastically. This happens especially in large applications containing many independent blocks.

Currently, the flat flow graph technique (see subsection 6.5.1) used has already proven its value in the past. But its speed becomes unacceptable large for huge loop bodies. Research is in progress to use dynamic programming and different type of schedulers to drastically speedup the original technique (see section 6.6 and [395]).

We have built a prototype tool which incorporates all the techniques of this section, to avoid the tedious labour for a designer.

6.5.6 Experimental SCBD results

The new prototype tool has been applied to drivers from multiple application domains to prove the effectiveness, as demonstrated by results in other recent papers [69, 66, 63, 524]. Here we will analyse the results and the detailed evolution of the SCBD tool for the Binary Tree Predictive Coder driver only. The experiment clearly shows the tradeoff between memory organisation cost (power

---

[3]The memory access is scheduled between the ASAP and ALAP time. Both the ASAP and ALAP are put close to the location of the previous ordering.

and area) and the memory subsystem cycle budget. All the results are obtained within reasonable tool execution time (several minutes) on a Pentium II-400. A Motorola library memory model is used to estimate the on-chip memory cost (see [93]). For the off-chip components, we have used an EDO DRAM series of Siemens.

Fig. 6.8 shows the tradeoff for the complete cycle budget range. This is obtained by letting the tool explore the cycle budget starting from the fully sequential budget, and then progress through the most interesting memory organisations (from a cost point of view) to reduce the cycle budget for multiple differently optimized implementations [63]. The number of allocated on chip memories is four for the entire graph shown in this figure. In the sequential budget (about 18M cycles), only single-port memories are employed. When a dual-ported memory has to be added, a clear discontinuity is present in the energy function. In order to reduce the budget below 8M cycles, dual-ported memories are needed though. The allocation of two dual-ported memories allows to decrease the cycle budget up to the critical path (6.5M cycles). The worst case needed "image" bandwidth can be guaranteed by inserting three intermediate on-chip memories (two dual port and one single port). These three intermediate memories can deliver up to five pixels per memory cycle.

The cycle distribution over the eight loops of the BTPC application is performed incrementally by the tool. The evolution of this distribution is shown in Fig. 6.20. At every step, the differently shaded bars represent the consumed cycle budget of every loop ("loop body time" × "number of loop iterations"). The lowest gray part represents *loop-1*, the next white part represents *loop-2*, the next one *loop-3* etc.. As can be seen in the figure, *loop-1* is dominant and consumes nearly 50% of the overall time in the fully sequential case (step-0). In addition, *loop-4* is so small that it only appears as a thin line. The dotted line and solid line represent respectively the estimated cost (internally estimated in SCBD without generating a memory organisation) and the actual power cost after memory allocation and assignment.

In the progressing steps of SCBD, the cycle budget is reduced and conflicts are added gradually. Step-0 is the fully sequential budget containing no conflicts. Therefore, the estimated cost is zero. However, the memory architecture will (of course) consume some power. In the first step *loop-2, 3* and *8* are reduced in cycle budget, decreasing also the global cycle budget just below 20M cycles. The (few) added conflicts increase the cost estimate. The actual power does not increase because the added conflict does not enforce changes in the optimal signal to memory assignment. The actual power cost does not increase until step-11. Then the cycle budget is reduced by 40% already. Due to an added selfconflict in step-16, both the estimated cost and actual cost make a large jump. For step-24 and step-25 the memory architecture constraints are demanding a 4-port memory. For the used memory library this is not feasible. Therefore, no valid solution exists any longer.

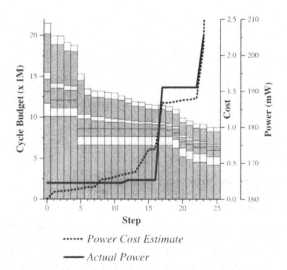

Figure 6.20.   The distribution of cycles over the loops.

The tool feedback is more then a memory architecture alone. Optimisation, for further cycle budget or cost, reduction can be based on the found schedule and conflict graph. A scheduled flow graph of a single loop is given in Fig. 6.21. Also the global conflict graph of this step is given in Fig. 6.22. The global conflict graph contains the conflict of the scheduled loop and of all the other loops. So in time-slot 2 the signals *in* and *glob_freq1* are scheduled together, and therefore it appears in the conflict graph. The gray nodes and edges in Fig. 6.21 show the critical path. Indeed, the dataflow of the critical needs to be broken to reduce cycle budget further. This is typically done with software pipelining techniques which are not discussed in this paper. Another approach to reduce the cycle budget further is partially loop unrolling. Then, a large parallelism becomes available. Note that this comes with a potential very high cost in both required bandwidth and code size. The cost of the memory architecture is reflected in the conflict graph. By analyzing the conflict graph at every step it can be located where a cost increase occurs and why. Counter measures can be taken in the form of basic group matching [166] and inserting intermediate arrays [154].

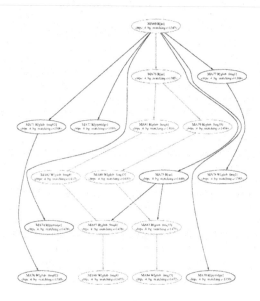

Figure 6.21.   Scheduled flow graph of one loop in step 23.

These large performance gains are only possible at moderately low cost when the memory architecture is designed carefully. At step-23 the cycle budget is reduced with 60% at a cost of 25% more power consumption, 5% more area and more complex interconnect compared to the fully sequential solution (step-0).

This experiment also proves the high usefulness of the tool. All these results can only be generated effectively by the introduction of our new approach and tool. It is clear that manual design would never allow to explore such a huge search space.

6.6   CONFLICT DIRECTED ORDERING TECHNIQUES

6.6.1   Basic principles

The idea of Force-Directed Scheduling (FDS) was first introduced by Paulin and Knight [425] to minimize the number of resources required in the high-level synthesis of high-throughput Application Specific Integrated Circuits (ASICs). FDS was later chosen as the basis of the Philips Research - PHIDEO silicon compiler, which resulted in an in-depth study of FDS implementation issues and

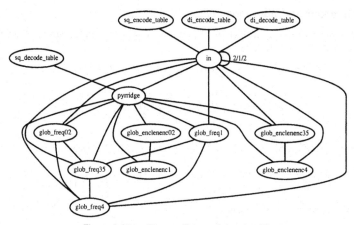

Figure 6.22.   The conflict graph for step 23.

improvements in [528]. State-of-the-art in FDS reports a runtime complexity of $O(m^2 n^2)$, where n is the number of operations and m is the average time window of each operation[4], and a feasibility of solving problems of n equals 100 in one minute using the computers of the mid 90s.

Because the requirement for the Conflict Directed Ordering (CDO) problem is based on global system-level analysis as opposed to the more local features of FDS, it can be proven that its straightforward implementation exhibits a runtime complexity of $O(n^4 m)$ [398]. This allows to solve problems for n equals 100 in only ten hours, even using today's workstation series. Moreover, the growing complexity of embedded multimedia systems results in embedded kernels exhibiting a number of memory operations well beyond n equals 100 hence necessitating the introduction of instruction-level parallelism to still meet stringent deadlines, which again contributes to growing time windows m. This trend for growing system complexity leads to (very) large-scale (access) scheduling problems, which necessitate novel, fast and effective solution techniques. To improve the designer interaction further, it is also desirable to have a controllable trade-off between user time and the quality of the solution (which directly influences the cost of the final solution).

In this section, we revisit the original idea of [425, 426] and show show that FDS can be treated efficiently as a *dynamic* graph algorithm [394, 395]. Based on this property, we introduce a Localized Conflict- Directed Ordering (LCDO) algorithm that instead of performing global system exploration at each optimization step relies on local system explorations, without any loss in the solution quality. This local reduction of the search space allows to reduce the overall complexity by one order of complexity, from $O(n^4 m)$ to $O(n^3 ln(n)m)$. Moreover, the large size of the remaining LCDO search space allows to also benefit from the latest scheduling and partitioning techniques ([200, 263, 218]) for a *k-parameterizable* computer-aided design runtime complexity of $O(kn^2 ln(n))$. Experiments of using the original CDO and LCDO on three real-life applications confirm that the design time complexity can be reduced by two orders of magnitude and more. Moreover, an interactive choice of the parameter k is shown to be an appropriate solution for designing large-scale and/or parallel data-dominated systems under stringent real-time constraints. These issues are discussed in more detail in the subsequent subsections.

---

[4]For a formal definition of n and m, please consult subsection 6.6.3.1

6.6.2 Illustration and motivation

Imagine that you have at your disposal two 1-port memories to compute $2n$ terms of the following four series[5] in $4n+2$ memory cycles.

$$s_1^{(0)} = 5, \quad s_1^{(2i-1)} = (s_1^{(2i-2)} - 2) \times 3 \quad s_1^{(2i)} = (s_1^{(2i-1)} - 1) \times 2$$

$$s_2^{(0)} = 5, \quad s_2^{(2i-1)} = (s_2^{(2i-2)} - 3) \times 4 \quad s_2^{(2i)} = (s_2^{(2i-1)} - 1) \times 2$$

$$s_3^{(0)} = 5, \quad s_3^{(2i-1)} = (s_3^{(2i-2)} - 3) \times 4 \quad s_3^{(2i)} = (s_3^{(2i-1)} - 2) \times 3$$

$$s_4^{(0)} = 5, \quad s_4^{(2i-1)} = (s_3^{(2i-2)} - 2) \times 3 \quad s_4^{(2i)} = (s_4^{(2i-1)} - 1) \times 2$$

You would probably start by writing the following C code:

```
float S1[2*N+1],S2[2*N+1],S3[2*N+1],S4[2*N+1];
S1[0]=S2[0]=S3[0]=S4[0]=5.;
for (int i=1; i<N+1; i++) {
  S1[2*i-1]=(S1[2*i-2]-2.)*3.;  S2[2*i-1]=(S2[2*i-2]-3.)*4.;
  S3[2*i-1]=(S3[2*i-2]-3.)*4.;  S4[2*i-1]=(S3[2*i-2]-2.)*3.;
  S1[2*i]=(S1[2*i-1]-1.)*2.;    S2[2*i]=(S2[2*i-1]-1.)*2.;
  S3[2*i]=(S3[2*i-1]-2.)*3.;    S4[2*i]=(S4[2*i-1]-1.)*2.;
};
```

The scheduler might come up with the following solution:

```
1: Load S1[0];  Load S2[0];    2: Load S3[0];  Store S1[1];
3: Store S2[1]; Store S3[0];   4: Store S1[2]; Store S2[2];
5: Store S4[1]; Store S3[2];   6: Store S4[2]; Store S1[3];
7: Store S2[3]; Store S3[3];   8: Store S1[4]; Store S2[4];
9: Store S4[3]; Store S3[4];  10: Store S4[4]; etc ...
...
2N+1: Store S4[2*N-1]; Store S3[2*N];
2N+2: Store S4[2*N];
```

This ordering does not violate data-flow dependencies, nor does it exceed a memory bandwidth of 2 and nor does it try to place two memory accesses to the same array within the same cycle, hence requiring expensive dual-port memories. However, because of array-to-memory placement constraints, this solution is not feasible using two 1-port memories. Hence, a third 1-port memory has to be added. In order to break this access bottleneck, one needs to find the following schedule:

```
1: Load S1[0];  Load S2[0];    2: Load S3[0];  Store S1[1];
3: Store S1[2]; Store S3[1];   4: Store S2[1]; Store S4[1];
5: Store S2[2]; Store S4[2];   6: Store S3[2];  etc ...
...
2N+1: Store S2[2*N]; Store S4[2*N];
2N+2: Store S3[2*N];
```

This solution is feasible using two 1-port memories by carefully assigning S1 and S4 to the first memory and S2 and S3 to the second memory. In our DTSE script, this post-assignment is the scope of the MAA step.

---

[5]Those series can for example arise in financial modeling. s_1 is a strategy consisting in alternating between medium and safe investment. s_2 is a strategy consisting in alternating between risky and safe. s_3 is a strategy consisting in alternating between risky and medium. s_4 is about leaving the most risky strategy s_3 for the safest strategy s_1 in the last round. In the near future, this type of application will be available in embedded decision support systems.

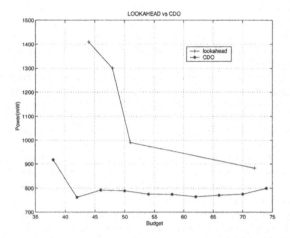

Figure 6.23.   List Scheduling with Look-Ahead (LS-LA) vs. our CDO

As a conclusion, schedulers targeted at embedded software **must** take into account *which* data is being placed in parallel in order to come up with effective solutions. As an illustration, we will now evaluate the well-known List Scheduling (LS) [359, 48] algorithm. LS is often cited by both the compiler and system synthesis communities as a reference ordering technique. We have applied a robust List Scheduling with Look-Ahead (LS-LA) to the design of a realistic Segment Protocol Processor module from an ATM network protocol application (see subsection 6.6.5.1 for a more detailed description of this module). We have measured the power consumption produced by LS-LA and compared it to the power-consumption of our reference, Conflict-Directed Ordering (CDO). The results presented in Fig. 6.23 clearly indicate three trends:

1. For soft real-time constraints (*i.e.* for a real-time budget higher than 50 memory cycles in our example), LS-LA provides a low-quality solution, already at a 10-20% distance from that of CDO;

2. For medium real-time constraints (*i.e.* for a real-time budget in the range of 44 to 50), LS-LA provides ineffective solutions, typically exhibiting twice the power consumption of CDO;

3. For hard real-time constraints (*i.e.* for a real-time budget below 44), LS-LA fails to avoid an extensive use of expensive multi-port memories while CDO can find a smart solution for a real-time budget as low as 38.

As a conclusion, conventional *scheduling-for-speed* algorithms (like LS-LA) are very ineffective for designing embedded systems under hard real-time constraints. In this context, as shown on our SPP example, conventional schedulers (like LS-LA) typically lead to solutions exhibiting twice the necessary power-consumption under real-time constraints relaxed by 10% and more. Moreover, those relaxed constraints are likely to generate an extra factor of 2 increase in cost in practice, by forcing the designer to duplicate the system to still meet his hard real-time requirement, for example using parallel computing techniques. Recent experiments [543, 410] estimate that these shortcomings in the quality of the memory to processor scheduling represent a typical 40% degradation of time performance. In the coming subsections, we will therefore develop novel algorithms based on the idea that schedulers **must** take into account *which* data is being placed in parallel.

6.6.3   Exploration of access orderings

Our problem is the problem of minimizing the cost of embedded multimedia systems, which can be partly defined as an optimization problem based on a flow-graph, an objective function to minimize and architectural constraints.

Using the example of subsection 6.6.2, placing `Load S1[0]; Load S2[0];` at memory cycle 1 and then `Load S3[0]; Store S1[1];` at memory cycle 2 and then again `Store S2[1]; Store S3[0];` at memory cycle 3 in parallel requires (at least) a 3-port memory architecture whereas only 2 accesses are being performed in parallel at each memory cycle. This is cost-ineffective by at least 50%.

In the following, we derive ordering techniques to remove this memory bandwidth bottleneck starting from ANSI-C input, without any restriction in the input format like many alternative techniques in the parallizing compiler domain have imposed.

6.6.3.1 Problem formulation.
Several definitions and theorems will be presented first, which define the context and the problem.

Definition 1. *EFG* An extended memory access flow graph $EFG = (V, E, O)$ is a node-weighted directed graph where V is the set of memory accesses, $E \subseteq V \times V$ is the set of precedence constraints, and $O \subseteq V \times B \times T^2$ specifies a time window (or partial ordering) $[t_1, t_2] \in T^2$ for each memory access $v \in V$ to a basic group of scalars in memory $b \in B$.

Remark 1. A basic group of scalars is an indivisible set of data w.r.t. its physical mapping onto memories. The choice of the basic groups B is a sensitive choice in the definition of an *EFG* [164].

Terminology 1. *ASAP and ALAP* t_1 is referred to as the *As Soon As Possible (ASAP)* timing for v, $ASAP(v)$. t_2 is referred to as the *As Late As Possible (ALAP)* timing for v, $ALAP(v)$. O is referred to as a *partial ordering* of *EFG*.

Definition 2. *acceptable* Given a memory access flow graph (V, E) to the *EFG*, an *acceptable* partial ordering O is a partial ordering of *EFG* meeting all the precedence constraints:

$$\forall e = (v_1, v_2) \in E, \quad t_1(v_2) \geq t_1(v_1) + 1 \quad t_2(v_1) \leq t_2(v_2) - 1$$

Definition 3. *maximal* Given a memory access flow graph (V, E) to the *EFG*, a *maximal* partial ordering O_{max} is an *acceptable* partial ordering of *EFG* minimizing $t_1 \in T$ and maximizing $t_2 \in T$ for all memory accesses $v \in V$.

Definition-theorem 4. Unicity of the *maximal* partial ordering Given a memory access flow graph (V, E) to the *EFG*, for any global timing constraint $T = [s, e]$, there exists a unique *maximal* partial ordering O_{max}.

Corollary to definition-theorem 4. The *maximal* partial ordering of *EFG* defines the unique maximum possible scheduling freedom for each memory access.

Theorem 1. Necessary inclusion into the *maximal* partial ordering Given a memory access flow graph (V, E) to the *EFG*, all *acceptable* partial orderings exhibit timing intervals within the range defined by the *maximal* partial ordering.

$$\forall o \in O, [(\forall e = (v_1, v_2) \in E, t_1(v_2) \geq t_1(v_1) + 1 \; t_2(v_1) \leq t_2(v_2) - 1)]$$
$$\Rightarrow (\forall (v, b, [t_1, t_2]) \in o, \exists (v, b, [t_{1min}, t_{2max}]) \in O_{max}, t_{1min} \leq t_1, t_2 \leq t_{2max})]$$

Terminology 2. *feasible (or schedulable)* An extended memory access flow graph *EFG* is *feasible* (or *schedulable*) if all *maximal* time windows are non empty:

$$\forall (v, b, [t_1, t_2]) \in O_{max}, \quad t_2 \geq t_1$$

and otherwise *infeasible*.

Definition-notation 5. *n and m* For any *feasible* extended memory access flow graph *EFG*, we denote $n = Card(V)$ as the *(total) number of memory accesses* and $m = \sum_{O_{max}} \frac{(t_2 - t_1 + 1)}{n}$ as the *(maximal) average time window* of each memory access.

Definition 6. *CG(EFG)* The basic group access conflict graph $CG = (B, C, p)$ of a *feasible* extended memory access flow graph *EFG* is an edge-weighted undirected graph, where $C \subseteq B \times B$ is the set of basic group access conflicts and p the corresponding probability of conflict. A basic group access conflict between two basic groups of scalars (*i.e.* an edge in the *CG*) means at least one occurrence of a parallel (or *concurrent*) access to those two basic groups of scalars within the extended memory access flow graph *EFG*:

$$\forall c = (b_1, b_2) \in C, \exists C(b_1, b_2) =$$
$$\{((v_1, b_1, [t_{11}, t_{12}]), (v_2, b_2, [t_{21}, t_{22}])) \in O^2, [t_{11}, t_{12}] \cap [t_{21}, t_{22}] \neq \emptyset\} \neq \emptyset$$

Terminology 3. *concurrent* b_1 and b_2 are referred to as two *concurrent* basic groups of scalars.

Definition 7. *uniform* p_u The probability of conflict p between two *concurrent* basic groups of scalars b_1 and b_2 is obtained by accumulating the probability of conflict of all occurrences of a *concurrent* access to b_1 and b_2. The most standard (and *local*) way of computing the individual probability of conflict between two accesses is by assuming that all conflict times in the two conflicting time windows $[t_{11}, t_{12}]$ and $[t_{11}, t_{12}]$ exhibit a *uniform* probabilistic repartition:

$$
\begin{aligned}
m(v_1) &= Card([ASAP,ALAP](v_1)) \\
&= t_{12} - t_{11} + 1 \\
m(v_2) &= Card([ASAP,ALAP](v_2)) \\
&= t_{22} - t_{21} + 1 \\
i(v_1, v_2) &= Card([ASAP,ALAP](v_1) \cap [ASAP,ALAP](v_2)) \\
&= \min(t_{12}, t_{22}) - \max(t_{12}, t_{22}) + 1 \\
p_u(c) &= \sum_{C(b_1, b_2)} \frac{i(v_1, v_2)}{m(v_1) \times m(v_2)}
\end{aligned}
\tag{6.1}
$$

Remark 2. Non-*uniform* techniques for computing $p(c)$ are discussed in subsections 6.6.3.5 and 6.6.3.6.

Pseudo-definition 8. Global (architecture-aware) cost model The basic group access conflict graph (CG) is a compact representation of the **global** memory architecture constraints. Whenever two basic groups of scalars conflict, the memory architecture must guarantee that the conflicting basic groups of scalars can be accessed in parallel in only one memory cycle. In practice, this is solved by multi-port, multiple or banked memory technologies. Based on the conflict graph representation, exploration and accurate cost estimation of memory architectures have been developed [524]. It is also possible to approximate this cost by a three-term weighted sum of the number of edges, number of self-edges and chromatic number of the conflict graph [561], which represents an implementation challenge in practice (see [399] section I-2.1.2). In the remaining of this paper, $cost(CG)$ is regarded as a plug-in function call used by our algorithms. The choice of a particular cost function has no major impact on the relative quality or runtime trends presented in the following subsections.

Terminology 4. *cost(CG)* **or** *cost(EFG)*   In the following, we will sometimes refer to the above cost model as $cost(CG)$ because it is based on the basic group access conflict graph (CG). For convenience of presentation, we will also refer to it as $cost(EFG)$ since Definition 6 defines a one-to-one match between an EFG and its CG(EFG).

Pseudo-definition 9. Exploration of memory access orderings Given a memory access flow graph (V, E) to the *EFG*, finding a low-cost memory architecture is the graph optimization problem consisting in identifying the cheapest *acceptable* partial ordering w.r.t the real technological cost of meeting the memory architecture constraints represented in the basic group access conflict graph (CG).

Definition 10. Time interval reduction of EFG Given a memory access $v \in EFG$ such that $O(v) = (b, [t_1, t_2])$ and a time interval $I \subset [t_1, t_2]$ containing (at least) t_1 or t_2, we denote $EFG \backslash t(v)$ the operator consisting in updating the partial ordering O such that I is removed from the time window of v. By definition, after applying the time interval reduction operator $EFG \backslash t(v)$, the new partial ordering of memory access v becomes $O(v) = (b, [t_1, t_2] \backslash I)$.

Theorem 2. Time interval reduction of EFG is strictly inclusive for partial orderings For any time interval reduction of EFG, the new partial ordering is strictly included in the previous one in the sense of the inclusion theorem 1.

$$
\forall v \in EFG, \forall t \in T^2(v), \qquad O(EFG \backslash t(v)) \subset O(EFG)
\tag{6.2}
$$

Remark 3. Greedy exploration by time interval reductions Algorithms presented in subsections 6.6.3.3 and following, are based on a greedy exploration of partial orderings. Starting from $O_0 = O_{max}$, a series of partial orderings are built such that each iteration of the exploration guarantees that the new partial ordering is strictly included in the former one *i.e* $O_{i+1} \subset O_i$ using the time interval reductions of Definition 10. Such algorithms converge in at most $n(m-1)$ and at least 0 iteration(s) while still spanning over the full range of *acceptable* orderings, provided that the *EFG* is *feasible* (*i.e* that the *maximal* partial ordering is at least *acceptable* and most likely *acceptable* and *reducible*). In the following, we always assume that EFG is *feasible*. In practice, unfortunately, there exists a class of real-time applications that are beyond today's technological capabilities, which of course are *infeasible* and cannot be solved by our techniques alone.

6.6.3.2 A 1D digital filter example. In this subsection, we present our Conflict-Directed Ordering (CDO) technique by the example of power-optimizing the one-dimensional digital filter $p(x) = (\alpha x + \beta \frac{1}{x}) + x^2$. This polynomial filter can be specified as follows using the C language:

```
float S[N]; AX[N]; BX[N]; AXBX[N]; X2[N];
for (int i=0; i<N+1; i++) {
  AX[i]=alpha*S[i]; BX[i]=beta/S[i]; X2[i]=S[i]*S[i];
  AXBX[i]=AX[N-i+1]+BX[N-i+1]; S[i]=AXBX[N-i+1]+X2[N-i+1];
}
```

The above specification is certainly sub-optimal in terms of memory usage and could be improved by blocking the i-loop so as to minimize the size of the intermediate AX, BX, AXBX and X2 arrays. However, in order to also minimize the power consumption, it is desirable to exploit as much as possible the underlying parallelism so as to potentially reduce the clock rate by concurrent memory to processing units data-paths. For this, the computation of AX and BX have to be done in parallel. The same applies for the computation of AXBX and X2. Moreover, the computations can be highly pipelined on the i-axis as shown in Fig. 6.24.

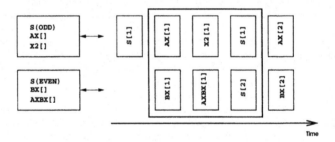

Figure 6.24.   Optimal low-cost ordering using two 1-port memories

The filter throughput is proportional to the inverse of the time required between two consecutive outputs of S[i] which corresponds exactly to the time required for one iteration of loop i. Using the above optimization, this is achieved in three memory cycles. Suppose we build a perfectly balanced memory-processor architecture in such a way that the processing units can deliver results in one memory cycle (even the division) when used in pipelined arithmetic and burst memory access modes while offering an instruction-level parallelism of at least 2. Based on the proposed loop pipelining scheme, such high-performance computing modes can be easily and fully exploited. As a result, the time required between two consecutive outputs of S[i] can be almost (including border effects) as low as three burst-mode memory cycles or three pipelined arithmetic processing cycles (which are equal on a perfectly balanced architecture). Therefore, the filter throughput can reach almost one third of the theoretical pipelined memory to processor throughput, which corresponds to an optimal exploitation of resources for the given application. This last statement remains valid even for a non-perfectly balanced memory-processor architecture offering an instruction-level parallelism of at least 2, which is always the case on modern general-purpose programmable platforms.

For any of the above architectural scenarios, the choice of the blocking factor remains a tradeoff between the pipelined memory to processor characteristics and the memory cost[6]. Last but not least,

---

[6]In pipelined computation theory, runtime performance improves with the number of computed elements *i.e.* in our case the blocking factor. Regarding our low-power objective, decreasing the number of memory cycles allows to reduce the

this tradeoff is a good example of a platform-dependent tuning that is best solved by a designer-in-the-loop use of the scheduler, as enabled by our IMEC - ATOMIUM/ACROPOLIS environment.

6.6.3.3 Conflict-directed ordering (CDO). In order to attain this optimal area vs. power solution, the original Conflict Directed Ordering (CDO) solution approach [562] works on a flow graph (FG) representation. In a first step, the as soon as possible (ASAP) and as late as possible (ALAP) scheduling interval is being initialized. From the ASAP-ALAP annotated FG, also called extended flow graph (EFG), a conflict graph (CG) incorporating the basic group-level pairwise probability of conflict is being constructed, as shown in Fig. 6.25. After this initialization phase, the set of

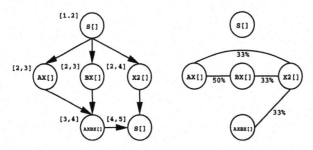

Figure 6.25.   Extended flow graph (EFG) and conflict graph (CG) for our 1D filter

time reducible nodes is being updated before entering the iteration process. The pseudo-algorithm presented below shows that each iteration of CDO consists in identifying the most conflicting data access using a single step look-ahead. The [ASAP,ALAP] scheduling interval of the identified data access is then reduced by one time interval.

begin
    $reducible := \{v \in EFG, ASAP(v) < ALAP(v)\};$ $O(n)$
    while ($reducible > 1$) $O(nm)$ outer-loops
        $cost(selected) := \infty;$
        for ($v \in reducible$) **do**   2.$reducible$ inner-loops
            if $(cost(EFG \backslash ASAP(v)) < cost(selected))$
                $selected := ASAP(v);$
            end
            if $(cost(EFG \backslash ALAP(v)) < cost(selected))$
                $selected := ALAP(v);$
            end;
        end;
        $EFG := EFG \backslash selected;$
        $reducible := \{v \in EFG, ASAP(v) < ALAP(v)\};$
    end;
end

Table 6.2.   Conflict-Directed Ordering (CDO).

Because of this unit reduction of the search space and because our 1D digital filter example exhibits a search space of size $n(m - 1) = 7$, CDO can in the worst-case reach the optimal so-

memory clock rate hence the power consumption. In memory optimization theory, performance decreases with the number of intermediate computation arrays *i.e.* in our case the blocking factor. Regarding our low-power objective, decreasing the number of intermediate arrays allows to come up with a smaller and less power-consuming memory architecture. Hence, for power-optimization purposes, there exists an optimal blocking factor.

lution in seven steps. Moreover, at each step, two look-ahead costs - $cost(EFG\backslash ASAP(v))$ and $cost(EFG\backslash ALAP(v))$ - have to be computed. In the worst-case again, each look-ahead can necessitate to recompute the Conflict Graph (CG) in full using $\frac{n(n+1)}{2}$ pairwise probability of conflict computation (PPCC). Hence, CDO exhibits a worst-case complexity (in PPCC) of $O(n^4m)$.

$$\sum_{i=1}^{n(m-1)} \frac{i}{m-1} \times n(n+1) \quad = \quad \frac{1}{2}n^4m + \dots \tag{6.3}$$

For $n(m-1) = 7$, we can estimate a complexity of 1,456 PPCCs.

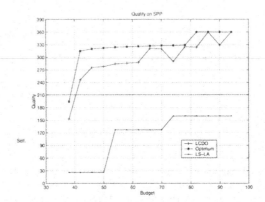

Figure 6.26.   LS-LA vs. CDO vs. near-optimal solution quality using our global architecture-aware cost model. Below 210, a self-conflict in the CG is introduced, requiring the use of expensive multi-port memories.

In Fig. 6.26, we present the results of applying Conflict-Directed Ordering (CDO) to the design of the Segment Protocol Processor (SPP) module. As already introduced in subsection 6.6.2, it should be first noticed that CDO exhibits a much higher quality of solution w.r.t to DTSE optimizations than conventional ordering techniques like List-Scheduling with Look-Ahead (LS-LA). This motivates the need for low-cost ordering techniques. In a second step, it should be stressed that the initial CDO algorithm is still within a 10-20% margin from the optimal solution. The above combined to an initial high polynomial complexity of $O(n^4m)$ motivates the need for both improvement in quality **and** reduction in design runtime for low-cost ordering techniques like CDO. In the following subsections, we will address both needs in detail.

6.6.3.4   Localized Conflict Directed Ordering (LCDO).   The complexity of CDO can be largely reduced by applying *dynamic* graph techniques [168, 242].
Indeed, when comparing $cost(EFG\backslash ALAP(v))$ to $cost(EFG\backslash selected)$, only a relative cost is required. Hence, it is possible to compare only the difference in cost w.r.t the initial cost(EFG) at the beginning of the current iteration. To do this, only the sons $s(v)$ (respectively only the fathers $f(v)$) of v are involved for an ASAP (respectively ALAP) look-ahead. This involves an update of the set of fathers and sons of the memory access $v \in EFG$, $sf(v) = s(v) \cup f(v)$, as opposed to a full update of the n nodes of EFG in the earlier version. Moreover, only the neighbours of those $sf(v)$ updated nodes have to be considered while updating the CG. This involves an average degree of CG nodes at step i, $d_i \leq n$, updates. Hence, a realistic upper-bound complexity (in PPCC) for a dynamic (or localized) CDO is $O(n^3 ln(n)m)$.

$$cplx_1(EFG) \quad = \quad \sum_{v \in EFG} [ALAP(v) - ASAP(v)] \times sf(v)$$
$$\equiv \quad O(n ln(n)m) \tag{6.4}$$

$$\sum_{i=1}^{n(m-1)} \frac{i}{m-1} \times 2 \times \frac{1}{n(m-1)} \times cplx_1(EFG) \times d_i$$

$$\leq \quad n^2(1 + \frac{1}{n(m-1)}) \times cplx_1(EFG)$$

$$\equiv \quad O(n^3 ln(n)m) \qquad\qquad\qquad (6.5)$$

The pseudo code then becomes:

begin
initialize EFG with the maximal ordering O_{max}
$reducible := \{v \in EFG, ASAP(v) < ALAP(v)\};$ $O(n)$
while $(Card(reducible) > 1)$ $O(nm)$ outer-loops
    $cost(EFG\backslash selected) := \infty;$
    for $(v \in reducible)$ **do**   2.reducible inner-loops
        if $(cost(EFG\backslash ASAP(v)) < cost(EFG\backslash selected))$
          $selected := ASAP(v);$
        end
        if $(cost(EFG\backslash ALAP(v)) < cost(EFG\backslash selected))$
          $selected := ALAP(v);$
        end;
    end;
    $EFG := EFG\backslash selected;$
    $reducible := \{v \in EFG, ASAP(v) < ALAP(v)\};$
    end;
end

Table 6.3. Localized Conflict-Directed Ordering (LCDO).

When applied to our digital filter example, this gives a complexity (in PPCCs) of:

$$6^2(1 + \frac{1}{7})(1.6 + 1.4 + 1.4 + 1.5 + 1.6 + 2.3) = 41, 14.31 = 1,275$$

Therefore, even on this simplistic example, LCDO can be $\frac{1,456-1,275}{1,456} = 12\%$ faster than CDO. In practice, it is possible to make LCDO even more dynamic by visiting only those fathers and sons that actually need an update. This data-dependent propagation combined to the strong approximation $d_i \leq n$ means that the above complexity formula (6.5) is very pessimistic. This also brings an end to all possible dynamic pairwise probability of conflict computation (PPCC) scenarios within one CDO iteration.

In Fig. 6.27, we compare the runtime of Localized CDO (LCDO) to that of CDO on a range of optimization problems arising from the SPP design. On this real-life example, we observe gains in runtime of order 100 which matches well the theoretical predictions. For low values of the real-time budget (*i.e.* for low values of m), real LCDO outperforms the theoretical prediction. This is mainly because the average degree of CG nodes at step i, $d_i \leq n$, decreases much faster in that case. Hence, almost another order of magnitude in the complexity can be gained (i.e. $\equiv O(nln(n)m)$). In simple terms, LCDO is even more dynamic on hard real-time problems.

Though LCDO can be very fast in practice, it should be stressed that it inherits *exactly* the quality of CDO results. Therefore. solutions are typically found within a 10-20% distance from the optimum. On most (and even simple) cases, CDO and LCDO are unable to find the optimal solution[7]. This motivates the need for higher-quality orderings. Provided *dynamic* techniques can

---

[7]For example, CDO does not find the optimal solution for the polynomial filter example presented at the beginning of this subsection.

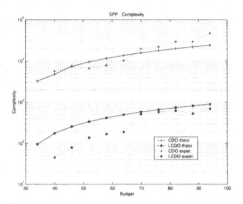

Figure 6.27.   Theoretical and real runtime for CDO and LCDO. Time is presented in seconds on the vertical logarithmic scale while the real-time budget $T = [0, B]$ is presented on the horizontal axis.

already speed CDO to a reasonable runtime, we will focus on the need for quality improvement in the coming subsections. Hopefully, this will inspire further runtime improvements.

6.6.3.5   LCDO with global constraint propagation (GCP).

When designing large systems, it is common practice to first divide the overall system into p smaller subsystems. When applying automatic design techniques such as LCDO to those subsystems, it becomes much easier to design each subsystem independently in $O(pn^3 ln(n)m)$ than to design and optimize the overall system in full in $O((pn)^3 ln(n)m)$. This divide and conquer mechanism is especially interesting in the case of local optimization techniques, typically found in *scheduling-for-speed* heuristics, because nothing more than just p fully concurrent designs are required in that case.

Unfortunately, as introduced in subsection 6.6.3.1, optimizing for making systems as cost-effective as possible, requires a *global* architecture-aware cost function. In the case of CDO for example, deciding *which* data can be put in parallel requires (at least) a *global* track of *which* data have already been put in parallel on other parts of the system. Hence, it remains possible to apply divide and conquer techniques to LCDO *only* if *global* constraints are propagated from one subsystem to another. This is for example the case in our data transfer and storage exploration (DTSE) methodology while optimizing over several nested loops [66]. Each nested loop is regarded as an individual system requiring an individual scheduling.

We propose to model global constraint propagation by introducing a $\delta(c)$ in the cost function. This δ means that the knowledge of prior decisions applied to other parts of the system guides to either take into account a cost factor ($\delta(c) = 1$) because it is new or to cancel it ($\delta(c) = 0$) because it has already been taken into account previously.

$$cost(CG) \quad = \sum_{c=(b_1,b_2)\in CG} \omega(c).p_n(c) \qquad (6.6)$$

$$p_n(c) \quad = \quad \delta(c).p_u(c)$$

By default, all $\delta(c)$ are equal to 1. In a first step, introducing δ in the LCDO algorithm impacts the cost of the individual CG edge cost computation, which remains a $O(1)$ process in PPCCs anyway. In a second step, added computation is also introduced when deciding to keep a conflict, hence necessitating to switch $\delta(c)$ to 0, again in $O(1)$ PPCCs provided the right data structure for the edge cost computation is put into practice[8]. Hence, constraint propagation exhibits the same upper-bound complexity as conventional LCDO. Some overhead can however be expected in practice.

---

[8] A two-level pointing technique to compute $\omega(c).\delta(c)$ "on the fly".

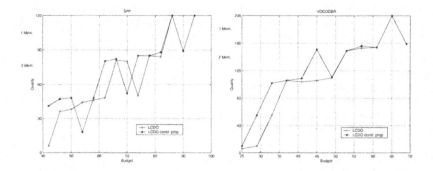

Figure 6.28.   Quality for LCDO with CP vs. CDO using our global architecture-aware cost model. For the SPP driver (left), only 1 memory is needed above a quality of 90, 2 concurrent memories above 60 and more below. For the VoCoder driver (right), only 1 memory is needed above a quality of 160, 2 concurrent memories above 120 and more below.

In Fig. 6.28, we show that global constraint propagation has a very positive impact on the quality of LCDO. Indeed, it really makes sense to re-use the knowledge of problems encountered on other parts of the system to guide new optimizations for new subsystems.

6.6.3.6   LCDO with deterministic GCP.   It is also possible to implement global constraint propagation on a deterministic conflict graph (CG), that is on a CG where edges are introduced only when a conflict is 100% certain. In that way, new LCDO optimizations leave out only those access nodes that are certain to generate a conflict. By introducing a deterministic CG, the LCDO upper-bound complexity is not affected.

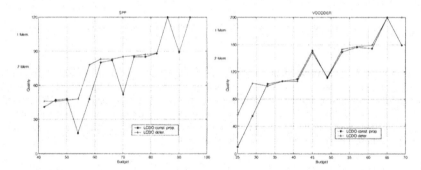

Figure 6.29.   Quality for LCDO with deterministic CP vs. CDO-CP using our global architecture-aware cost model. For the SPP driver (left), only 1 memory is needed above a quality of 90, 2 concurrent memories above 60 and more below. For the VoCoder driver (right), only 1 memory is needed above a quality of 160, 2 concurrent memories above 120 and more below.

Surprisingly, as shown in Fig. 6.29, a deterministic CG can contribute to further improve the quality of LCDO. Indeed, it really makes sense to leave out only those situations that cannot be solved as opposed to the many that are likely not to be solved. In simpler terms, exploration decisions based on crude facts are better than those based on higher-resolution predictions of future refinements.

6.6.4   Faster but still effective exploration

6.6.4.1   Multi-level Local CDO (LCDO(k)).   One obvious way to further enhance the LCDO solution techniques is to allow multiple reductions at each iteration. This technique corresponds exactly to that of Kernighan and Lin for fast graph partitioning [327]. Multiple reductions are also encountered in improved force-directed scheduling (IFDS) [528]. In the following, we introduce LCDO(k), the *clustered* (*i.e.* multiple reductions) version of CDO:

begin
  initialize EFG with the maximal ordering O_{max};
  for ($v \in EFG$) **do** $O(nmln(m))$
    $clusters(v) := k\text{-cluster}([ASAP, ALAP](v), k)$;
    $reducible(v) := \{cluster \in clusters(v), ASAP(v), ALAP(v) \in$
    $cluster\}$;
  end
  while ($Card(reducible) > 1$) $O(k)$ outer-loops
    $cost(EFG\backslash selected) := \infty$;
    for ($v \in EFG$) **do**   *clusters* inner-loops
      for ($cluster \in reducible(v)$) **do**
        if ($cost(EFG\backslash cluster) < cost(EFG\backslash selected)$)
          $selected := cluster$;
        end;
      end;
    end
    $EFG := EFG\backslash cluster$;
    for ($v \in EFG$) **do**
      $reducible(v) := \{cluster \in clusters(v), ASAP(v), ALAP(v) \in$
      $cluster\}$;
    end
  end;
end

Table 6.4.   (Multi-Level) Conflict-Directed Ordering (CDO(k)).

First, the time domain is decoupled into k clusters with as little inter-cluster conflicts as possible [231]. In the general case, this step can be automated by a k-clustering algorithm, for which state-of-the-art favors a k-way partitioning implementation on the complement graph [231] pp.359–360. In the case of CDO(k), this clustering is made very easy because two different times can never conflict. Therefore, the k-clustering has no constraint on the choice of the partitions. In practice, we therefore prefer to choose clusters in the following ad-hoc way [180]:

$$\bigcup_i C_i(v) = [ASAP, ALAP](v) \qquad (6.7)$$

$$l(v) = [0, ALAP - ASAP](v)$$

$$C_i(v) = ASAP(v) + [l(v).(1 - \alpha_k)^{i-1} \backslash l(v).(1 - \alpha)^i] \quad i \in [1, k-1]$$

$$C_k(v) = ASAP(v) + l(v).(1 - \alpha_k)^{k-1}$$

This ad-hoc choice has the nice property of making the reductions *adaptive* by being less and less aggressive as the optimization progresses if the reduction factor $\alpha_k = \alpha$ is chosen as a constant in [0,1]. We admit here that there exists a direct mapping between the reduction factor α and the number of clusters k. In practice, this ad-hoc clustering technique can be made even more adaptive by alternating the α-reduction from the ASAP and ALAP side dynamically at runtime.

    Second, those clusters can serve as input to the clustered version of CDO [218], which means that all times within a cluster are always reduced simultaneously.

Therefore, the complexity of LCDO(k) in pairwise probability of conflict computations (PPCC) is given by:

$$cplx_2(EFG) = \sum_{v \in EFG} k.sf(v)$$

$$\equiv O(kn\ln(n)) \qquad (6.8)$$

$$\sum_{i=1}^{n(k-1)} \frac{i}{k-1} \times 2 \times \frac{1}{n(k-1)} \times cplx_2(EFG) \times d_i$$

$$\leq n(1 + \frac{1}{(k-1)}) \times cplx_2(EFG)$$

$$\equiv O(kn^2 \ln(n)) \qquad (6.9)$$

Because the number of clusters k can be controlled by the user, it is possible to reduce the CDO(k) runtime to a minimum.

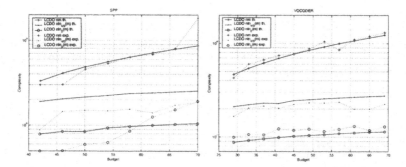

Figure 6.30.   Theoretical and real runtime for LCDO and LCDO(k). Time is presented in seconds on the vertical logarithmic scale while the real-time budget $T = [0, B]$ is presented on the horizontal axis.

In Fig. 6.30, we compare the runtime of various LCDO(k) choices to that of LCDO on a range of optimization problems arising from the design of the SPP. On this real-life example, we observe gains of order 10-15 which matches well the theoretical predictions. This result added to the gains observed in Fig. 6.30. means an overall speed-up of order 100-1,000 compared to CDO.

For low values of k, real LCDO(k) outperforms the theoretical prediction. This is mainly because the real number of steps can become much lower than $n(k-1)$ when aggressive clustering is used. In that case, the chances of observing a super-linear decline in the number of reducible clusters are much higher because of aggressive multiple reductions.

6.6.4.2   (Multi-level) Generalized CDO (G-CDO(k)).   Unfortunately, because it is always possible to select several nodes in graph partitioning and because IFDS is limited to the selection of a single access node for multiple time reductions, neither of the above multi-level approaches solve the general problem of multiple selections and reductions (MSMR) in the (*multi-dimensional*) *space–time* context of CDO. In order to allow MSMR in full, we propose to introduce a *space–time* conflict graph (ST-CG). As shown in Fig. 6.31, a ST-CG is a two-dimensional graph with accesses on the vertical axis and time on the horizontal axis. The freedom of scheduling access x at time t is (simply) represented by a two-dimensional (x, t) node. Provided this data model is initialised with a feasible schedule, there are two simple rules for solving the reducibility and feasibility problems:

1. **Reducibility**: never remove an access from the initial specification i.e. never empty a row;

2. **Feasibility**: always satisfy dependencies i.e. keep an access only after removing all concurrently scheduled fathers and sons.

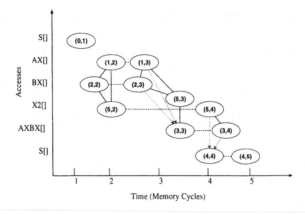

Figure 6.31.   Space-Time Conflict Graph (ST-CG) for our 1D filter

Based on those two rules (i.e. which provide the semantics for both the reducibility space *reducible* and the feasibility operator $EFG\backslash cluster$ in the generalized case), we propose a generalized conflict directed ordering (G-CDO(k)) using the following solution approach:

begin
    $clusters := k\text{-cluster}(\text{ST-CG}, k)$; $O(n^2.ln(n))$
    $reducible := \{cluster \in clusters, \exists v \in$
    $cluster, ASAP(v) < ALAP(v)\}$;
    while ($reducible > 1$) $O(k)$ outer-loops
        $cost(selected) := \infty$;
        for ($cluster \in reducible$) **do**   *clusters* inner-loops
           if ($cost(EFG\backslash cluster) < cost(selected)$)
           $selected := cluster$;
        end;
        end;
        $EFG := EFG\backslash cluster$;
        $reducible := \{cluster \in clusters, \exists v \in$
        $cluster, ASAP(v) < ALAP(v)\}$;
    end;
end

Table 6.5.   (Multi-Level) Generalized Conflict-Directed Ordering (GCDO(k)).

First, the *space–time* domain is decoupled into k clusters with as little inter-cluster conflicts as possible. In the general case, this step can be automated by a k-clustering algorithm, for which state-of-the-art favours a k-way partitioning implementation on the complement graph [231] pp.359–360. In practice, for pure acceleration purposes, we choose clusters in an ad-hoc way.

Second, those clusters can serve as the basis for a multi-level version of CDO [218], which means that all nodes within a cluster are always reduced simultaneously.

Because the number of clusters k can be controlled by the user, it is possible to reduce the G-CDO(k) runtime to a minimum. For example, using a greedy colouring technique for k-clustering (after isolating initially non reducible nodes in C1), our 1D digital filter ST-CG results in five clusters:

$$C1 = \{(0,1)\} \tag{6.10}$$
$$C2 = \{(1,2),(2,3),(5,4),(4,5)\} \tag{6.11}$$
$$C3 = \{(2,2),(1,3),(4,4)\} \tag{6.12}$$
$$C4 = \{(5,2),(5,3),(3,4)\} \tag{6.13}$$
$$C5 = \{(3,3)\} \tag{6.14}$$

Let us now assume a clever cost function to select and remove $C3$ from the first iteration. For reducibility, $C2$ has to be kept because $(4,5) \in C2$ is now the only node left in the 4th row. For feasibility, $C5$ has to be removed because $(3,3) \in C5$ is a concurrent son of $(2,3) \in C2$. For reducibility, $C4$ has to be kept because $(3,4) \in C4$ is now the only node left in the 3rd row. Hence, there is no reducible cluster left and the algorithm stops. Surprisingly enough, this aggressive use of G-CDO(k) still finds the optimal solution and we will see that this property will also hold for real-life demonstrators.

6.6.4.3 Five variants. In order to assess the flexibility of G-CDO(k), it is necessary to assess at least its two extreme *"mimic of CDO"* and *"aggressive"* variants. However, because the clustering is performed in an ad-hoc way in practice, it also proves necessary to separate clustering strategies. For example, the improved force-directed scheduling (IFDS) proposed in [528] experiments ways to select an access to reduce its scheduling interval to 1 in a single iteration. We call this *"single selection"* and *"multiple reduction"*. Because CDO can be relatively easily adapted for time adaptive reduction and space selection, we call this variant CDO(k). On the contrary, when full *space–time* selection is used, we refer to G-CDO(k). Depending on the level of selection and reduction, we obtain four variants as shown in Fig. 6.32. It should be stressed that there is still a big gap between

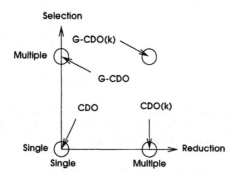

Figure 6.32.   G-CDO(k) prototype variants

the pseudo-algorithms presented in this section. and the actual implementation. Indeed, still a lot of improvement can be achieved by writing graph algorithms in a dynamic fashion [168], that is by maximizing the reuse of information from one iteration to another. We have built such a localised version for CDO and CDO(k), called LCDO and LCDO(k) respectively. For reference with previous work, we also keep the initial CDO implementation in mind, as the fifth variant tested in our benchmark. All five variants are presented in Table 6.6.

6.6.5 Experiments on real-life examples

6.6.5.1 Telecom networks - Segment Protocol Processor (SPP). In this subsection, we apply LCDO and LCDO(k) to the design of a large-scale system, namely the Segment Protocol Processor (SPP) module, which is a crucial part of an Asynchronous Transfer Mode (ATM) switching node [497]. SPP is a Telecom module involving the dynamic routing and buffering of ATM messages by successively decoding parts (also called segments) of its self-contained routing information. After

-	CDO	LCDO	LCDO(k)	G-CDO	G-CDO(k)
Iteration kernel					
Policy	static	dynamic	dynamic	static	static
Multi-level control					
Availability	no	no	yes	no	yes
k	nm	nm	$nln(m)$	nm	$nln(m)$
Look-ahead					
Availability	yes	yes	yes	no	no
Number of steps	1	1	1	0	0
Complexity					
	mn^4	$mn^3ln(n)$	$n^3ln(n)ln(m)$	m^2n^2	$n^2mln(m)$

Table 6.6.   Generalized conflict-directed ordering (G-CDO) typical variants

applying our Dynamic Memory Management (DMM) script to the ATM routing standard [565], SPP is refined to a specification exhibiting a complex conditional control flow with a total of 156 accesses to 14 arrays. In this context, the role of our storage-bandwidth optimization step is to multiplex the 156 accesses taking into account their exclusive patterns to minimize the required memory bandwidth and cost.

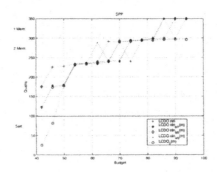

Figure 6.33.   LCDO quality on SPP using our global architecture-aware cost model. Only 1 memory is needed above a quality of 300, 2 concurrent memories above 250 and more below. Below 100, the introduction of expensive multi-port memories becomes unavoidable.

When designing under stringent real-time constraints, much parallelism has to be introduced in the data-path, involving little scheduling freedom (or, in other words, a small search space characterized by a low value of m and a low value of the real-time budget). By applying this stringent scenario to the design of SPP (and as already mentioned in subsection 6.6.2), we observe in Fig. 6.33 that LCDO is able to solve problems with a real-time budget lower than 44 memory cycles without introducing self-edges in the Conflict Graph (CG), which would turn up as prohibitively expensive memories. The optimal tradeoff between quality and runtime is met with $k = nln_{6/5}(n)$.

Moreover, the *clustered* reduction capability of LCDO(k) allows to improve the design runtime as illustrated in Fig. 6.34. This improvement factor reaches 300 for a real-time budget of 70, without any loss in the quality of the solution.

We observe however in Fig. 6.35 that LCDO is unable to solve problems with a real-time budget lower than 54 memory cycles while G-CDO can in Fig. 6.36 solve problems for a budget as low as 46. This proves that the *multi-dimensional* selection process of G-CDO can achieve better results than its *mono-dimensional* counterpart implemented in LCDO or the original CDO.

Moreover, the *multi-level* reduction capability of G-CDO(k) allows to further improve those results. For an extremely low budget of 42, we observe an optimal parameter tuning for $k = nlog_6(n)$. This proves that at least for large-scale problems like SPP, the choice of parameter k can help in

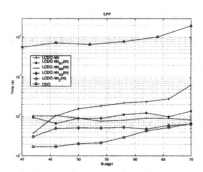

Figure 6.34.   Low-cost ordering runtime for SPP. Time is presented in seconds on the vertical loga-rithmic scale while the real-time budget $T = [0, B]$ is presented on the horizontal axis.

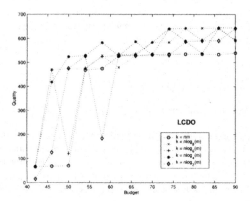

Figure 6.35.   LCDO quality on Alcatel's SPP

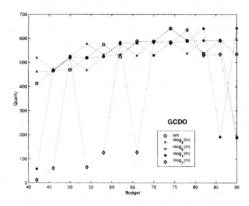

Figure 6.36.   G-CDO quality for Alcatel's SPP

finding the optimal granularity of the problem formulation w.r.t. the solving capabilities of the un-derlying heuristics.

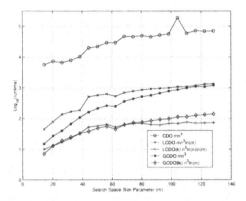

Figure 6.37.   Low-cost ordering runtime for Alcatel's SPP

Let us now study the SPP design runtime over a wide range of real-time constraints. As shown in Fig. 6.37, our algorithms are clearly classified in three categories from a runtime perspective:

- The traditional category, involving CDO;

- The fast category, involving LCDO and G-CDO;

- The ultra-fast category, involving CDO(k) and G-CDO(k).

Those results confirm our complexity analysis presented in Table 6.6. For a real-time budget of 104, G-CDO(k) can find a low-cost scheduling 1500 times faster than a straightforward implementation of CDO (which requires more than 35 hours of computation on an HP-PA1000 processor!). This runtime of around ten seconds for low budgets and one minute for a search space of size $mn > 15,000$ supports that G-CDO(k) is, to our knowledge, the fastest existing low-cost ordering technique for large-scale parallel programs.

6.6.5.2   Speech processing - Voice Coder (VoCoder).   In this subsection, we apply LCDO and LCDO(k) to the design of a Linear Predictive Coding (LPC) VoCoder. LPC provides an analysis-synthesis approach for speech signals [442, 296]. In the LPC algorithm, speech characteristics (energy, transfer function and pitch) are extracted from a frame of 240 consecutive samples of speech. Consecutive frames share 60 overlapping samples. All information to reproduce the speech signal with an acceptable quality is coded in 54 bits. It is worth noting that the steps LPC analysis and Pitch detection use autocorrelation to obtain the best matching predictor values and pitch (using Hamming window). The Schur's algorithm is used to obtain signal in baseband region in LPC analysis and to obtain high and low amplitudes in Pitch analysis. Both autocorrelation and Schur's algorithm have irregular access structures due to algebraic manipulations. Moreover, because of the manipulation of many samples concurrently, the LPC exhibits a highly parallel control flow involving many concurrent arrays. In this context, the role of our storage-bandwidth optimization step is to reduce as much as possible the number of memories by carefully multiplexing the accesses taking into account *which* data is being placed in parallel.

In Fig. 6.38, we compare CDO to various LCDO(k) choices on a range of optimization problems arising from the design of the VoCoder. When applying stringent real-time constraints, we observe that CDO is able to solve problems with a real-time budget as low as 25.

Moreover, the *clustered* reduction capability of CDO(k) allows to further improve those results. For an extremely low budget of 25, we observe an optimal parameter tuning for $k = nln_{6/5}(m)$. This proves that at least for very difficult problems like VoCoder, the choice of parameter k can help in finding the optimal granularity of the problem formulation w.r.t the solving capabilities of the underlying heuristics.

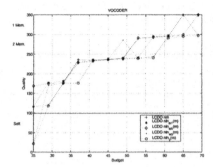

Figure 6.38.   LCDO(k) quality for the VoCoder using our global architecture-aware cost model. Only 1 memory is needed above a quality of 300, 2 concurrent memories above 250 and more below. Below 100, the introduction of expensive multi-port memories becomes unavoidable.

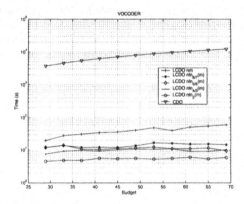

Figure 6.39.   Low-cost ordering runtime for the VoCoder module. Time is presented in seconds on the vertical logarithmic scale while the real-time budget $T = [0, B]$ is presented on the horizontal axis.

Let us now study the VoCoder design over a wide range of real-time constraints. As shown in Fig. 6.39, our algorithms are clearly classified in two categories from a runtime perspective:

- The conventional category, involving CDO;

- The dynamic category, involving LCDO;

- The fast category, involving LCDO(k).

Those results confirm our complexity analysis presented in subsection 6.6.3 for LCDO and CDO and subsection 6.6.4 for LCDO(k). For a real-time budget of 70, LCDO(k) can find a solution 2,000 times faster than a straightforward implementation of CDO (which requires 3 hours and 30 minutes of computation on an HP-PA1000 processor!). This runtime of around 5 seconds for a search space of size 10,000 supports that CDO(k) is, to our knowledge, one of the fastest existing low-cost ordering techniques for large-scale, real-time and parallel programs.

6.6.5.3   Image and video processing - Binary Tree Predictive Coding (BTPC).   In this subsection, we apply LCDO and LCDO(k) to the design of the BTPC. Because BTPC involves a complex algorithm with neighbourhood operators at high image rate, it is infeasible to meet stringent real-time constraints without accessing this data in parallel. In this context, the main role of our access ordering technique is to interleave loads, intermediate computations and stores such that only one access to the main image data is required at each cycle, hence avoiding the need for expensive multi-port memories for at least the main memory storage.

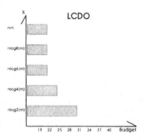

Figure 6.40. Quality results on BTPC for LCDO. For several choices of k on the vertical axis, we present the real-time budget region for which the use of expensive multi-port memories are unavoidable.

In Fig. 6.40, we indicate the circumstances under which LCDO is unable to find the desired low-cost ordering. Below a real-time budget of 25 memory cycles, LCDO is unable to find a good solution without introducing self-edges in the Conflict Graph (CG), which turn up as prohibitively expensive memories in practice. Among the LCDO(k) variants, the best choice is again $k = nlog_{6/5}(n)$.

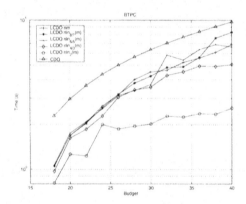

Figure 6.41. Low-cost scheduling runtime for BTPC. Time is presented in seconds on the vertical logarithmic scale while the real-time budget $T = [0, B]$ is presented on the horizontal axis.

In Fig. 6.41, we measure the runtime of our algorithms for designing the Binary Tree Predictive Coding (BTPC). Even on this small example, LCDO(k) is found to be the fastest solver in about two seconds, improving the CDO runtime by a factor of 4.

In Fig. 6.42, we indicate the circumstances under which LCDO and G-CDO are unable to find the desired low-cost ordering. Below a real-time budget of 25 memory cycles, none of the algorithms can find a good solution and it is likely that a good solution does not exist. From a budget of 25 to 37, LCDO clearly exhibits better results than the current version of G-CDO(k). This is because the nature of our problem is clearly *mono-dimensional*, involving the need to order carefully only one dataset. Among the LCDO(k) variants, the best choice is $k = nlog_6(n)$.

In Fig. 6.43, we measure the runtime of our algorithms for designing the Binary Tree Predictive Coding (BTPC). LCDO(k) is found to be the fastest solver in about two seconds whereas G-CDO(k) pays a *multi-dimensional* overhead of ten seconds. This difference may be important in practice when designing an even larger system consisting of many such subsystems. In that case, a global optimization strategy typically consists in optimizing the subsystems many times before arriving at a good real-time and cost balance among them.

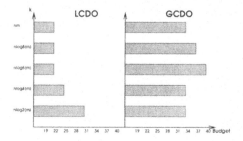

Figure 6.42.   Quality results on BTPC for GCDO

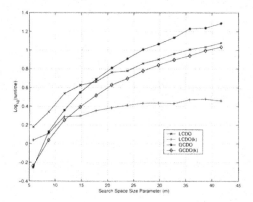

Figure 6.43.   Low-cost scheduling runtime for MPEG-4 BTPC

6.7 SUMMARY

Memory bandwidth is the limiting factor for performance of current multi-media systems. More bandwidth becomes available by adding multiport or more memories. These extra memories involve extra costs. The SCBD-MAA steps and tools described in this section optimize the memory architecture within the real-time constraint. Often the global system cost in terms of power/energy consumption, but also chip/board area or dollars, can be significantly reduced at the price of an acceptable increase in cycles spent in a particular submodule at the cost of less cycles for another less cost-sensitive submodule. Exploiting this tradeoff is only feasible for real complex systems if it can be systematically explored and when the useful tradeoff space is effectively visualized so that a designer can quickly evaluate the cost of different cycle budgets for a given submodule. The SCBD and MAA tools that implement the techniques of the previous subsections, explore different solutions in the performance, power and area space. In this section, we have shown a methodology to use these tools to effectively achieve this system-wide tradae-off. We believe that this opens up a whole new exploration space for industrial designers, that was always present but not practically accessible for real-life designs. It has been demonstrated for several real-life multi-media applications.

7 CACHE OPTIMIZATION METHODOLOGIES AND AUTOMATION

Architectural techniques for reducing cache misses are expensive to implement and they do not have a global view of the complete program which limits their effectiveness. Thus compiler optimizations are the most attractive alternative which can overcome both the above shortcomings of a hardware implementation. In this chapter, we will investigate the current state-of-the-art in compiler optimizations for caching. Afterwards, we propose a stepwise methodology which allows a designer to perform a global optimization of the program for a given cache organization. Then each of the main cache related steps will be studied in more detail including both problem formulations and techniques to solve them. The effectiveness of these automatable techniques will be substantiated by realistic demonstrators.

7.1 RELATED WORK IN COMPILER OPTIMIZATIONS FOR CACHES

In this section, we will discuss some of the commonly used compiler optimizations. The main goal of compiler optimizations is twofold; first, to increase the amount of array reuse in the cache - due to execution order and secondly, avoid conflicts between arrays - due to storage order. In literature, many frameworks and strategies for optimizing cache have been presented, some of the notable works are [18] [22] [192] [290] [363] [555]. Most of these only involve loop transformations for improving the access locality and these are linked only to the global loop transformation step of our methodology. We will list them here because they are crucial to enhance the opportunities for the cache usage. But their treatment will not be exclusive. Next, we will discuss each of the individual transformations in some detail.

7.1.1 Loop fusion

Loop fusion takes two loops, which are production and consumption of a particular function or an array respectively, that have the same iteration-space traversal and combines their bodies into a single loop [381] [555]. This results in much larger reuse of data with shorter reuse distances and hence improved array reuse in the cache.

An example of such a transformation is shown in Fig. 7.1. Loop fusion is legal as long as the loops have the same bounds and as long as there are no flow, anti-, or output dependences in the fused loop for which instructions from first loop depend on instructions from the second loop [381].

Fig. 7.1 presents an example illustration of loop fusion. Two loop nests exist with array $a[][]$ being produced in one and consumed in another. After loop fusion, only one loop nest remains and both the production and consumption are closer in time. This is illustrated in the address-time domain mapping in Fig. 7.1. The dark lines represent the reuse vector and the length of the vector along the time axis represents the lifetime of the array. The dark dots represent data access to the particular element. The distance between two accesses to the same element along time axis is also referred to as the reuse distance. Shorter reuse distances result in (potentially) larger reuse of the particular data element in the cache. Note that the mapping in Fig. 7.1 is for $N = M = 3$. Thus loop fusion contributes to shorter reuse vectors, plus smaller life times and hence enhances the ability to cache the array $a[][]$ in our example.

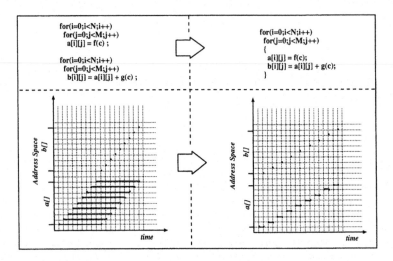

Figure 7.1.   Example illustration of loop fusion.

Loop fusion is a powerful transformation, which is used for enhancing data locality, increasing instruction-level parallelism and reduction of memory accesses. Many approaches have been proposed in the past for performing this transformation [305] [555] based on different cost functions [185] [293] [361] [554].

7.1.2 Loop tiling/blocking

Loop tiling is a transformation, which transforms an existing iteration space traversal into a tiled iteration space traversal[1]. This results in smaller and less complex iteration spaces for further partitioning and can improve the cache performance by improving the amount of variable reuse [554]. In particular, loop tiling helps in reducing capacity misses for an algorithm.

An example of loop tiling is shown in Fig. 7.2 on a matrix multiplication. Here we have tiled only the j and k loops. The term B in the source code represents the tile/block size. From the address-time domain mapping we make two important observation for the initial code namely that the reuse distances for arrays $b[][]$ and $c[][]$ are large and the access to array $c[][]$ are column-wise. Note that the storage order is row-wise. The address-time domain mapping in this example is for $N = 2$ and $B = 1$. For the loop tiled code the reuse distances for array $b[][]$ and $c[][]$ have reduced significantly (from 4 to 2 - since $B = 1$). Also the accesses to array $c[][]$ are not column-wise but tiled. These two issues contribute significantly to a enhanced cache utilization due to increased temporal locality. Thus the critical component in the above example is the size of tile B.

---

[1]Iteration space traversal is also referred to as execution order of an algorithm.

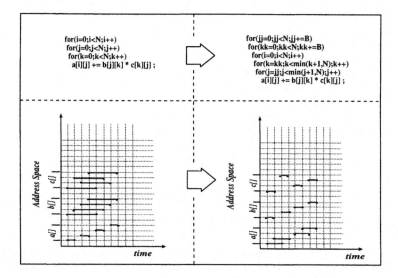

Figure 7.2.   Example illustration of loop tiling.

In the past much work has focussed on loop tiling. In particular the work in the group of Monica Lam at Stanford [306], the group of Ken Kennedy at Rice [77] - who call it "unroll and jam" and Michael Wolf at University of Illinois at Urbana [555] - who call it "strip mine and interchange". In an compiler context, most approaches assume that the loops to be transformed are perfectly nested, fully permutable [554] and the loop bounds are affine functions of the surrounding loops iteration variables which is too limiting for our target domain. However recently, loop tiling has also been investigated, with promising results, for reduction in energy consumption [240] in an embedded context [481].

7.1.3 Multi-level blocking

Multi-level blocking is an extension of the classic loop tiling technique. This technique extends the single level of loop tiling, intended for one level of memory hierarchy, to multiple levels of memory hierarchy namely registers, cache levels, TLB and virtual memory [250]. A standard procedure in this technique involves the use of Multilevel Orthogonal Block (MOB) forms [390] to compute the tiles in each level of the memory hierarchy. The MOB form provides maximum reuse of data in all levels simultaneously and their orthogonality property provides a simple method to optimize the form and the size of the tiles at each level. Another popular approach for multi-level blocking is from the group of Pingali at Cornell University [278].

An example illustration of multi-level blocking is shown in Fig. 7.3. The notable differences between typical blocking and multilevel blocking are more than one size of blocks, here $B1$ and $B2$. Indeed the code here is for a two level memory hierarchy as proposed in [278]. Note that the initial code is the same as used for illustrating loop tiling in Fig. 7.2. Thus for example, for $N = 4$, L1 cache size of 8, L2 cache size of 16 - we can reduce most capacity misses with tile sizes of $B1 = 1$ and $B2 = 2$. This is because for $B1 = 1$ we will need $(2 \times 3 =)$ 8 data elements in L1 cache, and for $B2 = 2$ we will need $(4 \times 3 =)$ 12 data elements in the L2 cache.

7.1.4 Array padding

Array padding is a data-reorganization technique that involves the insertion of dummy elements in a data structure for improving cache performance namely to reduce (self-) conflict misses [405] [407]. Array padding [340] increases the size of the array dimension aligned with the storage order, and is most effective in reducing the occurrence of self conflicts when the array dimensions are power of

```
for(i3=0;i3<N;i3+=B2)
 for(j3=0;j3<N;j3+=B2)
  for(k3=0;k3<N;k3+=B2)
   for(k2=k3;k2<min(k3+1,N);k2+=B1)
    for(j2=j3;j2<min(j3+1,N);j2+=B1)
     for(i2=i3;i2<min(i3+1,N);i2+=B1)
      for(i=i2;i<min(i2+1,N);i++)
       for(j=j2;j<min(j2+1,N);j++)
        for(k=k2;k<min(k2+1,N);k++)
         a[i][j] += b[j][k] * c[k][j] ;
```

Figure 7.3.   Matrix multiply blocked for two levels of memory hierarchy.

two. However, the amount of padding which minimizes the occurrence of conflicts between arrays is difficult to determine directly when data from different arrays must remain cached for reuse. Hence, arbitrary use of padding cannot guarantee the best performance.

An example illustration of array padding is provided in Fig. 7.4. In our example, the cache size is 8 bytes and cache line size of 1 byte. The initial code has self-conflicts at every iteration of i, whereas the code after array padding has no self-conflicts but only compulsory misses. Thus for $N = 12$, the initial code has 5 self-conflict misses and the array padded code has 2 compulsory misses and no self-conflicts. In our example, there is only one reuse vector with a distance of 8. If there were many reuse vectors or multiple arrays it becomes very difficult to evaluate the pad values. Thus for calculating pad values between arrays requires precise knowledge of base addresses, which until now nobody has exploited. Also, only affine storage orders have been investigated in the literature until now. Nevertheless this is the only true storage order transformation available for conflict misses in the literature.

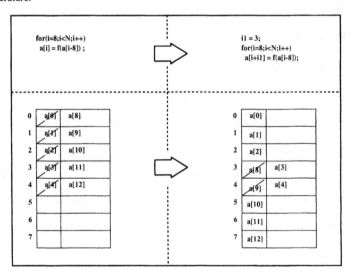

Figure 7.4.   Example illustration of array padding.

[448] propose an algorithm to carry out array padding for both individual arrays and between arrays. However their reduction of conflict misses are quite modest (35%) as compared to those of

[405] and [340] (up to 60%). Apart from [340], [405] and [448] to our knowledge no work has been carried out on reducing conflict misses by means of storage order transformation.

7.1.5   Combination of loop and data transformations

Apart from just loop transformations or data (layout) transformations individually, a combination of the two has also been shown to have many benefits for reducing capacity misses. Cierniak and Li [112] have presented a unified approach to optimize locality using both data and control (loop) transformations. Similarly Kandemir et al [259] have proposed an algorithm to optimize locality under data layout constraints. Kulkarni and Stumm [292] propose a technique to decompose iteration space into finer computation spaces and then linearly transform each of the computation spaces. [23] propose a similar approach for multi-processors.

The advantages of combining loop and data transformations in the context of cache optimization are: (1) cache misses related to the execution order namely capacity misses are reduced and (2) cache misses related to storage order namely conflict are also reduced. However, some issues need further investigation namely, most approaches currently explore a only particular loop transformation for a particular subset of data layouts. For example, [259] explores only loop tiling for either row-wise or column-wise data layouts. Thus, approaches which perform a much more global exploration of many different execution order transformations, not just one particular loop transformation, for different data layouts need to be investigated. This can be achieved even at the cost of large compilation times, which are allowed in our application domain, as long as the results are beneficial.

7.1.6   Software prefetching

It is increasingly common for designers to implement data prefetching by including a fetch instruction in the (micro-) processor instruction set. A fetch specifies the address of a data word to be brought into the cache. Upon execution, the fetch instruction passes this address to the memory system, which forwards the word to the cache. Because the processor does not need the data yet, it can continue computing while the memory system brings the requested data into the cache.

An example illustration of software prefetching is shown in Fig. 7.5. Let us assume that the cache size is 2KB and the cache line size is 4 bytes. Also for simplicity, we assume that we need one cycle to fetch a data from the main memory and all the elements are integers. Thus almost all the elements required in the algorithm fit in the cache. So it is likely that only compulsory misses will occur. In the given algorithm, we need 150 accesses to array $a[][]$ and 300 accesses to array $b[][]$. Out of the 450 accesses we will have $150 + 51 = 201$ misses. Note that array $b[][]$ causes only 51 misses since it reuses the $(j + 1)$th element as well as the complete j elements for succeeding iterations of i. The code with prefetch instructions shows that we now have a trade-off to make, since for avoiding 201 cache misses we have to execute 201 additional instructions. Depending on the miss penalty of the machine, this trade-off can be a good one or not. For example if the miss penalty is 20 cycles, then this trade-off becomes attractive, since these misses can potentially degrade the performance quite a lot.

Software initiated prefetching involves an overhead in additional instructions hence trade-offs are involved as to determine the best possible solution for an algorithm [546]. Combining loop unrolling and loop pipelining [163] with prefetching has been shown to yield better results [379].

7.1.7   Scratch pad versus cache memory

Scratch-pad memory refers to data memory residing on-chip [408], that is mapped into an address space disjoint from the off-chip memory, but connected to the same address and data buses. Both the cache and scratch-pad SRAM allow fast access to their residing data, whereas an access to the off-chip memory (usually DRAM) requires relatively longer access times. The main difference between the scratch-pad SRAM and data cache is that, the SRAM guarantees a single-cycle access time, whereas an access to the cache is subject to compulsory, capacity and conflict misses.

In a compiler context, the contents of the scratch-pad memory are decided at compile-time and are completely user controlled. Whereas, for the data cache, the cache controller manages the data

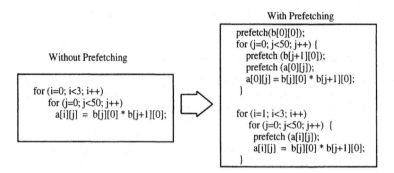

Figure 7.5.   Example illustration of software prefetching.

transfers from/to the cache at run-time. Thus for applications where full compile-time analysis is not feasible a scratch-pad SRAM is not preferred. Indeed for such cases, it is preferred to use a lockable cache which combines the benefits of both a scratch pad type SRAM and a hardware controlled cache [507] wherein the non-static parts of the application code are managed by the hardware controller and the statically analyzable part being locked in the cache by the compiler.

7.2   THE GLOBAL METHODOLOGY

The cache related issues in our global DTSE methodology are situated in Fig. 7.6.

We will deal with both (1) the sub-steps needed in our methodology to improve cache utilization for lower power and reduced bandwidth, and (2) the additional constraints on different steps due to the caching issues. Next to the execution order transformation steps that were applied up to now, in this chapter the focus lies on the storage order transformation steps. Most of these will be fixed in the code by means of source code transformations. However, some of them require the definition of *pragmas*, like locking status in the TriMedia TM1 processor.

***Running example:*.**   We will use an autocorrelation function which is found in a voice coder to illustrate the steps of our methodology [440]. The autocorrelation function measures the similarity between samples of the same signal as a function of the time delay between them. The initial algorithm is shown below in Fig. 7.8. Note that we have two loop nests and two important signals[2] namely "ac_inter[]" and "Hamwind[]" with respective sizes of 26400 and 2400 integer elements. Observe that "Hamwind[]" is produced and consumed in two different loop nests and the second loop nest has non-rectangular access patterns dominated by accesses to the temporary variable "ac_inter[]".

Thus we have two objectives to achieve for reducing memory accesses : first, bring the production and consumption of "Hamwind[]" closer in time and secondly reduce the number of accesses to "ac_inter[]". The first objective is achieved in the global loop transformation step of the DTSE script. The result is shown in Fig. 7.9.

7.2.1   *Data reuse and cache bypass decisions*

In this step the various signals exhibiting data reuse identified in the first step are arranged in order to match the target memory hierarchy. This is done by introducing temporary signals which satisfy the size limits of smaller memories in the memory hierarchy. Note that this step is not limited to the software prefetching methods as described in subsection 7.1.6. A more detailed discussion about data reuse decisions and the corresponding exploration using data reuse trees is available in [154] [563].

---

[2]This implies that these two signals are responsible for most memory accesses and are dominant in size.

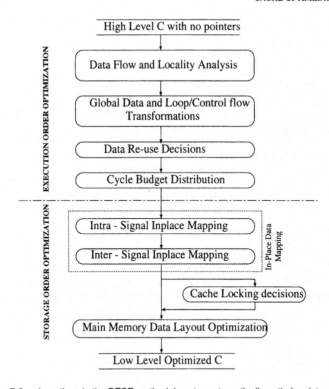

Figure 7.6.   Locations in the DTSE methodology to systematically optimize data caches.

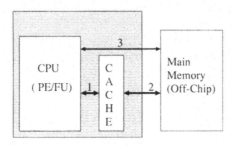

Figure 7.7.   Various points in typical processor organization where power consumption can be reduced.

Apart from introducing temporary signals to exploit data reuse from the data reuse exploration trees, we also fix the signals which will be transferred directly to the CPU. This is possible only for software controlled caches, where the programmer or the compiler can pass directives to bypass the cache . This process of cache bypassing is done along with memory hierarchy layer assignment step since we have an estimate of the power requirements for different hierarchy considerations at this stage. This information is used to identify signals with the least amount of data reuse and bypass them so that they do not interfere with the existing elements of the cache. It is also important to note that apart from just the reuse of data, it is as important that we take in to account the additional delay/latency due to this bypassing of signals. Bypassing large signals (with no reuse at all), can have adverse effect on performance and needs to be taken in to account. This step influences the reduction in the number of accesses at points 2 and 3 in Fig. 7.7.

```
for(i5=0;i5<2400;i5++)
{
  if (pr_max2[180] < 0.00781250)
       Hamwind[i5] = (Ham_int[i5] << 5 );
  else  Hamwind[i5] = (Ham_int[i5] >> 5 );
}

for(i6=0;i6<11;i6++)
{
 ac_inter[i6][i6] = Hamwind[0] * Hamwind[i6];
 ac_inter[i6][i6+1] = Hamwind[1] * Hamwind[i6-1];
 for(i7=(i6+2);i7<2400;i7++)
   ac_inter[i6][i7] = ac_inter[i6][i7-1] + ac_inter[i6][i7-2] +
                      (Hamwind[i7-i6] * Hamwind[i7]);

 AutoCorr[i6] = ac_inter[i6][2399];
}
```

Figure 7.8.   Pseudo code of the autocorrelation function found in the voice coder algorithm.

```
for(i57a=0;i57a<40;++i57a)
 for(i57b=0;i57b<60;++i57b) {
   i57 = i57a*60 + i57b;

   if (pr_max2[180] < 0.00781250)
        Hamwind[i57] = (Ham_int[i57] << 5 );
   else Hamwind[i57] = (Ham_int[i57] >> 5 );

    if (i57 < 11){
       ac_inter[i57][i57] = Hamwind[0] * Hamwind[i57];
       ac_inter[i57][i57+1] = Hamwind[1] * Hamwind[i57-1];
    }

    for(i6=0;i6<11;i6++)
    if (i57 >= i6+2)
       ac_inter[i6][i57] = ac_inter[i6][i57-1] +
                           ac_inter[i6][i57-2] +
                           (Hamwind[i57-i6] * Hamwind[i57]);

    if (i57 == 2399)
     for(i6=0;i6<11;i6++)
       AutoCorr[i6] = ac_inter[i6][2399];
 }
```

Figure 7.9.   Pseudo code of the loop transformed autocorrelation function.

Running example:. In our example, we have two large signals which are responsible for most of the memory accesses, namely "ac_inter[][]" and "Hamwind[]". We illustrate in brief the process of exploring different alternatives for data reuse copies using the data reuse tree for the signal "Hamwind[]". Note that in the Fig. 7.10, the levels L0, L1, L2 and L3 indicate abstract (potential) levels of memory hierarchy. The corresponding power consumption for each copy of data is also

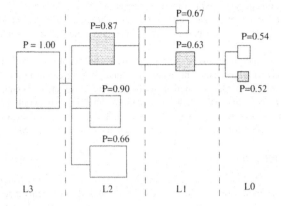

Figure 7.10.   Data reuse trees for signal Hamwind[].

indicated. Note that the copy in $L3$ has the reference power consumption of 1.0, which means that all the memory accesses are made to this single memory. Similarly for copies of different sizes in each level the power consumption changes with respect to the reference value of 1.0. For example, in level $L2$ the case with shaded copy shows that introducing this copy reduces the power consumption by 13%. Thus for each of these abstract levels of hierarchy we explore the possible copies of "Hamwind[]" which can be created. Thus for three potential levels of hierarchy we have the different possibilities as illustrated in Fig. 7.10.

7.2.2 Storage cycle budget distribution, write back and replacement policy management

The next step after fixing the data reuse and cache bypass decisions is to optimize the algorithm for a given cycle budget and the associated timing constraints. In this step, the memory accesses are ordered under the constraints of memory size and the number of ports and the associated bandwidth. Hence this step is also termed as conflict directed ordering (CDO). Also the cycle budget is distributed over the individual loop nests. Different issues pertaining to the storage cycle budget distribution are available in [66] [395] [483] [564] [566].

Before actually performing the CDO sub-step, we have to take some additional decisions namely write back decisions and replacement policy decisions based on the given execution ordering defined in all previous steps and sub-steps. These decisions are valid only for software controlled caches and imposed using the high-level directives.

The write back decision identifies the various signals and instances in the algorithm wherein the next level of memory hierarchy needs to be updated. Evidently, this decision introduces additional transfers to the main memory (as compared to no transfers at all without these decisions). The presence of a write buffer (in the write back path) and the amount of cycles required to completely transfer all the elements of the write buffer to the next level of memory need to be considered. This consideration is important, since in many applications, if the dependencies are not short we will require elements produced in one function immediately in the next function, but due to cache size constraints we will have to write back many elements. This increases the time criticality of such applications and also influences the amount of write backs (and hence power). Also the replacement policy decision pertains to the issue of which element to be replaced on a cache miss. For a hardware controlled cache, typical replacement policies like LRU are present but with prior knowledge of the data flow and dependencies of the algorithm we can decide to retain signals which are actually reused further in the execution order. For instance, if several signals are reused further on, their local cost (in power and/or performance) when replacing them can easily be calculated to decide. Thus write back and replacement policy decisions cause additional transfers to the main memory. Moreover, these issues need to be decided before the cycle budget is fully distributed that is before the final CDO step. Thus this step reduces the number of accesses at point 2 in Fig. 7.7 and it enforces the timing constraints, taking into account the number of memories and their bandwidths.

Running example:.   In the data reuse decision step we have only explored the different possibilities of placing the copies of larger data. The actual placement of these copies in the code based on a memory hierarchy is performed in this step. Thus, for example, if we had a machine with four levels of hierarchy then we will choose the shaded alternatives in the Fig. 7.10. Similarly, if we had only three levels of memory hierarchy then we will choose the last three levels of shaded hierarchy. The power consumption values for each copy are indicated in the corresponding levels.

7.2.3 Data layout optimizations

The *storage order* of a program refers to the order of storage of the different data elements in a given program. Thus any modification to the storage order of these data elements is termed as storage order transformation. Examples of storage order transformation include in-place data mapping and array padding. In this section, we will discuss the steps of the DTSE methodology which mainly impact the storage order of a program. They modify the final data layout of the arrays so they will be termed data layout optimizations in our script.

7.2.3.1 In-place data mapping.   Once the execution order is fully fixed for the memory accesses we now focus on the storage order of the various signals. It is important to note that the cache performance and utilization is highly influenced by the storage order of these signals. A more detailed discussion of this step and its impact on caching is presented in Chapter 5.

The first sub-step is the optimization of storage order inside individual signals. This process is termed as intra-signal in-place mapping. More details about this step is available in [148] [147]. Intra-signal in-place mapping reduces storage size and hence contributes to a reduction of the capacity misses. In the context of our methodology, it is also important to ensure that the copies in the cache and the main memory are compatible in terms of their storage order at various levels of the memory hierarchy. This is important because we have decided in the CDO step the various instances of write backs and this in turn constrains the in-place mapping. This implies that we cannot perform valid write backs if the copies in different levels of the memory hierarchy are inconsistent. Thus the intra-signal in-place mapping step needs to be applied to the individual data reuse copies at different levels of hierarchy. This step reduces the amount of traffic between levels of memory hierarchy (that is point 2 in Fig. 7.7) and influences point 3 indirectly.

In the previous sub-step we have optimized the storage order inside a signal. Next, we concentrate on optimizing the storage order between different signals. This process is termed as inter-signal in-place mapping. More details about this technique are available in [148] [145]. If the exact location of mapping these signals in the main memory is not considered, that is they are placed as and where free locations are available, a degradation in cache performance can occur. Thus once the inter-signal in-place mapping is performed we ensure that the various signals are mapped taking the cache size and the cache line size in to account.

Running example:.   Now that we have brought the production and consumption closer in time for the "Hamwind[]" signal, the remaining bottleneck is the large memory size required by the temporary signal "ac_inter[]". Note that "ac_inter[]" has dependency only on two of its earlier values, thus only three (earlier) integer values need to be stored for computing each of the autocorrelated values. Thus by performing in-place data mapping, as shown in Fig. 7.11, we are able to drastically reduce the size of this signal. Thus initially "ac_inter[]" could not have been accommodated in the cache due to cache size limitation[3] but now we have removed this problem. This transformation has large benefits in the cache locking stage shown next.

---

[3]Thus most cache misses caused by "ac_inter[]" were due to capacity misses and this is now reduced to a large extent.

```
LOCK(&Hamwind, 2400);
LOCK(&ac_inter, 33); /* Initially this was 11*2400 = 26400 */
for(i57a=0;i57a<40;++i57a)
 for(i57b=0;i57b<60;++i57b) {
   i57 = i57a*60 + i57b;

   if (pr_max2[180] < 0.00781250)
        Hamwind[i57] = (Ham_int[i57] << 5 );
   else Hamwind[i57] = (Ham_int[i57] >> 5 );

   if (i57 < 11) {
       ac_inter[i57][i57%3] = Hamwind[0] * Hamwind[i57];
       ac_inter[i57][(i57+1)%3] = Hamwind[1] * Hamwind[i57-1];
   }

   for(i6=0;i6<11;i6++)
    if (i57 >= i6+2)
       ac_inter[i6][i57%3] = ac_inter[i6][(i57-1)%3] +
                            ac_inter[i6][(i57-2)%3] +
                            (Hamwind[i57-i6] * Hamwind[i57]);

   if (i57 == 2399)
    for(i6=0;i6<11;i6++)
       AutoCorr[i6] = ac_inter[i6][2];
}
WRITEBACK(&Hamwind, 2400);
```

Figure 7.11.   Pseudo code of the in-place data mapped autocorrelation code.

7.2.3.2   Cache locking for software controlled caches.   After all the above transformations we have to decide on locking various signals in a software controlled cache at particular instances. Thus this step is valid only for software controlled caches. These decisions need to address the following two questions : (1) which consecutive or individual lines in the main memory should be locked? and (2) whether signal unrolling is needed since many signals will be part of the working set and only a part of these signals is alive at any instance. A detailed knowledge of software controlled cache architectures is assumed. This means we must know whether multiple parts (this could be statements or loop nests, in general, parts of main memory) of program can be locked in the cache or only a single continuous portion of the program is allowed to be locked. This also includes issues like, whether partial lines are allowed to be locked or only complete lines can be locked. These features determine the amount of gain we can achieve in power and performance using the cache locking feature. A more detailed discussion of this step and its impact on cache related issues is presented in Chapter 5.

Running example:.   Now that we have removed the data locality and memory size bottlenecks, we can perform the data cache locking step - which reduces the required bandwidth of the given application. This is necessary because signals which are locked in the cache are not written back to the next higher-level of memory. The task of maintaining cache coherency is now with the programmer or the compiler and not with the hardware anymore. To perform cache locking we identify signals which have more write backs (memory accesses) to the main memory and then perform cache locking and the associated write backs at the necessitated time instances in the code. Thus for an 8KB (8192 integer elements) lockable data cache, we can easily lock both the (memory access) dominant signals in the cache for our example as shown in Fig. 7.11. We have observed a reduction in the bandwidth by a factor 10 after these transformations on a TM1000 simulator. This demonstrates the very large gains that are achievable by the combination of sub-steps in section 7.2.3.1 and section 7.2.3.2.

7.2.3.3 Main memory data layout organization. The last step once all the code transformations are over is the organization of data in the main memory. The main objective of this step is to place the data at particular base addresses in this memory taking into account the cache size[4] so as to remove most of the conflict cache misses. Typically the compilers allocate single contiguous memory locations to signals, which causes conflict misses since the execution order does not match the storage order. Only array padding has been able to reduce conflict misses in the past, but as stated earlier it also has many limitations. In our approach, we partition the signals and interleave partitions of different signals in the main memory. This reduces the instances of mapping signals with similar lifetimes onto same cache locations. This step has significant effect on cache utilization and reduces to a large extent the conflict misses. A more detailed discussion of this step is presented in Chapter 6 and Chapter 7.

Running example:. Fig. 7.12 shows the initial and optimized data organization in the main memory for the LPC analysis function. For a cache size of 32 bytes, we have the data organization as shown in Fig. 7.12. Note that there is a recursive allocation of all the data where each unit of allocation is equal to the cache size. For the initial organization all the signals are allocated contiguously in the main memory and get mapped to all the locations in the cache. In contrast, after data organization all the data which are alive simultaneously get mapped to mutually exclusive cache locations. For a direct mapped cache of 32 bytes, we note that there are 2641 compulsory misses and 21768 capacity misses and 10050 conflict misses. Thus to avoid 10050 conflict misses we need an overhead in memory size of 656 locations or 23% as shown in Fig. 7.12. This is a trade-off that we have to make globally across the application.

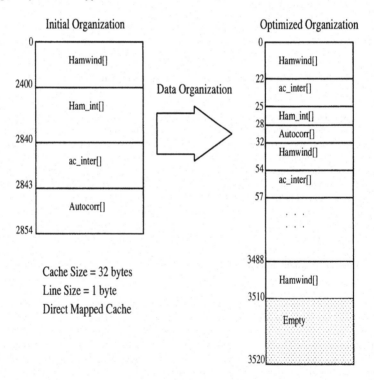

Figure 7.12.   Illustration of data organization on LPC analysis.

[4]All along we have been doing relative placement of data at a higher-level and not the actual data placement in the main memory.

7.3 IN-PLACE DATA MAPPING FOR CAPACITY MISS REDUCTION

Two issues are central to efficient cache utilization, namely maximal reuse of data in the cache and minimal write backs to the next level of memory in a memory hierarchy. Most compiler optimization try to achieve the first goal, that is to maximize the reuse of data in the cache but they usually sacrifice the second goal, since this goal is not in the critical path of the CPU accesses. However to achieve a cost efficient implementation both the issues need to be addressed. Aggressive in-place data mapping addresses both the issues and cache locking allows for further reduction in the write backs. In this section, first, we will introduce the existing in-place data mapping technique and then, show the correlation between in-place data mapping and cache miss as well as write back reduction. Based on these discussions we will derive some constraints which need to be applied to the existing in-place data mapping technique to obtain better cache utilization. Lastly, we will present the experimental results of applying our complete methodology on three real-life applications.

7.3.1 Introduction to in-place mapping technique

Many models have been proposed in the past for representing a given program [9] [112]. In this work, our main concern is the analyses of memory occupation and the related storage order(s). Hence traditional models which focus solely on modeling the execution order of a program are not suitable for performing storage order optimizations like in-place data mapping. In the literature, approaches which enable memory reuse have also been reported [178] [214] [549]. A technique to estimate memory size early in the design flow (even) when the execution order is not completely fixed is available in [275]. A more detailed discussion about different models and the model presented here is available in [148].

7.3.1.1 Geometrical model for memory storage analysis. We use the concept of *binary occupied address/time domains* (BOATD's) to model the detailed execution and storage order of a program [148]. Each point with integer coordinates in this domain represents an occupied address/time tuple that is an address that is (possibly) being occupied at that time. The binary OATD (BOATD) domain of each value based flow dependency in the program is described using the expression below.

$$
\begin{aligned}
D_{ijklm}^{BOAT} = \ & \{ [a,t] \mid \exists s \in D_m^{var}, i \in D_i^{iter}, j \in D_j^{iter}, x, y, w \ s.t. \\
& a = O_m^{addr}(s,w) \wedge C_m^{addr}(w) \geq 0 \wedge \\
& M_{ikm}^{def}(i) = \mathbf{s} = M_{jlm}^{oper}(j) \wedge \\
& t_{init}' \geq O_{ikm}^{wtime}(O_i^{time}(i,x)) \wedge C_{ikm}^{time}(x) \geq 0 \wedge \\
& t_{final}' \leq O_{jlm}^{rtime}(O_j^{time}(j,y)) \wedge C_{jlm}^{time}(y) \geq 0 \}
\end{aligned}
\tag{7.1}
$$

In equation 7.0, the first two subscripts of D_{ijklm}^{BOAT}, namely i and j refer to the statements where the concerned array is present (either defined or as an operand). The third and fourth subscript, namely k and l refer to the position of the definition or the operand in statements i and j respectively (there can be multiple definitions or operands in the same statement), while the last index (here m) refers to the array to which the domain belongs.

In the above equation, D_i^{iter} refers to the iteration domain for statement i. Each integral point inside this domain corresponds to the execution of one statement. The set of points in this domain thus represent all the executions of this statement. Similarly, D_m^{var} represents the variable domain of an array m and is a mathematical description of an array of variables. Each integral point inside this domain corresponds to exactly one variable in the array. The definition and operand domains of a statement describe which variables are being accessed during all possible executions of that statement and are represented by D_{ikm}^{def} and D_{jlm}^{oper} respectively. Each integral point inside these domains correspond to exactly one variable that is being written (in case of a definition domain) or read (in case of an operand domain).

The relation between the executions of the statements and the variables that are being written or read, are represented by means of mathematical mappings between the dimensions of the iteration domains and the dimensions of the definition or operand domains. We refer to these mappings as the definition mappings (M_{ikm}^{def}) and operand mappings (M_{jlm}^{oper}) respectively.

The time order function of D_i^{iter} is represented by $O_i^{time}()$. Evaluating this function for a point in D_i^{iter} results in the "execution date" of the corresponding operation. The meaning of "execution date" is context dependent as it may refer to an absolute execution date or to a relative execution date. The time offset function of D_{ikm}^{def} is represented by $O_{ikm}^{wtime}()$. The resulting offset is an offset relative to $O_i^{time}()$. A write access corresponding to a point in D_{ikm}^{def} occurs at this offset in time relative to the execution of the corresponding operation. The time offset function of D_{jlm}^{oper} is represented by $O_{jlm}^{rtime}()$. A read access corresponding to a point in D_{jlm}^{oper} occurs at this offset in time relative to the execution of the corresponding operation. These two time offsets allow us to evaluate the first time instance when the array was accessed (becomes alive), represented by t_{init} and the last time instance when the array is accessed, represented by t_{final}, after which the array ceases to exist[5].

The storage order function of D_m^{var} is represented by $O_m^{addr}()$. Evaluating this function for a point in D_m^{var} results in the storage address at which the corresponding variable is stored. In general, each of these functions may have (extra) hidden[6] or auxiliary dimensions as arguments and may be accompanied by extra constraints on these extra dimensions (for example, a symbolic constant), represented by $C_i^{time}() \geq 0$ and $C_m^{addr}() \geq 0$.

Thus, for a given execution and storage order, the above equation contains all the information available at compile time about the (potential) memory occupation due to a flow-dependency between two statements. In fact a BOAT-domain is nothing more than a mapping of a dependency on an address/time space.

Now, we can easily find the complete memory occupation of an array if we know the memory occupation due to *each* value-based flow dependency related to that array. The memory occupation of the array is then simply the *union* of the memory occupation due to each of the flow dependencies. Consequently, we can also model the memory occupation of an array by a geometrical domain, which we call the *occupied address/time domain* (OAT-domain) of that array.

$$D_m^{OAT} = \bigcup_{ijkl} D_{ijklm}^{BOATD} \qquad (7.2)$$

Further more, we can now derive an expression for the occupation of a memory itself. It is again simply the union of the memory occupation of all arrays *assigned to that memory*. The result is again a geometrical domain, which we call the *collective occupied address/time domain* (COAT-domain) for that memory.

$$D_m^{COAT} = \bigcup_m D_m^{OAT} \qquad (7.3)$$

A more detailed description of the above model and the associated issues can be found in [148] [93].

Example:.   An example of a B(C)OAT-domain for a non-manifest program and a memory containing 3 arrays is given next.

```
    int A[10], B[5], C[10];
    for ( i = 0; i < 10; ++i )
S1:     C[i] = f1(...);

    for ( j = 0; j < 5; ++j )
S2:     A[j] = f2(...);
```

---

[5]In the equation above we know the time instances for each dependency, hence the superscript of ′ is used. By evaluating this function for all the dependencies for the particular array we obtain the actual t_{init} and t_{final}.

[6]Hidden dimensions are not really part of the domains, but are used to model non-manifest behaviour.

```
    for ( k = 5; k < 10; ++k )
S3:     if ( C[k-5] > 0 )
S4:         A[k] = f3(...);
        else
S5:         B[k-5] = f4(...);

    for ( l = 0; l < 10; ++l )
S6:     if ( l >= 5 && C[l] <= 0 )
S7:         ... = f5(B[l-5]);
        else
S8:         ... = f6(A[l]);
```

The variable, iteration, definition and operand domains for this example are the following:

$$
\begin{aligned}
D_A^{var} &= \{\, s_a | 0 \le s_a \le 9 \wedge s_a \in \mathbb{Z} \,\} \\
D_B^{var} &= \{\, s_b | 0 \le s_b \le 4 \wedge s_b \in \mathbb{Z} \,\} \\
D_C^{var} &= \{\, s_c | 0 \le s_c \le 9 \wedge s_c \in \mathbb{Z} \,\} \\
D_1^{iter} &= \{\, i | 0 \le i \le 9 \wedge i \in \mathbb{Z} \,\} \\
D_2^{iter} &= \{\, j | 0 \le j \le 4 \wedge j \in \mathbb{Z} \,\} \\
D_3^{iter} &= \{\, k | 5 \le k \le 9 \wedge k \in \mathbb{Z} \,\} \\
D_4^{iter} &= \{\, k | 5 \le k \le 9 \wedge p = F(k) \wedge p > 0 \wedge k \in \mathbb{Z} \,\} \\
D_5^{iter} &= \{\, k | 5 \le k \le 9 \wedge p = F(k) \wedge p \le 0 \wedge k \in \mathbb{Z} \,\} \\
D_6^{iter} &= \{\, l | 0 \le l \le 9 \wedge l \in \mathbb{Z} \,\} \\
D_7^{iter} &= \{\, l | 0 \le l \le 9 \wedge q = F(l) \wedge l \ge 5 \wedge q \le 0 \wedge l \in \mathbb{Z} \,\} \\
D_8^{iter} &= \{\, l | 0 \le l \le 9 \wedge q = F(l) \wedge (l < 5 \vee q > 0) \wedge l \in \mathbb{Z} \,\} \\
D_{11C}^{def} &= \{\, s_c | \exists i \in D_1^{iter} \text{s.t. } s_c = i \,\} \\
D_{21A}^{def} &= \{\, s_a | \exists j \in D_2^{iter} \text{s.t. } s_a = j \,\} \\
D_{31C}^{oper} &= \{\, s_c | \exists k \in D_3^{iter} \text{s.t. } s_c = k - 5 \,\} \\
D_{41A}^{def} &= \{\, s_a | \exists k \in D_4^{iter} \text{s.t. } s_a = k \,\} \\
D_{51B}^{def} &= \{\, s_b | \exists k \in D_5^{iter} \text{s.t. } s_b = k - 5 \,\} \\
D_{61C}^{oper} &= \{\, s_c | \exists l \in D_6^{iter} \text{s.t. } s_c = l \,\} \\
D_{71B}^{oper} &= \{\, s_b | \exists l \in D_7^{iter} \text{s.t. } s_b = l - 5 \,\} \\
D_{81A}^{oper} &= \{\, s_a | \exists l \in D_8^{iter} \text{s.t. } s_a = l \,\}
\end{aligned}
$$

We assume the following execution and storage orders (assuming that arrays A, B and C all have the same element size):

$$
\begin{aligned}
O_1^{time}(i) &= i & O_{11C}^{wtime}(x) &= x \\
O_2^{time}(j) &= j + 10 & O_{21A}^{wtime}(x) &= x \\
O_3^{time}(i) &= 2(k-5) + 15 & O_{31C}^{rtime}(x) &= x \\
O_4^{time}(i) &= 2(k-5) + 16 & O_{41A}^{wtime}(x) &= x \\
O_5^{time}(i) &= 2(k-5) + 16 & O_{51B}^{wtime}(x) &= x \\
O_6^{time}(i) &= 2l + 25 & O_{61C}^{rtime}(x) &= x \\
O_7^{time}(i) &= 2l + 26 & O_{71B}^{rtime}(x) &= x \\
O_8^{time}(i) &= 2l + 26 & O_{81A}^{rtime}(x) &= x
\end{aligned}
$$

$$
\begin{aligned}
O_A^{addr}(s_a) &= s_a \\
O_B^{addr}(s_b) &= s_b + 5 \\
O_C^{addr}(s_b) &= s_c + 10
\end{aligned}
$$

¿From this, we can derive the following BOAT-domains (after simplification):

$$D_{1311C}^{BOAT} = \{ [a,t] | 10 \le a \le 14 \land a - 10 \le t \le 2a - 5 \land a \in \mathbb{Z} \}$$
$$D_{1611C}^{BOAT} = \{ [a,t] | 10 \le a \le 19 \land a - 10 \le t \le 2a + 5 \land a \in \mathbb{Z} \}$$
$$D_{2811A}^{BOAT} = \{ [a,t] | 0 \le a \le 4 \land a + 10 \le t \le 2a + 26 \land F(a) > 0 \land a \in \mathbb{Z} \}$$
$$D_{4811A}^{BOAT} = \{ [a,t] | 5 \le a \le 9 \land 2a + 6 \le t \le 2a + 26 \land F(a) > 0 \land a \in \mathbb{Z} \}$$
$$D_{5711B}^{BOAT} = \{ [a,t] | 5 \le a \le 9 \land 2a + 6 \le t \le 2a + 26 \land F(a) \le 0 \land a \in \mathbb{Z} \}$$

The corresponding graphical representations can be found in Fig. 7.13. The resulting OAT-domains and COAT-domain can be found in Fig. 7.14. Note that the OAT-domains of arrays A and B seem to overlap graphically. This is caused by a virtual overlap between D_{4811A}^{BOAT} and D_{5711B}^{BOAT}. One can verify that this overlap is non-existent in reality:

$$D_{4811A}^{BOAT} \bigcap D_{5711B}^{BOAT} = \{ [a,t] | 5 \le a \le 9 \land 2a + 6 \le t \le 2a + 26 \land$$
$$F(a) > 0 \land F(a) \le 0 \land a \in \mathbb{Z} \} = \phi$$

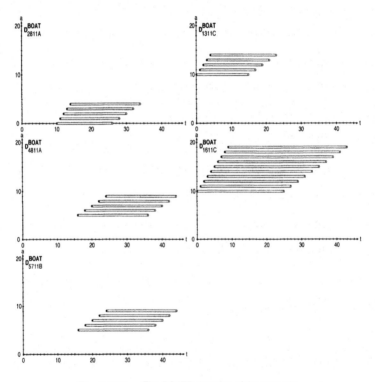

Figure 7.13.   The BOATD domains of the example.

So, the memory occupation of arrays A and B is partly conditional, and the conditions for A and B are complementary.

In practice, this means that some of the elements of arrays A and B can share storage locations, simply because they can never occupy those locations at the same time.

If we would not have modeled this exclusive condition accurately, we could not have been sure that this storage order was valid, and we would have to make sure that the address ranges of arrays A and B were disjoint. This would require extra storage locations. So this example already clearly illustrates how a more accurate modeling of non-manifest programs may result in better solutions, compared to the ones obtained through traditional worst-case modeling techniques [114] [115].

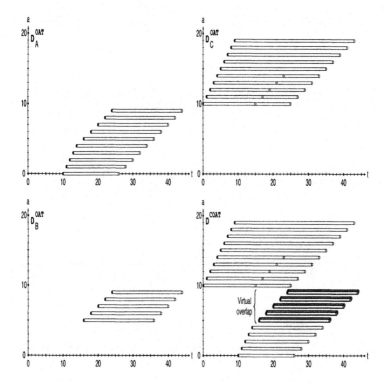

Figure 7.14.   The OAT and COAT domains of the example.

7.3.1.2 Intra-signal inplace mapping for custom memory organizations. The inplace data mapping step consists of two major sub-steps. In a first sub-step, we optimize the internal organization of each array separately, resulting in a partially fixed storage order for each array.

¿From the discussion of the storage order strategies in section 7.2.3, we can derive that we have to try to find the intra-array storage orders that result in occupied address-time domains that are as "thin" as possible, that is storage orders which have the smallest address reference window. The number of possible storage orders is huge however, even if we restrict ourselves to the affine ones. A more detailed discussion about how this step is automated in a custom memory organization context is available in [149]. Here we provide a brief discussion of the material necessary to understand our context.

The intra-signal inplace mapping step comprises calculation of the address reference window, as shown in Fig. 7.15(a), for every array and identification of an (optimal) intra-array storage order as shown in Fig. 7.15(d) [147] [149]. In Fig. 7.15 the term C_x represents an offset in the physical memory space. The real address is obtained by combining the abstract address with an offset in the physical memory space. The address reference window represents the maximal distance between two addresses being occupied by the array at the same time. The BOATD descriptions allow us to calculate this distance by solving a finite number of integer linear programming (ILP) problems. More in particular, for each pair of BOATDs $(BOATD_1, BOATD_2)$, we have to calculate $max(|a_1 - a_2|)$ for $(a_1,t) \in BOATD_1$ and $(a_2,t) \in BOATD_2$. The overall maximum plus one is then equal to the size of the address reference window. This step is also referred to as *windowed allocation* in Fig. 7.15(d).

7.3.1.3 Inter-signal inplace mapping for custom memory organizations. The next step, inter-signal inplace data mapping, optimizes the storage order between different arrays. This step involves re-using the memory space between different arrays, both for arrays with disjoint and partially over-

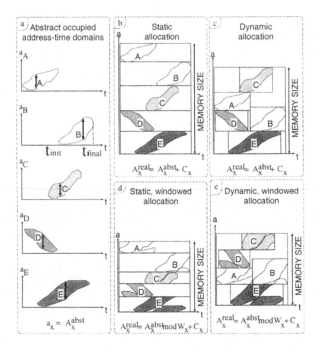

Figure 7.15.   Intra-signal and inter-signal inplace data mapping.

lapped lifetimes. The criteria which identifies (or checks) arrays with disjoint life times is called compatibility and the criteria which identifies arrays with partially overlapped life times which can be merged is called merge-ability [145]. Alternatively, those arrays whose memory space can be reused since all the read/writes to particular addresses of one array are over just before the start of second one to those addresses are considered to be merge-able. This step is illustrated in Fig. 7.15(c) and (e).

The actual array placement algorithm is as follows :

1. We start by setting up two sets of arrays: arrays that have been placed (that is whose offset has been fixed) and arrays that still have to be placed. Originally, all arrays belong to the second set, and we assume that all their offsets are zero.

2. We select from the arrays-to-place set the array that currently has the lowest offset. If two arrays have the same offset, we select the one with the largest window size. If they also have an equal window size, we select the one which has the smallest number of compatible and merge-able arrays.

3. We then compare the selected array with those that have already been placed, and check for conflicts. Two arrays have *no* conflicting positions if: (1) they are compatible, or (2) they have non-overlapping address ranges, or (3) they are merge-able and aligned at the same offset. We then check for conflicts between the selected array and the already placed ones. Two arrays have *no* conflicting positions if they are compatible, or they have non-overlapping address ranges, or they are merge-able and aligned at the same offset. If there is no conflict with any of the already placed arrays, we place the selected array at its current position, and transfer it to the placed-array set. If a conflict exists however, we move the array upwards until we place the selected array at its current position. Otherwise we move the array upwards until its base crosses the top of an already

placed array, and *leave the moved array in the arrays-to-place set.* In both cases, we return to step 2, unless no more arrays have to be placed.

7.3.2 Impact of inplace data mapping on caching

In this section, we will discuss the impact of inplace data mapping on caching. Three main issues are of relevance in this context; first, the reduction in capacity misses due to inplace mapping, secondly, the reduction in write backs to the main memory and lastly, the improvement of cache reuse due to inter-signal inplace mapping. We also discuss the impact of inplace mapping for varying cache line size and associativities.

7.3.2.1 Relating capacity misses to inplace data mapping. Capacity misses occur due to inefficient temporal reuse of data and due to larger working set of data in a given loop nest. Thus to reduce capacity misses one has to either increase the temporal reuse of data or reduce the working set of that data in the particular loop nest namely reduce the size of the particular data.

Intra-signal as well as inter-signal inplace mapping helps in reducing the memory size occupied by the individual array variables and hence helps in reducing the associated capacity misses. In practice, it has been observed that loop transformations (like loop fusion and loop tiling) help in improving the temporal locality of data (see section 7.1). Note that an array has better temporal locality if the occupied address-time domains are as thin as possible. And we know that intra-signal inplace data mapping is able to reduce the memory size of the data which has thin OATD's. This results in array data which has good temporal, due to thin OATD's, as well as good spatial locality, due to the smaller size of data which is accessed close in time. Thus inplace data mapping can potentially reduce the capacity misses to a large extent due to enhanced spatial locality of data.

Example:.   An example illustration of capacity miss reduction due to inplace mapping is illustrated here. Fig. 7.16 shows the initial code and the corresponding cache states for different iterations of j. Note that the cache is fully associative, cache size used is 8 bytes and cache line size is 1 byte, where each element of $a[][]$ and $b[][]$ needs 1 byte of memory space. For $W = 1$ and $H = 8$, we observe that we need 30 data accesses. For completing these 30 data accesses we have 13 cache misses. Out of these, 10 are compulsory misses and the remaining 3 are capacity misses due to elements $b[0][3]$, $b[0][4]$ and $b[0][5]$ respectively.

Fig. 7.17 shows the inplace mapped code and the corresponding cache states for different iterations of j. All other assumptions are similar to those stated above. The real mapping now is a modulo 4 mapping hence array $a[0][0]$ and array $a[0][4]$ occupy the same memory space. Similarly array $a[0][1]$ and array $a[0][5]$ occupy the same memory space and so on. Note that the arrays in italics indicate the coefficients of arrays as in the initial code whereas in reality values greater than 4 get folded due to the modulo mapping. We now observe that to complete 30 data accesses we have only 8 cache misses and all of these misses are compulsory. Thus for the inplace mapped code we are able to eliminate the 3 capacity misses, because all data fit in the cache size, and to reduce the compulsory misses because the same memory locations are reused by different values.

7.3.2.2 Write backs and intra-signal inplace mapping. Write backs occur due to a cache miss or replacement of a data from the cache which has been modified (written in to). Thus to reduce write backs the memory space in the cache has to be reused efficiently namely data has to be kept current by reading/writing to it often enough and writing to as few as possible distinct memory locations in the cache.

Intra-signal inplace mapping reduces the memory size by a windowed allocation, which in turn results in memory accesses closer in space as well as closer in time for a single array. Inter-signal inplace mapping reuses memory space between two arrays closer in time. Thus both the intra- and inter-signal inplace mapping try to reuse memory space closer in time which results in enhanced cache reuse. This in turn results in fewer write backs to the main memory.

Note that loop transformations like loop fusion or loop tiling will enhance locality by reusing the data in the cache as much as possible but they cannot reduce the number of writes to distinct memory locations. In contrast, inplace data mapping is able to reduce number of writes to distinct

Example Algorithm :

```
for (i=0; i<W; i++)
  for(j=3; j<H; j++) {
  b[i][j]  = g() ;
  a[i][j]  = a[i][j-2] + a[i][j-3] + f(b[i][j], b[i][j-2], b[i][j-3]) ;
  }
```

Cache States for different iterations (i=0, j) :

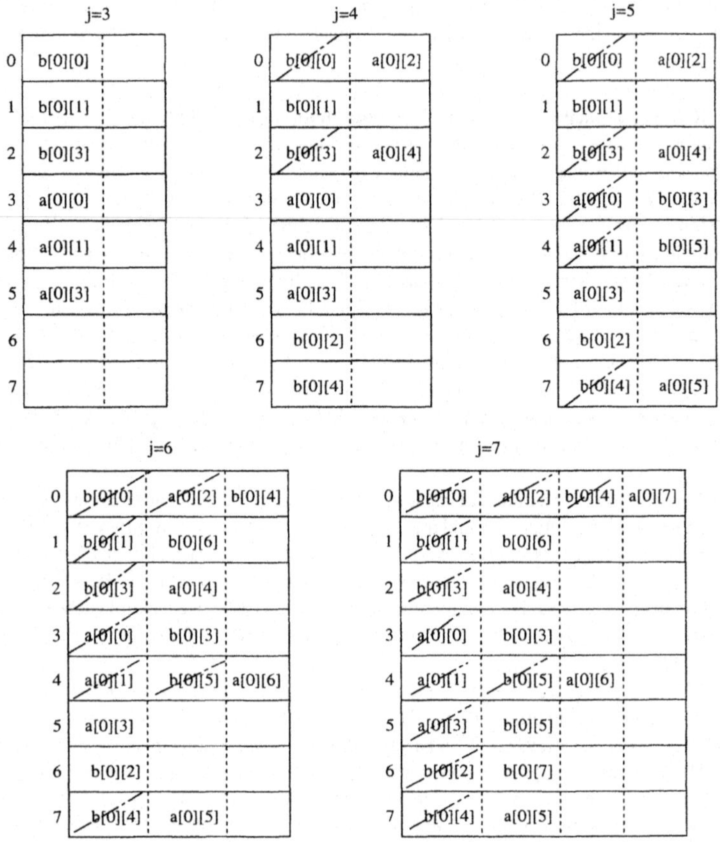

Figure 7.16.   Example illustration of capacity miss reduction with inplace mapping - initial code.

memory locations in the cache, which results in fewer write backs as compared to the conventional loop transformations.

Example:.   Fig. 7.16 and Fig. 7.17 illustrate the reduction in capacity misses due to inplace data mapping. We now focus on another aspect of cache miss related memory traffic namely the write backs.

In our example we write in to both array $a[][]$ and array $b[][]$. Thus only those elements of array $a[][]$ and $b[][]$ which have been written into will be written back to the main memory on a cache miss/replacement. In Fig. 7.16 we observe that this is the case for element $a[0][3]$, $b[0][3]$, $b[0][4]$ and $b[0][5]$. Also observe that the second cache miss due to element $b[0][4]$ is not written back. Thus we have 4 write backs in our example. Note that in Fig. 7.17 we have writes to all the elements but no cache misses or replacements as yet hence there are no write backs. If we assume that our program ends at this point then for the initial code we will have to write back elements $a[0][4]$ to $a[0][7]$ and

Example Algorithm :

```
for (i=0; i<W; i++)
  for(j=3; j<H; j++) {
    b[(i*H+j)%4] = g() ;
    a[(i*H+j)%4] = a[(i*H+j-2)%4] + a[(i*H+j-3)%4]
                 + f(b[(i*H+j)%4], b[(i*H+j-2)%4], b[(i*H+j-3)%4]) ;
  }
```

Cache States for different iterations (i=0, j) :

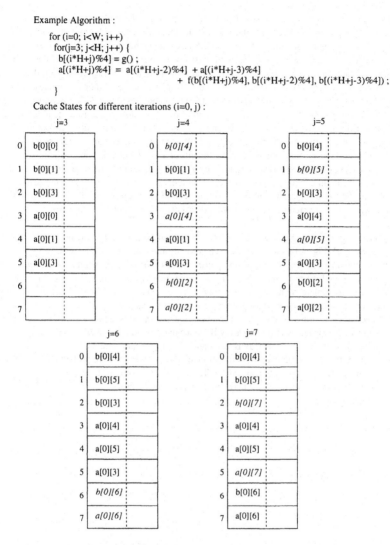

Figure 7.17.   Example illustration of capacity miss reduction with inplace mapping - inplace mapped code.

$b[0][6]$ to $b[0][7]$, which implies that we will have $(4 + 6 =) 10$ write backs now. In contrast, for the inplace mapped code we will have only 8 write backs.

In summary, for the initial code, irrespective of whether we have performed loop tiling [306] or not, we will have $2 \times W \times (H - 3)$ write backs for any value of W and H. In contrast, for the inplace mapped code we will always have 8 write backs. Thus to reduce the memory traffic due to write backs, inplace data mapping is a very effective technique.

7.3.2.3   Inter-signal inplace mapping to improve cache reuse.   Improving cache reuse is critical to achieving an overall better cache utilization. Two issues are central to ensuring a better cache reuse : (1) data must be reused closer in time, which increases the probability of finding the data in the cache and (2) reuse data with more space (addresses) closer in time, which ensures that more cache lines will be reused.

As we have seen, inter-signal inplace data mapping optimizes the storage order between different arrays namely it determines the array placement in the memory. Originally this was only employed in a fully custom memory organization context [145]. To extend this step in a programmable systems context, we need to take into account the caching issues. As mentioned above, one of the key issues is a maximal cache reuse. To use inter-signal inplace mapping in a caching context, trade-off's are involved as to which signals are in-placed which differs from [147]. Based on the above discussion, we use two additional criteria in array placement apart from those used in [147] and discussed earlier. Both the criteria are applied when arrays are found to be compatible and not merge-able. The first criteria evaluates the closeness in time between two arrays which are compatible. In fact, we search for arrays which will result in shorter reuse vectors, thus improving the temporal reuse and locality. The second criteria determines the number of elements of a particular array which are most probable to be in the cache. Here, we are searching for arrays which will result in most number of contiguous addresses closer in time, hence improving the spatial locality. Thus first we reduce the search space based on the first criterion and then based on the second criterion we determine the arrays with similar address-time characteristics which will potentially result in better cache reuse.

Example:. In our example in Fig. 7.15(e), array B can reuse the complete memory space of either array A or array D. Fig. 7.18 shows the two scenarios for the placement of array B. In Fig. 7.18(a), the complete memory space of array D and partial space of array A has been reused. This results in a partially unused memory space for array A, represented as a_1 in Fig. 7.18(a). In contrast in Fig. 7.18(b), the complete memory space of array A is used but only a partial use is made of memory space of array D. This results in unused memory space represented as d_1. In a caching context, our steering decision is to reuse the complete memory space of array A (see Fig. 7.15(e)) because of two reasons : (1) array A is closer in time to array B, hence there is a large probability of finding array A in the cache when the life time of array B starts and (2) array A has more elements (a_1) alive closer in time to array B than that of array D, hence more elements of array A are probable to be (already) present in the cache than those of array D (d_1). Thus in general our two steering criteria allow us to determine the best inter-signal inplace mapping for better cache reuse.

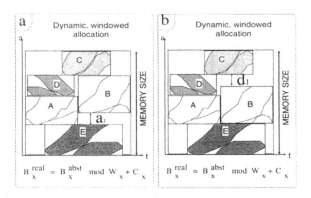

Figure 7.18.   Cache conscious array placement for inter-signal inplace data mapping.

Example:. Fig. 7.19 presents an example of how inter-signal inplace mapping affects caching. We assume that we have a fully associative cache of size $4N$. In the example, arrays $a[]$, $b[]$ and $c[]$ are not temporary and cannot be removed. All other arrays are temporary and can be removed. For the initial code in Fig. 7.19(a) we have $4 \times (N-2) + (N-1)$ compulsory misses and $(N-1)$ capacity misses[7]

---

[7]The capacity miss estimation is highly approximated for the sake of simplicity, as total number of elements required in the loop nest needed minus the cache size.

and we have $6N$ write backs. In Fig. 7.19(b) we show the effect of applying array contraction [381], as done in conventional compilers, on the code. Now we have $4 \times (N-2) + (N-1)$ compulsory misses and no capacity misses and we have $5N$ write backs. Fig. 7.19(c) shows the inter-signal inplace mapped code, where we reuse the unused space of cache for a temporary signal. For this code we have only $4 \times (N-2)$ compulsory misses and $4N$ write backs. Thus we have removed $N-1$ compulsory misses and N write backs by inter-signal inplace mapping as compared to the code with array contraction.

```
tmp1[0] = a[0];
tmp1[1] = a[1];
for(i=2;i<N;i++) {
  tmp1[i] = f(tmp1[i-2],a[i],b[i]);
  c[i] += tmp1[i] * C0;
}

for(i=1;i<N;i++) {
  tmp2[i] = f(c[N-i],a[i],b[i]);
  c[i+N] = tmp2[i] * C1;
}
```

(a) Initial code

```
tmp1[0] = a[0];
tmp1[1] = a[1];
for(i=2;i<N;i++) {
  tmp1[i] = f(tmp1[i-2],a[i],b[i]);
  c[i] += tmp1[i] * C0;
}

for(i=1;i<N;i++) {
  tmp2 = f(c[N-i],a[i],b[i]);
  c[i+N] = tmp2 * C1;
}
```

(b) After array contraction

```
c[0+N] = a[0];
c[1+N] = a[1];
for(i=2;i<N;i++) {
  c[i+N] = f(c[i-2+N],a[i],b[i]);
  c[i] += c[i+N] * C0;
}

for(i=1;i<N;i++) {
  tmp2 = f(c[N-i],a[i],b[i]);
  c[i+N] = tmp2 * C1;
}
```

(c) Inter-signal inplace mapped code

Figure 7.19.   Example illustration of inter-signal inplace mapping and its impact on caching.

7.3.2.4   Hardware controlled cache issues: block size and set associativity.   In this section, we investigate the impact of inplace mapping on caching for different cache line sizes and different set associativities. This is essential to show that inplace mapping is not only valid for a single cache configuration.

We have used an autocorrelation function, which is a part of a voice coder algorithm [440]. Fig. 7.20 shows the reduction in memory bandwidth for different cache line sizes - also called a block size. Note that the number of main memory accesses represents the sum of number of cache misses and the number of write backs to the main memory. We observe that the inplace mapped code always performs better and the trend is similar to the variations in the initial code.

Fig. 7.21 shows the reduction in memory bandwidth for different set associativities. Here too, we observe that the inplace mapped code always performs better than the initial code. In fact, we are always close to 50% better than the initial code. Note that beyond 2-way associative, there is no appreciable reduction in the main memory accesses. This shows that not enough temporal locality is present in this application, whereas after inplace mapping we are able to increase the cache reuse and reduce the write backs which results in a significant reduction in the number of main memory accesses.

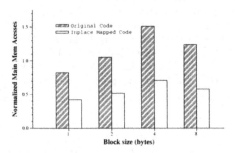

Figure 7.20.   Memory bandwidth reduction due to inplace mapping for varying cache line sizes.

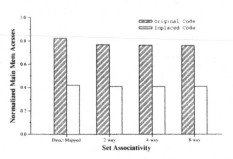

Figure 7.21.   Memory bandwidth reduction due to inplace mapping for varying set associativities.

7.3.3   Experimental results and discussion

This section presents the experimental setup and the results for two applications namely a voice coder and a quad-tree structured differential pulse code modulation (QSDPCM) algorithm. A brief discussion on the algorithms is provided first.

7.3.3.1   Voice coder and QSDPCM algorithm description.

1. **Voice coder algorithm:**
   Most of the communication applications like GSM etc. require digital coding of voice for efficient, secure storage and transmission. We have used a Linear Predictive Coding (LPC) vocoder which provides an analysis-synthesis approach for speech signals [440]. In our algorithm, speech characteristics (energy, transfer function and pitch) are extracted from a frame of 240 consecutive samples of speech. Consecutive frames share 60 overlapping samples. All information to reproduce the speech signal with an acceptable quality is coded in 54 bits. The general characteristics of the algorithm are presented in Fig. 7.22. It is worth noting that the steps LPC analysis and Pitch detection use autocorrelation to obtain the best matching predictor values and pitch (using hamming window). The Schur algorithm is used to obtain signal in baseband region in LPC analysis and to obtain high and low amplitudes in Pitch analysis. Both autocorrelation and Schur's algorithm have irregular access structures due to algebraic manipulations and they form a crucial form of the total algorithm. A more detailed discussion about the various parts of the algorithm is available in [294].

2. **QSDPCM algorithm:**
   The Quadtree Structured Difference Pulse Code Modulation (QSDPCM) [489] algorithm is an interframe adaptive coding technique for video signals. The algorithm consists of two basic parts: the first part is motion estimation and the second is a wavelet-like quadtree mean decomposition of the motion compensated frame-to-frame difference signal.

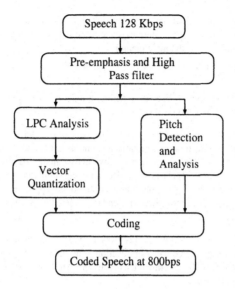

Figure 7.22.   The different parts of the voice coder algorithm and the associated global dependencies.

A global view of the QSDPCM coding algorithm is given in Fig. 7.23. Most of the submodules of the algorithm operate in a cycles with an iteration period of one frame. For the first step of the algorithm (motion estimation in the 4×4 subsampled frames), the previous frame must already have been reconstructed and 4×4 subsampled. The only submodules that are not in the critical cycle (path) are both the 4×4 and 2×2 sub-sampling of the current (input) frame. It is worth noting that the algorithm spans 20 pages of complex C code. A detailed discussion on the algorithm is left out since it is not crucial to understanding this work. A more detailed explanation of the different parts of the algorithm is available in [489].

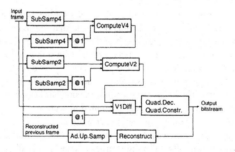

Figure 7.23.   The different parts of the QSDPCM algorithm.

7.3.3.2 Experimental set-up. The experimental setup comprises five different system organizations described below:

1. **Software controlled cache:** For simulation of a software controlled cache we have performed all the data transfer in/out of the cache manually and we have added C++ based counting functions to count the number of reads and writes to the particular arrays. For this setup we have varied the cache size from 32 words to 512 words, depending on the data size in the algorithms, which are indicated for every figure. We have made the following assumptions:

(a) On-chip level one cache, dual-ported, write back and write through updating policy[8] and an off-chip shared main memory. There exists a cache bypass for accessing arrays with little data locality, directly from the main memory.

(b) We make no assumptions on the type of CPU, namely it can be either a RISC or superscalar or VLIW processor.

(c) Additional hardware support like pre-fetching, branch prediction etc. is not present, since they consume more area and power.

2. **Dinero III simulator:** To simulate a hardware controlled cache we have used the dinero III cache simulator [228]. The architectural assumptions are similar to the one made above, but the CPU is defined using the DLX RISC instruction set. We chose smaller data sets for simulations on DLX simulator due to (extremely) large trace sizes and long simulation times[9]. This nevertheless does not compromise the behaviour or the characteristics of the application, since we chose cache sizes such that the ratio between total data size and cache size for typical embedded processors is maintained.

3. **ARM-6 emulator:** The ARM-6 emulator [27] was used to obtain the number of instructions and the code size values for the voice coder application. The main reason is that the code size is a problem especially on RISC-based embedded processors. Moreover, ARM cores are widely used in the embedded application domain.

4. *HP/SUN workstations:* We have used HP-777/879 workstations to execute the optimized codes and to obtain the execution times to measure the performance. We have also measured performance on a SUN machine running SUNOS 4.2.

5. **4-way SMP machine:** We have used a 4-way SMP machine, where each node is Power PC 604 with clock frequency of 112 MHz, to investigate the impact of DTSE oriented transformation on a performance oriented parallelizing compiler. A state-of-the-art optimizing and parallelizing compiler (OPC) from IBM research (xlc for AIX 4.2) was used with this system.

7.3.3.3 Experimental results for hardware cache. In this subsection, we discuss the experimental results for a hardware controlled cache obtained using the Dinero III cache simulator on the voice coder as well as the QSDPCM algorithm.

Fig. 7.24(a) gives the normalized number of cache misses for the different functions in the voice coder observed using the dinero III cache simulator. Note that the inplace data mapped and the globally optimized case has almost half the number of cache misses for the three (main) functions. Similarly, Fig. 7.24(b) shows that the reduction in power is quite significant for both the inplace data mapped and the globally optimized case. This has a dominant effect on the whole algorithm.

Fig. 7.25(a) shows the number of cache misses for each function in the QSDPCM algorithm for initial and optimized versions. The locally transformed version has been optimized for caching with loop blocking, whereas the entire methodology presented in section 7.2 is applied for the globally transformed case. We observe a gain of, on average, 26% for all the concerned functions. It is interesting to note that for this application we have concentrated mostly on improving cache performance for power as compared to looking for reducing overhead due to power oriented transformations on regular code and we found that this benefits both power and performance as seen in Fig. 7.25(b). This shows that our methodology is not only beneficial for power but also for CPU performance.

We observe that the initial specification consumes a relatively large amount of power. The functions V4, motion estimation with four pixel accuracy and calculation of difference between current and previous frame, and V2, motion estimation with two pixel accuracy, have almost similar gains due to the different steps in the methodology. Whereas for the QC, quadtree construction, and UPSA (up-sampling) functions we observe that we gain more due to global transformations than due to just

---

[8]Since the cache is managed by software there is no need for a replacement policy. The user has to ensure the coherency of the main memory.
[9]For a frame size of 128×196 we had trace of size 90MB.

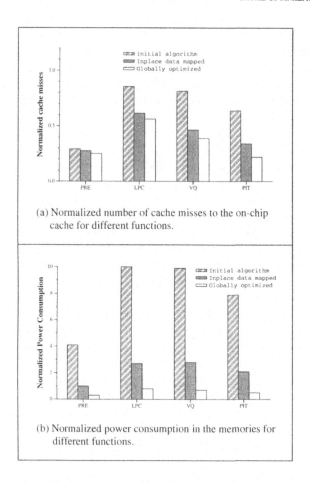

(a) Normalized number of cache misses to the on-chip cache for different functions.

(b) Normalized power consumption in the memories for different functions.

Figure 7.24.   Experimental results for voice coder algorithm with a hardware controlled cache.

loop blocking (initial case) but after in-place mapping (cached and locally transformed) we again have a large extra gain. This indicates that apart from gains due to improving locality by means of global transformations and data re-use (bar 3), there are equal gains due to only applying the aggressive in-place mapping and caching step (bar 2). The various sources of power consumption are described below. In general we observe that the gain in power for the above methodology is quite significant for hardware controlled caches (approximately a factor 10).

7.3.3.4   Experimental results for software cache.   In this subsection, we discuss the experimental results for a software controlled cache obtained on the voice coder and the QSDPCM algorithms.

Fig. 7.26(a), shows the number of read/writes to the on-chip cache. Note that the number of accesses to the cache is much larger for the globally optimized case, whereas it is much less for the initial algorithm. Similarly, the inplace data mapped algorithm has significantly larger number of accesses to the cache. In Fig. 7.26(b), we observe a similar phenomenon but now the number of memory accesses are reduced for the optimized cases. The presence of a cache bypass allows us to observe an interesting feature in both the above results. Namely, for the initial algorithm we observe that we have more accesses to the main memory than to the cache. In contrast, for the globally optimized case we access the main memory only a bit more than an average once for different cache

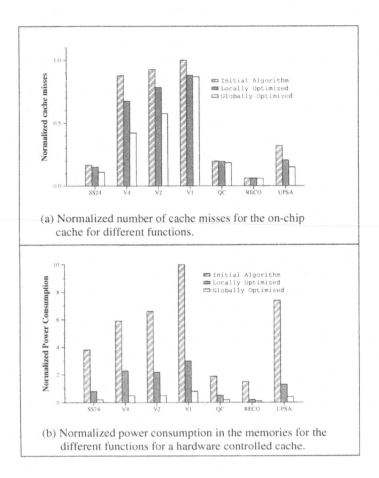

(a) Normalized number of cache misses for the on-chip cache for different functions.

(b) Normalized power consumption in the memories for the different functions for a hardware controlled cache.

Figure 7.25.   (a) Normalized cache misses and (b) Normalized power consumption for the QSDPCM algorithm with a hardware controlled cache.

sizes. The rest of the accesses are to the cache. This is due to the increased data locality which is brought about by the execution order optimization(s). In addition, we obtain reduced working set sizes which is due to the storage order optimization(s). Indeed the hardware would have generated the same amount of accesses to the main memory for the optimized algorithm but it would also have increased the number of cache accesses, since all the accesses which were bypassed would now be through the cache and hence would incur a large penalty in power and system bus load. Fig. 7.26(c) gives the total power consumption in the memories for the different cache sizes and optimization levels. We observe two issues here: (1) the globally optimized algorithm always performs better and (2) the performance based conclusion that increasing cache size always increases performance does not hold true for power, since the lowest power consumption for the globally optimized case is for a cache size of 128 words and not for any higher size.

Fig. 7.26(d) gives the results of applying the methodology for a cache with a write through updating policy. Here too we illustrate that the globally optimized algorithm always performs better. The gain though are not as large as those with a write back updating policy. Fig. 7.26(e) presents the comparison of number of instructions and the code size for different cases. We observe that the overhead in code size and instructions is less than 15% for the most optimized case (64 words

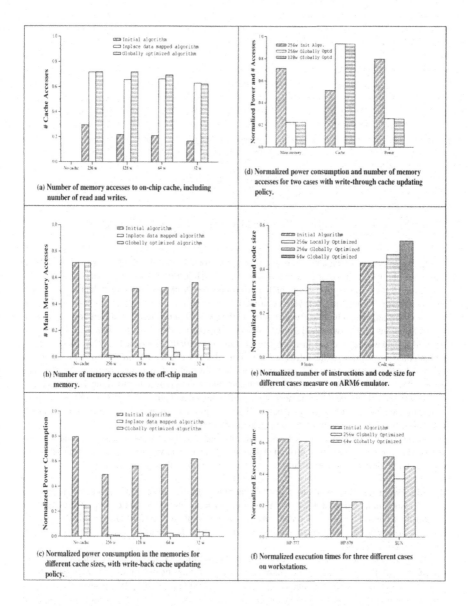

Figure 7.26.   Experimental results for voice coder algorithm with a software controlled cache.

globally optimized[10]), which is very acceptable for these application kernels because typically also a larger amount of additional purely control dominated code is present in the program memory [354]. To conclude, we have executed the different algorithms on three different workstations to observe the performance gains. We observe an increase in performance by 20% on the average, which is obtained apart from the large gains in power consumption, as shown in Fig. 7.26(f).

---

[10] As the cache size becomes smaller and smaller we try to apply more and more complex transformations to achieve the best solution for power, hence a increase in the code size.

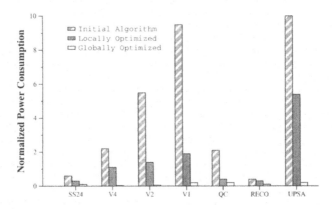

Figure 7.27.   Experimental results for QSDPCM algorithm with software controlled cache.

Fig. 7.27 gives the normalized power consumption for the different functions in the QSDPCM algorithm obtained using a software controlled cache. Here too we observe a large reduction in power for all the functions for the globally optimized case. Note that the inplace data mapped case clearly performs better for a software controlled cache than for a hardware controlled cache. This motivates the addition of more software control to the cache organization. The difference in values of UPSA functions for hardware and software controlled cache are due to the fact that using a software controlled cache we could not fit the entire working set initially in the cache. Thus these signals which could not be put in the software cache were directly accessed from the main memory and hence a large penalty in power consumption is paid. In contrast, once we perform the data and control flow transformation along with inplace data mapping to reduce the working set, the number of accesses directly to the main memory are heavily reduced and now the software controlled cache is able to manage the data cache much better. Hence the globally optimized algorithm requires much less power for a software controlled cache than for the hardware controlled cache.

This motivates the addition of software controlled caches. The hardware controlled variants are still needed for the parts of the algorithm that are not statically analyzable since our current approach cannot handle these cases. Research is ongoing to partly reduce these restrictions [343].

7.3.3.5   Experimental results on parallel machine.   In this subsection, we discuss the experimental results obtained on a 4-way SMP machine. The main goal of this particular study was to demonstrate that DTSE oriented transformations do not inhibit the performance oriented optimization of a parallelizing compiler. Issues related to the relation between DTSE and parallelization are discussed in [128] [130].

Fig. 7.28 gives the execution times for different optimization levels for the parallelized code on different platforms. Fig. 7.28(a) shows the execution time for the initial and DTSE optimized code using parallel virtual machine (PVM) on HP PA-8000 processors. We observe that, for the initial code there is a speed-up of factor 2.2 whereas for the DTSE optimized code the speed-up is an almost ideal factor 3.9. Similarly Fig. 7.28(b)-(d) give execution times on a PowerPC 604 based 4-way machine. Here also the total execution time decreases consistently after the DTSE optimizations, indicating that these transformations do not inhibit parallelization. Note that for larger frame sizes (as in (c) and (d)) the original algorithm performs even more poorly when the number of processors increases, due to increased communication overhead. In contrast, the DTSE-optimized case performs quite well for a larger number of processors. Thus apart from reducing the power cost, DTSE heavily enhances the performance for data dominated applications by reducing the potential inter-processor communication.

Fig. 7.29 gives the execution times for different optimization for the QSDPCM algorithm. We observe that there is quite a significant gain in total execution time for the DTSE case and a further gain due to the OPC related transformations. Also the DTSE-transformed parallelized code has

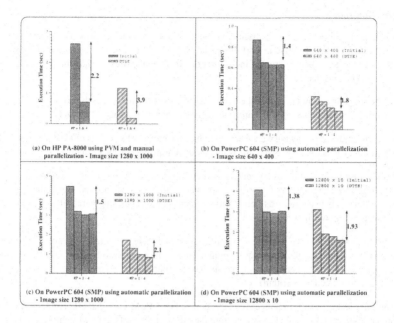

Figure 7.28.   Performance of parallelized cavity detection algorithm for different frame sizes using a HP PA-8000 based machine and a Power PC 604 based 4-way SMP machine. Here P represents the number of processors and the execution times are in seconds.

better execution times compared to that of the original parallelized code. Note that the execution time for the original case reduces by a factor 3 with four processors, whereas for the DTSE transformed case a reduction in execution time is present by an almost ideal factor 3.9. This once again shows that DTSE is complementary and is even augmenting the effect of the performance and parallelization related optimizations.

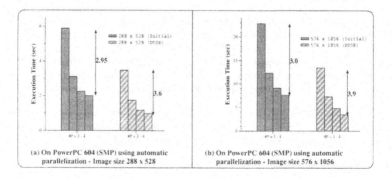

Figure 7.29.   Performance of parallelized QSDPCM algorithm for different frame sizes using a Power PC 604 based 4-way SMP machine. Here P represents the number of processors and the execution time is in seconds.

7.3.4 Summary

In this section we have presented a discussion about the impact of inplace data mapping on caching and also presented a technique to perform cache locking. Experimental results on two real-life applications have also been presented in this section. In summary, we conclude the following from the experimental results:

1. Inplace mapping allows to reduce the temporal memory space for data with partially overlapping life times and hence has a positive influence on both cache misses and write backs.

2. By performing cache locking we are able to reduce the memory bandwidth of the cavity detection algorithm by a factor 3 and due to the whole methodology by a factor 12 in total.

3. DTSE optimizations allow us to reduce power consumption for voice coder as well as QSDPCM algorithms to a large extent, factor $6 - 20$ in each of the cases without sacrificing the real-time constraints for both hardware as well as software controlled caches. In general software controlled cache allow a lower power implementation.

4. The conclusion based on performance optimization that increasing cache size always increases performance does not hold for power minimization, since the lowest power consumption for our globally optimized algorithms typically have a minimum for intermediate cache sizes (for example, voice coder algorithm).

5. The performance results of both cavity detection and QSDPCM show that DTSE-transformations are complimentary to and even augmenting the effect of the performance oriented parallelizing compiler. For both the algorithms we observe a consistent decrease in the required execution times after each optimization step. This comes on top of the power and bus load reduction effects of the DTSE step for individual processor nodes.

7.4 STRATEGY FOR PERFORMING CACHE LOCKING

Cache locking is achieved by switching-off the replacement policy of the hardware cache. Thus the elements in the locked part of the cache are not replaced on cache misses and hence are not written back to the main memory. This results in reduced memory traffic. This in turn reduces the memory bandwidth and the corresponding power consumption.

The cache locking step involves identifying memory locations which will be placed in the cache lockable region of main memory (eg. SDRAM) by the linker at link time. The hardware in turn ensures that any data which is placed in this particular memory area (of main memory) will be present in the L1 data cache also.

Identifying these memory locations involves two phases: (1) defining single contiguous[11] abstract address ranges for which we need to evaluate the reduction in memory accesses and (2) actual evaluation of a cost function to identify the best candidate for cache locking namely the address range which has the highest value of the cost function. This implies that the particular address range has more write backs to the main memory and hence locking this address range will result in more reduction of the data transfers. The two phases are explained below :

1. In the first phase of this step, we choose individual arrays and hence their corresponding abstract address ranges. This is done since current state-of-the-art media and DSP processors do not support cache locking of non-contiguous memory space or partial arrays.

2. In the second phase, we evaluate for each of these arrays the following cost function:

$$W_k = (\#memory\ accesses_k \times t_{diff}) \tag{7.4}$$

---

[11]Currently, we are allowed only to lock single contiguous (main) memory space in to the cache. Also since this task is performed by the linker, we need to specify all the data separately at link-time. This imposes a restriction that we cannot lock part of arrays, unless we convert these arrays into scalars. For large arrays (larger than 10-20 elements) this is not feasible.

$$t_{diff} = t_{final} - t_{init} \qquad (7.5)$$

This equation allows to identify the array(s) with most memory accesses and maximum potential for reducing memory accesses by means of cache locking. The t_{diff} for every array is obtained from the BOATD's as shown in Fig. 7.15(a) and equation 7.0.

The weight W_k evaluates the total number of memory accesses, both reads and writes, made to the particular address range (which here is an array k) over a period of time. Thus the array variables which have the most memory accesses and are alive for the largest amount of time (t_{diff}) throughout the program execution will be chosen to be locked in the cache. Note that we can choose more than one array depending on the size limit of the cache lockable region (for example, in TM1000 processor this size is 8KB [507]). Thus in such a case we will chose arrays with their weights in ascending order. Thus note that if more arrays can be inplace mapped (both intra- and inter-signal) then the cache locking step will have a larger impact. Due to this dependence the cache locking step succeeds the inplace data mapping steps.

Example:.   For example, in Fig. 7.15(e), arrays A and part of array B (which reuse the same abstract addresses) have the highest weight. Array D has the next highest weight, followed by array E and array C. Thus here we will start by locking array A and array B first. Since array B reuses (part of) memory space of array D, we have to lock array D with array A and array B. We then check for available (lockable) space for array E and if array E is larger than available space we try to match the array with next highest weight, here array C and so on.

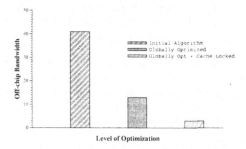

Figure 7.30.   Reduction in system-bus load for cavity detection algorithm on a TriMedia TM1000 processor.

Fig. 7.30 shows the reduction in system bus loading due to our methodology using the cache locking feature of the TM1000 processor on a cavity detection algorithm used in medical imaging. We have achieved a reduction in bandwidth by factor 3 because of the global optimizations. By performing cache locking we reduce the bandwidth further by a factor of 4 (hence a total reduction of a factor 12). As a side effect of this system bus load reduction also the CPU on the processor has been improved with almost a factor 2.

7.5 MAIN MEMORY DATA LAYOUT ORGANIZATION FOR CONFLICT MISS REDUCTION

The main source of stalls in the cache organization are the cache misses. Three types of misses namely compulsory, capacity and conflict misses exist (see [424]). We had to partly redefine the original definitions however to arrive at a fully unambiguous for all the applications and cache organizations.

Compulsory and capacity misses are related to the execution order of a program. In the past most of the work has concentrated on optimizing the execution order of the programs [381]. The conflict misses on the other hand are solely due to the layout of data in the main memory. Thus to remove conflict misses, data layout organization techniques need to be investigated. In this section, first, we will discuss the correlation of conflict misses with data layout of a program. Next, we will present two detailed example illustrations to show that by modifying the data layout of a given program majority of the conflict misses can be removed. Lastly, we will present preliminary experimental results for the proposed data layout organization technique. In the final subsections, we will present the formalization of the problem and the corresponding solution(s).

Then we will present a cache miss model which is used to estimate the number of conflict misses needed to drive the optimization process. Next, we will present a complete problem formulation as well as solutions to the main memory data layout organization problem. Lastly, we will demonstrate the data layout organization technique using a prototype tool on real-life applications that we are indeed able to reduce a large amount of conflict misses than before.

7.5.1 Revisiting the conflict miss problem

Conflict misses occur due to the limited associativity of the cache memory. Thus a direct mapped cache will have a higher number of conflict misses as compared to any other cache configurations. In this subsection we will revisit the conflict miss problem but look at it from the perspective of compilers and what compilers can do to eliminate these misses. Two issues are central to this understanding, namely what is the main cause of conflict misses and how does this relate to the compilation process.

7.5.1.1 Correlating base addresses and conflict misses. We know that a fully associative cache has no conflict misses and a direct mapped cache has the most conflict misses. The main difference between a fully associative cache and a direct mapped cache is the (hardware) protocol used to determine the exact cache locations for a particular data from the main memory.

The process of transferring data elements from the main memory to the cache is termed as *mapping*. The fundamental relation which governs the mapping of data from the main memory to a cache is given below:

$$(Block\ Address)\ MOD\ (Number\ of\ Sets\ in\ Cache) \tag{7.6}$$

In this equation, the block address is actually the index part of the block address and the number of sets in the cache varies depending on the type of the cache.

A direct mapped cache obeys the above protocol whereas a fully associative cache does not. This is also the reason why a fully associative cache has the largest freedom for data placement, namely any data from main memory can be placed at any location in the fully associative cache.

Thus the main cause of the conflict miss lies in equation 7.6. In the above equation, we observe that the term *number of sets in cache* is determined by the hardware and hence the compiler has no control over this aspect of the equation. The remaining term *block address* (in the main memory) can be modified by the compiler though. It is clear that, if we arrange the data in the main memory so that they are placed at particular block addresses depending on their lifetimes and sizes, we can control the mapping of data to the cache and hence (largely) remove the influence of associativity on the mapping of data to the cache.

Example:. We present a small example based on the above discussion. In Fig. 7.31, we present two instances of data layout in the main memory and the corresponding mapping to a direct mapped and a 2-way associative cache. Note that for the initial layout, as in Fig. 7.31(a), there are two (cross-) conflicts due to arrays $a[]$ and $b[]$ for the direct mapped cache. For the 2-way associative cache there are no conflict misses.

Now, we modify the data layout in the main memory, as in Fig. 7.31(b), so that all the four elements get mapped at different cache lines. Thus we see that for the direct mapped cache there are no conflicts anymore. In fact, the 2-way associative and direct mapped cache have similar cache mapping now. Thus we have managed to obtain a performance of a 2-way associative cache using a direct mapped cache by modifying the data layout in the main memory.

7.5.1.2 Role of compiler and linker. In this subsection we briefly present the different steps involved in determining the base addresses of data structures in the main memory for programmable processors.

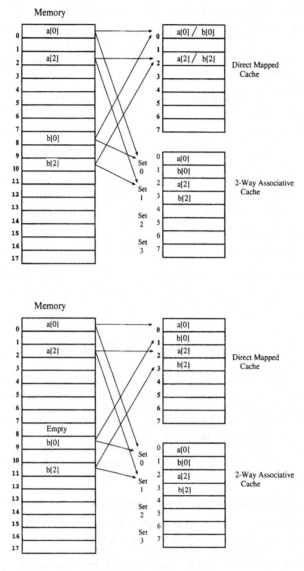

Figure 7.31. Correlating base addresses in the main memory with conflict misses in the cache (a) Initial layout with conflict (b) Modified layout with no conflicts.

The two main steps involved in the determining the base addresses of data in the main memory are as below [159].

1. A program in a high-level language (like C) is first compiled using a compiler. The compiler in turn generates a equivalent program in the *assembly language* of a particular machine. The assembly program is then converted to the *machine language* by an assembler. This machine language program is called as *object code*. This process is termed as *compilation*.

2. The object code is then converted into an executable code by a *linker* or *loader* as illustrated in Fig. 7.32. A linker/loader performs the following functions.

(a) Allocate space in the memory for the programs (*allocation*).

(b) Resolve symbolic references between object codes (*linking*).

(c) Adjust all address dependent locations, such as address constants, to correspond to the allocated space (*relocation*).

(d) Physically place the machine instructions and data into the memory (*loading*).

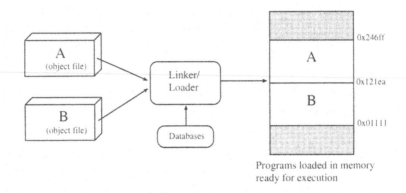

Figure 7.32.   General principle behind linking and loading schemes.

So in the third step of linking process, we determine the base addresses of the data structure in a program. Thus if we want to modify the base addresses of data in the main memory, we have to modify the linking/loading process which is quite difficult since we will have to perform all the book keeping tasks of a linker also. A better alternative to this process is to modify the high-level source code to reflect the modified data layout in the main memory. This is possible if we merge all the concerned data in a single address space (single name) at higher level, which in turn imposes restrictions on how the data must be allocated in the main memory. The latter approach has been used in this work.

7.5.2 Example illustration of data layout organization

In this subsection we will present in detail an example illustration of the proposed data layout organization technique. Thus we have a reference for comparing the data layout organization technique.

Fig. 7.33 shows the transformed algorithm and the corresponding modified main memory data layout. Note that the main transformation in the algorithm has been the modification in the addressing, which reflects that we have first split the arrays $a[]$ and $b[]$ and then merged them together in a single array called $tmp[]$. The storage order inside $tmp[]$ array is a repetitive allocation of 8 bytes (cache size = 8 bytes), where array $a[]$ gets 4 bytes and array $b[]$ gets the remaining 4 bytes as shown in Fig. 7.34. Thus the data layout is not single contiguous any more but depends on the algorithm characteristics and the cache size. A more detailed explanation of this process follows in the following subsections. Here we will concentrate on how the cache behaviour is impacted due to modified data layout.

Fig. 7.34 illustrates the detailed cache states for a direct mapped cache for the data layout organized algorithm. This modified algorithm requires only 14 cache misses for a direct mapped cache. Note that a fully associative cache would require also 14 cache misses. Thus we have the best of both worlds namely we have a low complexity hardware and a software implementation which is able to achieve the performance of a complex hardware. This indeed, is one of the main goals of the data layout organization technique.

for (i=3 ; i<11 ; i++)
    tmp[i%4 + i/4 * 8] = tmp[(i-3)%4 + (i-3)/4 * 8] + tmp[(i-1)%4 + (i-1)/4*8 + 4] + tmp[(i-3)%4 + (i-3)/4*8 + 4] ;

Let, Cache Size = 8 bytes; Cache line Size = 1 byte;
    Direct Mapped Cache

We need (8 x 4 =) 32 accesses to complete above task.

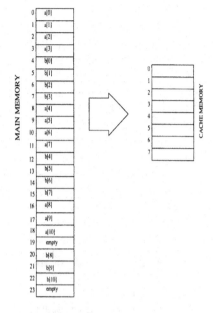

Figure 7.33.   Data layout organized algorithm and the corresponding data layout in the main memory.

7.5.3 Prerequisites

In this subsection we present some issues which are prerequisite to understanding the following sub-sections. These mainly comprise the basic definitions and terms used in this work, the assumptions made (be)for applying this technique to any program and the metrics used to quantify the results obtained after applying the data layout organization technique.

7.5.3.1 Basic terms and definitions.   The following terms are used in this work and they are interpreted (and defined) as below:

1. **Effective Size:** The total number of elements of an array accessed in a loop nest represents the effective size (ES_k) of the array in the particular loop nest.

2. **Tile Size:** The total number of contiguous lines of (main) memory allocated to an array in the cache is termed as tile size[12] (x_k) of the particular array.

3. **Reuse Factor:** The ratio of total number of accesses to an array in a particular loop nest and the effective size of that array in the loop nest is termed as the reuse factor (RF_i) of that array in the particular loop nest.

7.5.3.2 Assumptions.   The following assumptions are made (be)for applying the data layout organization technique to any program:

---

[12]The term tile size here should be differentiated from that used in the context of loop blocking. We are concerned primarily with the storage order and not the execution order of the program. Indeed, we assume that loop blocking is already performed.

Example Algorithm :

for (i=3 ; i<11 ; i++)
    tmp[i%4 + i/4*8] = tmp[(i-3)%4 + (i-3)/4*8] + tmp[(i-1)%4 + (i-1)/4*8 + 4] + tmp[(i-3)%4 + (i-3)/4*8 + 4] ;

Cache States for different iterations (i) :

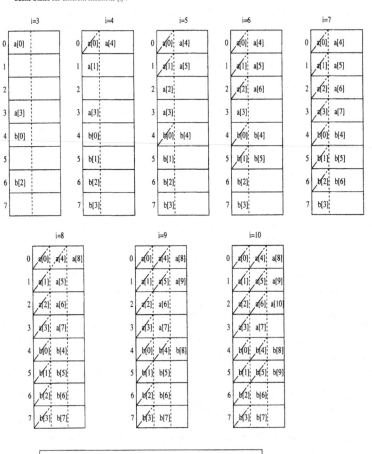

We have 14 cache misses for the above example (14/32 = 44 % miss rate).
Note that we are able to achieve the same miss rate as a fully associative cache.

Figure 7.34. Data layout organized algorithm and the corresponding cache states for different iterations of the algorithm for a direct mapped cache.

1. Only the storage order of the program can be modified and all transformations which modify the execution order (like loop transformations [293] [306]) have already been applied.

2. Loop blocking [306] has been applied successfully and so is in-place data mapping [149] which ensure that the number of capacity misses is minimal. This is indeed the case, as observed in the experimental results.

7.5.3.3 Metrics used. The following two (main) metrics have been used in this thesis:

1. **Miss rate:** The miss rate refers to the ratio of number of data cache misses to that of the sum of total number of hits and misses for the particular data cache.

2. **Memory bandwidth:** The memory bandwidth refers to the sum of number of cache misses and the number of write backs from the data cache to the off-chip main memory. This term quantifies the number of memory accesses made to the off-chip memory and hence provides a useful estimate of the amount of power consumed in data transfers, the required bandwidth and the performance penalties due to off-chip memory accesses.

3. **Power:** The power consumption in the memories is evaluated using the cost models of chapter 1.

4. **Other metrics:** We also measure the total execution time of the concerned program on a given processor or simulator. The total number of instructions needed to complete the execution of the concerned program are also measured.

7.5.4 Qualitative explanation of data layout organization method

We now illustrate our data organization methodology on a compact but still representative real-life test vehicle namely a full search motion estimation kernel with four pixel accuracy [56]. This illustration is mostly qualitative, a more formal and quantitative approach is presented in the last subsections.

The main loop nest of the kernel is shown by the pseudo-code in Fig. 7.35. The two important signals are *Current*[][], which is the current frame, and *Previous*[][], which is the previous frame, having overlapping lifetimes. The size of each frame is $N \times M$ integer elements of one byte each (for sake of simplicity) and the default block size is given by Nb. In this example illustration, we assume a direct mapped cache of size 512 bytes and cache line size of 16 bytes as indicated in Fig. 7.37. We will now illustrate how reorganizing the data in the main memory, as shown by the pseudo-code in Fig. 7.36, improves the cache performance.

```
int current[50176],previous[50176];
int v4x[3136],v4y[3136];

main()
{
  for(x=0; x<N/Nb; x++)        // N = 784, Nb = 16
   for(y=0; y<M/Nb; y++)       // M = 1024, Nb = 16
   {
     for(v4x1=0; v4x1<9; v4x1++)
      for(v4y1=0; v4y1<9; v4y1++)
      {
        dist=0;
        for(x4=0; x4<4; x4++)
         for(y4=0; y4<4; y4++)
         {
           p1=f1(current[(4*x+x4)*256+(4*y+y4)]);
           p2=f2(previous[(4*x+v4x1-4+x4)*256+4*y+v4y1-4+y4]);
           dist+=abs(p1-p2);
         }
        v4x[x*64+y]=g1(v4x1,dist);
        v4y[x*64+y]=g2(v4y1,dist);
      }
   }
}
```

Figure 7.35.   Pseudo-code for the initial full search motion estimation kernel.

Typically, the memory allocation in traditional compiler/linkers is single contiguous and no cache parameters are taken into account for this process. As shown in Fig. 7.37, the initial data organization is similar to the typical data organization performed by a linker, wherein the memory is allocated according to the data declarations in the program. The second case in Fig. 7.37, indicated as improved initial, exhibits a modified data layout incorporating the cache size. We have modified the base addresses of the arrays taking into account their life-times and cache size[13]. This technique is called base offset modification and is performed by some advanced compilers/linkers. The third case in Fig. 7.37 is the data layout optimized, where we first split the existing arrays and then merge them into groups based on the cache size and the line size, as explained in the subsequent subsections. This data organization can be imposed on the linker by carefully rewriting the code as shown in Fig. 7.36.

In the present example, we observe that initially variables $Current[][]$ and $Previous[][]$ are mapped using the relation in equation 7.6. Thus whenever elements of $Current[][]$ and $Previous[][]$ are separated by a distance of a multiple of the cache size, they will conflict with each other[14] and cause additional cache misses. But for the data layout optimized case, variable $Current[][]$ and $Previous[][]$ can be mapped only to (mutually exclusive) cache locations 0 to 240 and 240 to 480 respectively. Thus we have eliminated the cross conflict misses altogether. A similar explanation holds for variables $V4x[][]$ and $V4y[][]$. In addition, since the number of partitions, due to tile sizes, of $Current[][]$ and $Previous[][]$ does not match those of $V4x[][]$ and $V4y[][]$, we have to let some of the locations, corresponding to the multiple of base address(es) of $V4x[][]$ and $V4y[][]$, remain empty so as to avoid the mapping of other variables on to those locations in cache. Thus we have some overhead in memory locations, in the present example 1%, but this is very reasonable and acceptable in our target domain as motivated earlier. Note that the added complexity in addressing can be removed nearly fully using the address optimizations in the ADOPT approach of IMEC [37] [201] [216], which is used as a post processing stage at the end of DTSE script.

We note that by performing the data layout organization as illustrated above, we are able to decrease the miss rates by up to 82% and reduce the write backs by up to 50%.

The three sub-steps comprising the main memory data layout organization will now be discussed.

7.5.4.1 Array splitting. The first step in the above example is the process of array splitting. The following observations are made regarding this step.

1. The arrays are split-up as a multiple of the cache line size, here 16 bytes, into many sub-arrays.

2. The size of each sub-array depends on the overall effective size of the array and the number of accesses to the array in the complete program; that is the larger the *reuse factor* and the *effective size* the more continuous spaces (or lines) it gets and vice-versa.

3. The sub-array size for arrays with the same effective size are equal but rounded to the nearest possible multiple of the cache line size.

7.5.4.2 Array merging. The second step in the above example is the process of array merging or array interleaving. The following observations are made regarding this step.

1. All the sub-arrays which are simultaneously alive are now merged (or interleaved) so as to form a larger chunk. The size of each chunk is equal to the cache size.

2. During the evaluation of sub-array sizes, we have to ensure that the total size of all the sub-arrays which are simultaneously alive does not exceed the cache size. Thus, this step provides a constraint to the array splitting step.

3. Each of these chunks, with a size equal to cache size, is repetitively allocated in the main memory until all the arrays in the program are allocated in the main memory.

---

[13] By doing so we have eliminated some possibilities of cross conflict misses for arrays which are having simultaneous life-times placed on base addresses which are a multiple of the cache size.

[14] Since they are simultaneously alive at any given instance.

```
int tmparray[114688];

main()
{
 for (x=0; x<N/Nb; x++)        // N = 784, Nb = 16
  for (y=0; y<M/Nb; y++)       // M = 1024, Nb = 16
  {
    for (v4x1=0; v4x1<9; v4x1++)
     for (v4y1=0; v4y1<9; v4y1++)
     {
      dist = 0;
      for (x4=0; x4<4; x4++)
       for (y4=0; y4<4; y4++)
       {
        kk1 = (4*x+x4)*256+4*y+y4;
        p1 = f1(tmparray[(kk1 % 240)+((kk1 / 240) * 512)]);
        kk2 = (4*x+v4x1-4+x4)*256+4*y+v4y1-4+y4;
        p2 = f2(tmparray[(kk2 % 240)+((kk2 / 240) * 512)+ 240]);
        dist += abs(p1 - p2);
       }
      kk3 = x*64+y;
      tmparray[(kk3%16)+((kk3/16) * 512)+ 480] = g1(v4x1,dist);
      tmparray[(kk3%16)+((kk3/16) * 512)+ 496] = g2(v4y1,dist);
     }
  }
}
```

Figure 7.36.   Pseudo-code for the data layout organized full search motion estimation kernel.

7.5.4.3 Dependence on cache and algorithm characteristics.
The following dependencies on cache and algorithm characteristics are observed from the example illustration in Fig. 7.37.

1. The data layout organization technique depends on the cache size and the cache line size. This is seen in the array merging and array splitting step respectively. Thus for a given algorithm increasing the cache size for constant cache line size increases the freedom for performing data layout organization. Similarly, increasing associativity also increases the freedom for performing data layout organization for constant cache size and cache line size.

2. The initial data organization requires less space compared to the optimized data organization. This happens due to the mismatch in the number of sub-arrays due to the initial size of different arrays and also due to the smaller signals which are used permanently in cache. Thus the algorithm characteristics determine the trade-off in the memory size versus the reduction in cache miss.

7.5.5   Initial experimental results and discussion

In this subsection we present the results of manual experiments on some smaller yet representative kernels encountered in several real-time multimedia applications. The data layout transformations were done manually and the values obtained are all estimations compared with the DLX simulator.

We have used a 2-dimensional convolution kernel, a 2-dimensional discrete cosine transform (DCT) using row and column decomposition method kernel, a full search motion estimation kernel, a successive over relaxation algorithm (SOR) algorithm and the linear predictive coding (LPC) analysis part of a voice coder. The experimental results are discussed next.

Table 7.1 shows the different functions with different optimization levels, number of hits, number of misses, overhead in memory locations due to data organization and the reduction in misses due to

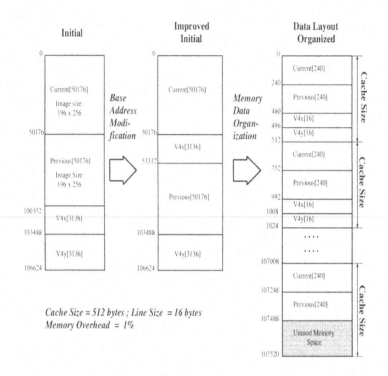

Figure 7.37.   The initial and final data organizations for the motion estimation kernel.

data organization for the respective functions. As mentioned earlier, the different "Models" represent the amount of area being traded-off to reduce the number of conflict misses. This also means that the versions with larger overhead in area have more arrays partitioned and allocated in the main memory. The optimized version indicated as "OModelX" represents a source code-transformed version of the original specification which is implemented assuming a direct mapped (DM) or fully associative (FA) cache. Also note that the size of the cache is 256 bytes with a cache line size of 16 bytes for all these experiments[15].

We observe from table 7.1 that all the functions except SOR have good inherent cache performance (hit rates greater than 90%). Also source level loop transformations reduce the total misses significantly. But observe that the data organized specifications (OModelx), are able to further reduce the misses by almost 48% for convolution, 7% for DCT, 21% for motion estimation and 3% for LPC analysis. For SOR where initial platform independent source-level transformations were not fully exploited due to single loop nest, the total reduction in misses was 93%. Note that both convolution and motion estimation have regular access patterns, hence its easy to obtain low overhead in area, whereas for DCT, which acts on the rows as well as on the columns, more area has to be traded-off. Observe that for the LPC test-vehicle we also gain in the number of memory locations for the optimized specification (hence the "+" sign). This is because of the heavy impact of the in-place mapping on this function. Also table 7.1 shows that due to the gain in conflict misses, there is a reduction in traffic between on-chip and off-chip memory, we gain both in average memory access time and the power required for these additional transfers for all the test-vehicles. In summary, con-

---

[15]The choice of a relatively small cache size is due to the data sets in many of our illustrative examples which are relatively small compared to actual data sets in complete multimedia applications. Thus we expect no change in the relative amount of gain due to our methodology for larger cache sizes when a complete RMP algorithm is addressed.

Case	# Hits	# Misses	Mem Size Overhead	Reduction in Misses (%)	Relative Power	Relative Time
			Convolution			
Direct Map.	112900	5645	0	0	1.0	1.0
Fully Asso	117480	1065	0	81.1	0.52	0.76
OModel1	114227	4318	0	23.5	0.85	0.93
OModel2	116149	2396	5.6	57.6	0.66	0.83
OModel3	116600	1945	6.04	65.6	0.61	0.81
OModel4	116943	1602	33.52	71.6	0.57	0.78
			2-D DCT			
Direct Map.	573000	57344	0	0	1.0	1.0
Fully Asso	620282	10502	0	81.6	0.39	0.65
Model1	598016	32768	31.3	42.8	0.68	0.83
Optimized (DM)	620544	10240	1.4	82.7	0.39	0.65
Optimized (FA)	625036	5746	1.4	89.9	0.34	0.62
OModel1	624640	6144	33	89.3	0.35	0.62
			Motion Estimation			
Direct Map.	838660	41933	0	0	1.0	1.0
Fully Asso	865287	15306	0	63.49	0.62	0.81
Model1	875150	5443	22.05	87.02	0.48	0.74
Model2	874040	6553	24.63	84.37	0.50	0.77
Optimized (DM)	868497	12096	0.12	71.12	0.58	0.79
Opt. 2-way Asso.	871373	9220	0.12	78.01	0.54	0.77
Optimized (FA)	877711	2882	0.12	93.13	0.44	0.72
OModel1	876964	3629	16	91.34	0.46	0.73
OModel2	877185	3408	15	91.87	0.46	0.72
OModel3	877418	3175	23.4	92.43	0.45	0.72
			SOR			
Direct Map.	17425	3485	0	0	1.0	1.0
Fully Asso	20698	212	0	93.92	0.22	0.44
OModel1	20668	242	12.8	93.05	0.22	0.44
			LPC Analysis			
Direct Map.	23532	2120	0	0	1.0	1.0
Fully Asso	24872	780	0	63.21	0.55	0.73
Optimized (DM)	24932	720	+91.6	65.7	0.53	0.71
Optimized (FA)	25058	594	+91.6	71.98	0.47	0.68
OModel1	24992	660	+89.0	68.87	0.50	0.68

Table 7.1.   Reduction in misses, memory location overhead, the corresponding reduction in power consumption and the average memory access time for all the test vehicles.

sidering the reduction of conflict misses for all the test-vehicles, we are able to achieve a close-to fully associative cache performance using a direct mapped cache by performing main memory data layout organization.

Fig. 7.38 is a plot of the amount of overhead in memory locations compared to the reduction in misses for three different test-vehicles. We observe that the curve saturates beyond a certain point, which clearly illustrates that we can trade-off between main memory space overhead and gain in misses to a certain extent, which is very interesting to designers. Thus the motion estimation curve, which has a larger slope, is the preferred one due to lower overhead in area. The ripples in the curve for motion estimation represent the boundary effects wherein the conflict misses (self conflicts) increase but they are removed by allowing a slight increase in memory overhead space.

7.5.6   Main memory data layout organization for multi-level caches

Until now we have considered a memory hierarchy with only one level of cache memory. However many modern microprocessors, used mostly in desktop computers and workstations, also have a level two cache. In this subsection, we will briefly discuss as to how this affects the main memory data layout organization technique that we have proposed.

Fig. 7.39 shows how a block of data gets mapped from the main memory to the L2 cache and then to the L1 cache. Note that, the mapping protocol stated in equation 7.6 is now applied repetitively at both the levels of memory hierarchy. Let C_1 and C_2 be the total number of cache lines in L1 and

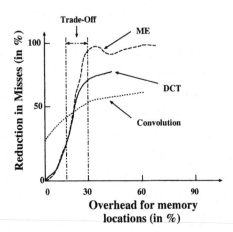

Figure 7.38.   Reduction in Misses vs Overhead in memory locations.

L2 caches and x be an (line) address in the main memory which gets mapped to the L1 cache. In the example in Fig. 7.39 the L2 cache size is $512KB$, the L1 cache size is $8KB$ and the cache line size is 32 bytes. Thus in our example, we have $C_1 = \frac{32KB}{8bytes}$ and $C_2 = \frac{512KB}{32bytes}$ respectively. Also the value of the example case for an address in the main memory is 8224. For the sake of simplicity, if we assume that both L1 and L2 are direct mapped cache, then we see that the following relation is used to map a data from the main memory to the L1 cache via L2 cache by the hardware: $x \% C_2 \% C_1$, which reduces to $x \% C_1$ since $C_1 < C_2$. Thus, if we repetitively (due to the modulo) map data in the main memory with each repetition equal to the L1 cache size, we will obtain the same result even when there is an intermediate memory between the main memory and the L1 cache, as long as the size of intermediate levels are multiples of the level one cache. In practice, we have observed that the size of the higher levels of cache memories are multiples of the lower level cache memories [427]. Thus as long as the cache size of L1 cache is a multiple of the L2 cache size, our data layout organization technique can be applied as is to a level two cache also[16].

7.5.7 Cache miss model

We now present the cache miss model used to drive the optimizations in this work. The cache miss model presented here is used to estimate the number of conflict misses. The conflict miss estimation has two main components namely the number of cross-conflict misses and the number of self-conflict misses.

In the next sub-subsections we will first explain how we evaluate the number of cross-conflict misses between two arrays for an instance of these two arrays in the program. To evaluate the total number of cross-conflict misses we will have to calculate these misses for all the instances of different arrays in the complete program, which is explained in a subsequent subsection. A similar explanation holds true for estimating the self-conflict misses. We will now briefly discuss how we estimate these two components in the conflict miss estimation.

7.5.7.1 Evaluating Cross-conflict Misses. A conflict miss is said to be a cross-conflict miss when the conflicting elements in the cache belong to two different arrays or in general different data structures. Note that in Fig. 7.31(a), both the conflict misses are cross-conflict misses.

In our model, we first evaluate the *effective size* as well as the *number of accesses* for all the arrays in every loop nest in the program. This implies that an array can have more than one effective

---

[16]Since $x \% C_{n-1} \% ... \% C_1$ is reduced to $x \% C_1$ and the cache sizes of memories in higher levels are multiples of those in lower levels of the memory hierarchy.

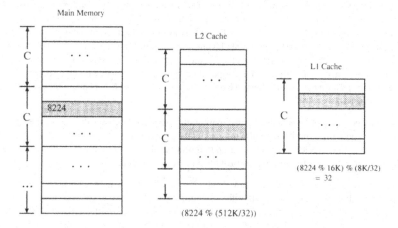

Figure 7.39.   Example illustration of mapping in multi-level caches.

size, depending on the number of loop nests it is alive in, and correspondingly also many different number of access counts.

Now based on these two characteristics we evaluate the number of cross-conflict misses. Let C be the cache size and ES be the effective size of any given array. In this discussion, our strategy is to estimate conflict misses for individual arrays (represented by k) for different cases of other remaining arrays (represented by j), which are simultaneously alive with this particular array (k). Then we sum all these estimated cross-conflict misses for individual arrays over all the loop nests for the complete program.

Fig. 7.40 illustrates the concept of a sliding window which is used to evaluate the cross-conflict misses. The cross-conflict misses are calculated as the intersection between the number of elements of an array with a particular effective size (ES_k) and all other arrays (j) which are alive along with the array k with a particular effective size (ES_j). Since we do not have the knowledge of base addresses of the individual arrays at compile-time we need to evaluate all the possibilities for the different base addresses of the particular array (j).

To evaluate the intersection between the effective size of two arrays which are simultaneously alive we perform the following steps :

1. First we fix the base address of one of the array (k) to zero. Thus we reduce the number of possibilities for different base addresses of array k as well as the combination between addresses of array k and addresses of array(s) j.

2. Now we fix the base address of arrays j to zero. Next we evaluate the intersection between the effective size of array k and array j.

3. We now slide the base address of array j to 1, using the *offset*, and evaluate the intersection between the effective sizes. This step of sliding (or incrementing) the base address of array j is carried out until the base address is equal to the cache size ($C - 1$) at which point we have passed through all the possible positions in the cache. To average the results, we then divide the total number of estimated cross-conflict misses by the cache size (C).

4. To account for the reuse of elements in the cache, we multiply the above estimated cross-conflict misses by the reuse factor of array j. By doing so, we assume that all the elements which are estimated as *misses* in the above intersection will always be misses - even during the subsequent accesses, when they are (supposedly) reused. This assumption holds true if we have a uniform reuse namely if all the elements of the array are reused the same number of times and the reuse distance between any two successive accesses to the individual array element(s) is greater than

the cache size. This limitation results in an overestimation of the cross-conflict misses, which are illustrated in subsection 7.5.7.4. Note that this limitation can be overcome by modeling the reuse in more detail namely using either the BOATD's, introduced earlier, or detailed array data flow analysis, as used by [203]. In our data layout organization context however we believe that the added complexity of this extension can be quite large and hence needs to be investigated. This extension is needed specifically if the application has iteration spaces with holes in them. For the applications presented in this thesis we do not have this case and hence the model presented here can be used.

5. At this point we have obtained the estimation of cross-conflict misses due to conflicts between array k and array j for the particular instance in the program. Hence to estimate the cross-conflict misses for the complete program we need to repeat the above process for the different arrays and loop nests and for different cases as explained in the subsection 7.5.7.3.

The above step is represented mathematically in Fig. 7.43 and the term evaluating cross-conflicts is present in both the cases. In the algorithm in Fig. 7.43, we have two different cases. The first case is when the effective size of array k is greater than the cache size - in this case, we have both the cross-conflict misses as well as the self-conflict misses. In the second case, when the effective size of array k is less than the cache size, only cross-conflict misses occur because all the elements of the particular array (k) fit in the cache and hence do not cause any self-conflict misses. However, in the context of our main memory data layout organization, after array splitting, we can still have self-conflict misses if the tile size of the particular array is smaller than the effective size of that array. This results in a special case which is explained in the next sub-subsection. Hence, in Fig. 7.47, the cost function used to drive the data layout organization step has both the terms corresponding to cross-conflict miss as well self-conflict miss estimation - even though the effective size of array k will be less than the cache size.

In the above process we note that the array(s) j are pushed through all the base addresses for a given cache size. Thus if we consider the effective size of array(s) j to be a *window* as in Fig. 7.40 then we have an effect similar to that of sliding a window through the whole cache.

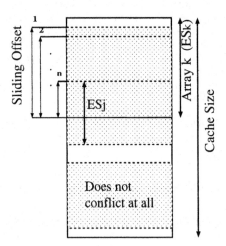

Figure 7.40.   Sliding window concept for evaluating cross-conflict misses.

7.5.7.2 Evaluating self-conflict misses.   A conflict miss is said to be a self-conflict miss when the conflicting elements in the cache belong to the same array or in general the same data structure.

Similar to the cross-conflict miss estimation, we use the effective size and the number of accesses as the base information for our estimation. The conventions remain similar to those used in subsubsection 7.5.7.1. Fig. 7.41 illustrates the main concept of how the self conflicts are evaluated. The self-conflict misses are calculated as the intersection between the number of elements needed for a particular array (ES_k) in the particular loop-nest and the cache size. Thus for arrays with effective size less than the cache size we do not have any self-conflicts since all the required elements fit in the cache. In contrast, for cases when the effective size is larger than the cache size, the elements which are in excess of the cache size will conflict with the elements of the same array. Here too, we multiply the reuse factor to account for the reuse of array elements in the cache. The limitations of using a coarse reuse factor are similar to those stated earlier. The above step is presented mathematically in Fig. 7.43, where the second term in the case of $ES_k > C$ represents the self-conflict misses.

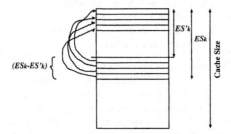

Figure 7.41.   Evaluating self conflict misses.

Note that in Fig. 7.41 and Fig. 7.42, we depict the special case of a self conflict miss which needs to be evaluated during the main memory data layout organization process, as stated earlier. Let the initial effective size of the particular array be ES_k and the tile size of the particular array be ES'_k. The tile size of an array is always less than or equal to the effective size of an array ($ES'_k \leq ES_k$) (see subsection 7.5.8). For the case when the tile size is strictly less than the effective size, we have a scenario where even though the effective size is less than the cache size, we will have self conflict misses.

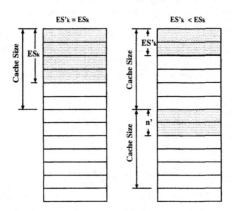

Figure 7.42.   Special case for evaluating self-conflict misses when $ES_k < C$.

If our tile size was equal to the effective size then all the elements, with size equal to the effective size, would have been stored single contiguous in the main memory and hence would have been mapped in single contiguous locations in the cache. When we have a case wherein the tile size is less than the effective size, we have ES'_k number of elements stored single contiguous in the main memory. The remaining elements of the same array required in this loop nest namely ($ES_k - ES'_k =$) n' elements are stored at a base address (in the main memory) which is separated from the earlier ES'_k elements by (a window equal to the) cache size. From the main memory to cache mapping protocol of equation 7.6, we will have these remaining n' elements of the array k conflicting with the elements ES'_k in the cache. The storage of elements of array k, in the main memory, for these two cases is shown in Fig. 7.42. We will now discuss how the total number of conflict misses is calculated for a given program based on the cross-conflict and self-conflict miss estimations.

7.5.7.3   Evaluating total conflict misses.   The total number of conflict misses is equal to the sum of total number of cross-conflict misses and the total number of self-conflict misses for the entire program. Fig. 7.44 gives the complete algorithm used to estimate the conflict misses. In this algorithm, the outermost summation is for the loop nests - representing all the loop nests in a program. The next summation is for the variable with fixed base address (k) and the innermost summation for the conflicting arrays (j). Additional summations due to the arrays with effective sizes with non-multiples of the cache size are also present[17] (represented by $kpart$ and $jpart$ respectively).

In the algorithm in Fig. 7.44 we have two main cases:

1. The first case is when at least one of the conflicting array has an effective size larger than the cache size. Note that earlier we have assumed that $ES_k > C$ but not stated what happens when for example the $2C < ES_k < 3C$ [18] namely if the effective size is not an exact multiple of cache size and greater than (twice) the cache size. In such a scenario we will have to calculate the cost function (at least) $kpart = \frac{Effectivesize}{Cachesize}$ times more. One should view this case as having accesses to $kpart$ number of caches, whereas in reality we are accessing a particular cache, with cache size C, (at least) $kpart$ number of times. This is because the effective size of the array k is equal to $kpart \times C'$ where $C' \geq C$. The term C' is greater than C because $kpart$ is an integer and for cases where ES_k is not a multiple of the cache size C we will have additional misses due to $ES_k \bmod C$ elements, next to the $kpart \times C$ elements. The term $kpart'$ represents these additional misses due to the arrays with effective sizes larger than the cache size but not a multiple of the cache size. A similar explanation holds for the term $jpart$, which accounts for the (non cache size multiple) effective size of the array under consideration namely array j.

2. The second case is when neither of the conflicting arrays have an effective size larger than the cache size. In this case, we evaluate only the cross-conflict misses for the given cache size. Since the accesses are confined, in space to just the cache size, the overhead due to the earlier case is absent here.

In the algorithm in Fig. 7.44 we have four cases, which are derived from the two main cases discussed above. Four cases exist since at any given instance in our estimation two arrays are conflicting with each other and each of these arrays has two cases as described above.

$if\ (ES_k > C)\ then$

$Cost\ Function = \sum_{offset=0}^{C-1} \frac{ES_j - ((ES_j + offset) - ES_k)}{C} \times \frac{\#\ accesses\ to\ j}{Effective\ size\ of\ j} +$
$$(ES_k - C) \times \frac{\#\ accesses\ to\ k}{Effective\ size\ of\ k}$$

$Else$

$Cost\ Function = \sum_{offset=0}^{C-1} \frac{ES_j - ((ES_j + offset) - ES_k)}{C} \times \frac{\#\ accesses\ to\ j}{Effective\ size\ of\ j}$

Figure 7.43.   Pseudo code for the cost function used to evaluate the conflict misses in a given algorithm.

7.5.7.4   Estimation versus simulation.   In this subsection we briefly present the comparison of cache misses obtained using estimation using the above method and simulated values obtained using

---

[17]This term will be zero if the effective size is a multiple of the cache size.

[18]In general, for the case when $C < ES_k < nC$.

$If\ (ES_k > C)\ then$
$\quad If\ (Effective\ size\ (of\ j) \leq C)\ then$
$$\| \ \textstyle\sum_{i=1}^{\#loops} \sum_{k=1}^{n} (\sum_{k'=1}^{kparts} \sum_{j=1}^{\#c-vars} (Cost\ Function) +$$
$$(Cost\ Function)_{ES_k = kpart'}) \ \|$$
$\quad Else$
$$\| \ \textstyle\sum_{i=1}^{\#loops} \sum_{k=1}^{n} (\sum_{k'=1}^{kparts} \sum_{j=1}^{\#c-vars} (\sum_{part=0}^{jparts} (Cost\ Function) +$$
$$(Cost\ Function)_{ES_j = jpart'}) +$$
$$(Cost\ Function)_{ES_k = kpart'}) \ \|$$
$Else$
$\quad If\ (Effective\ size\ (of\ j) \leq C)\ then$
$$\| \ \textstyle\sum_{i=1}^{\#loops} \sum_{k=1}^{n} \sum_{j=1}^{\#c-vars} (Cost\ Function) \ \|$$
$\quad Else$
$$\| \ \textstyle\sum_{i=1}^{\#loops} \sum_{k=1}^{n} \sum_{j=1}^{\#c-vars} (\sum_{part=0}^{jparts} (Cost\ Function) +$$
$$(Cost\ Function)_{ES_j = jpart'}) \ \|$$

$Where,$
$$kparts = \frac{Effective\ size\ of\ k}{C}$$
$$kpart' = (Effective\ size\ of\ k)\ \%\ C$$
$$jparts = \frac{Effective\ size\ of\ j}{C}$$
$$jpart' = (Effective\ size\ of\ j)\ \%\ C$$

Figure 7.44.   Pseudo code for evaluating the total number of conflict misses in a given algorithm for the general case.

cache simulation. The SimpleScalar tool set [74] was used for cache simulation. A row-filtering algorithm with three loop nests and four arrays was employed for this comparison.

Fig. 7.45 gives the comparison of the estimated cache conflict misses versus the simulated values for different cache sizes. We observe that the estimated values behave in a monotonic fashion whereas the simulated values do not. Also, the estimation always overestimates the cache misses as compared to the simulation. However, the relative difference (or variance) of cache misses obtained using estimation and those obtained using simulation for different cache sizes is quite close.

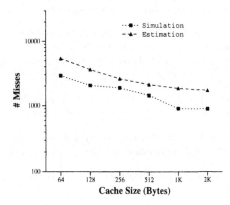

Figure 7.45.   Comparison of estimated versus simulated values for conflict misses for different cache sizes

Fig. 7.46 gives the comparison of the estimated and simulated cache conflict misses for varying reuse factors. Here we increased the outer loop of the filtering algorithm, which increases the number of misses. Now we observe that the estimated and the simulated values have quite similar behaviour. Here too, the estimation overestimates the number of misses as compared to the simulation values. Hence for performing optimizations, which require that relative values for different cache sizes and algorithm characteristics namely the reuse factor and the effective sizes, are consistent, we can use this cache miss model for our data layout organization technique.

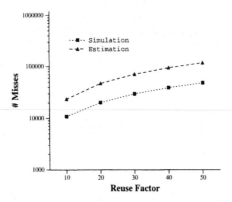

Figure 7.46.   Comparison of estimated versus simulated values for conflict misses for varying the outer loop bounds for a test-vehicle.

7.5.7.5   Limitation of the model and state-of-the-art.   In the context of compile-time based cache miss analysis to our knowledge only [203] have presented a complete analysis framework. However their framework has certain limitations at the moment namely,

1. Array subscript expressions are assumed to be affine. Hence in the context of DTSE no modulo or divisions are allowed in the array subscripts.

2. All loops are normalized namely have a step of one. This though is not a hard condition but can cause problems with applications where the steps of loop are data dependent.

3. No conditions are allowed inside loop nests or alternatively no multiple loop nests[19]. This indeed is a major restriction for performing global analysis, which is needed in a context like ours for performing data layout organization based on global constraints.

Our model compromises the absolute accuracy of conflict miss estimation but is able to handle a larger class of code namely nested loops with conditionals and nested loops with steps of non-unity unlike the framework presented by [203]. However our model has some limitations, which do not allow for accurate cache miss estimation:

1. We use a coarse estimation of the reuse namely we do not have detailed information about reuse of each array element. Hence we (always) overestimate the total number of conflict misses.

2. We do not model the execution order in detail namely at the level of individual array elements. This is also one of the reasons for overestimation of the total misses in our model.

3. Our model does not have a concept of replacement policy. This is important for estimating misses in set associative caches.

---

[19]However irregular loop nests, with single basic blocks are allowed.

The first and second issues above are interrelated namely we need detailed execution order for obtaining detailed reuse information. These two issues can be modeled by using the BOATD's, presented earlier. The third issue has been modeled by [203] by means of Diophantine equations. Thus the benefits of our model, namely utilizing all possible base address information in the main memory, and those of [203], namely accurate modeling of reuse and replacement of elements in real caches, need to be combined to obtain a model which can be applied to realistic applications and also provides accurate estimation.

Apart from [203] at Princeton, [495] have proposed a technique to evaluate the number of cache interferences (namely conflict misses) but their technique has more limitation than the technique proposed by [203]. Also cache miss analysis based on abstract interpretation [177] have been proposed, however this technique works fine for instruction cache and for data cache with only scalars. The main use of above technique was to provide an upper bound and a lower bound for worst case execution time (WCET). Similarly [322] present a technique for instruction cache modeling for obtaining WCET. Hence they cannot be (directly) employed in our context.

7.5.8   Problem formulation

The general main memory data layout organization problem for conflict miss reduction can be stated as,"For a given program with m loop nests and n variables (arrays), obtain a data layout which has the least possible conflict misses". This problem has two sub-problems. First, the tile size evaluation problem and secondly the array merging/clustering problem.

7.5.8.1   Cost function based on cache miss estimation.   In the context of main memory data layout organization problem, we use a modified form of cache miss estimation presented earlier in subsection 7.5.7. The modified cost function is shown in Fig. 7.47. The main differences with the general conflict miss estimations as compared to the one used here are: (1) we have replaced ES_k with an unknown x_k, which is the tile size in our case as will be explained in the next subsubsection, (2) the definition of two terms in the algorithm to estimate total conflict misses have been modified: $kparts = \frac{x_k}{C}$ and $kpart' = x_k \% C$ and (3) we now evaluate the special case for self conflict miss when $x_k < ES_k < C$, as explained in subsection 7.5.7.2. Hence there are no separate cases for the cost function.

$$\sum_{offset=0}^{C-1} \frac{ES_j - ((ES_j + offset) - x_k)}{C} \times \frac{\# \; accesses \; to \; j}{Effective \; size \; of \; j} +$$
$$(ES_k - x_k) \times \frac{\#}{Ef}$$

Figure 7.47.   Pseudo-code for evaluating the cost function to calculate the number of conflict misses as used in the data layout organization technique.

7.5.8.2   The tile size evaluation problem.   The problem of tile size evaluation refers to the evaluation of the size of sub-array(s) for a given array (as discussed in subsection 7.5.4.1). Let x_i be the tile size of the array i and C be the cache size. For a given program we need to solve the m equations below to obtain the needed (optimal) tile sizes. This is required because of two reasons. Firstly, an array can have different effective size in different loop nests. The second reason is that different loop nests have different number of arrays which are simultaneously alive.

$$L_1 = x_1 + x_2 + x_3 + ... + x_n \leq C$$

$$\cdots \cdots$$

$$L_m = x_1^{(m-1)} + x_2^{(m-1)} + x_3^{(m-1)} + ... + x_n^{(m-1)} \leq C \qquad (7.7)$$

The above equations need to be solved so as to:

1. Minimize the conflict misses estimated using the algorithm in Fig. 7.47,

2. Ensure that $0 < x_i \leq max(ES_i)$. This means that the tile size cannot be larger than the maximum effective size of the array, else there will be inefficient cache utilization and

3. Ensure that $x_i \bmod L = 0$, where $i = 1 \dots n$. This ensures that all the tile sizes are multiples of cache line size, hence no misses due to cache line size and execution order mismatch[20].

The optimal solution to this problem comprises solving ILP problem, which requires large CPU time [392]. Also, note that we can ensure an optimal solution only by imposing a strict equality to C in above equations but for $n < m$[21], the strict equality does not guarantee a result and hence we use an inequality. Therefore we have developed heuristics which provide good results in a reasonable CPU time.

7.5.8.3 The array merging problem. We now further formulate the general problem using the loop weights for the heuristic approach. The weight in this context is the probability of conflict misses calculated based on the simultaneous existence of arrays for a particular loop-nest i.e. sum of effective sizes of all the arrays as given below :

$$L_{wk} = \sum_{i=1}^{n} ES_i \qquad (7.8)$$

Hence, now the problem to be solved is, which variables to be clustered or merged and in what order that is from which loop-nest onwards so as to minimize the cost function. Note that we have to formulate the array merging problem this way because, we can have many tile sizes for each array[22] and there can be different number of arrays alive in different loop nests. In the example illustration in subsection 7.5.4 we have only one loop nest and hence we did not need this extension. Using the above considerations, we can identify loop nests which can potentially have more conflict misses (and assign corresponding weights) and focus on clustering arrays in the highest weighted loop nests (first).

7.5.9 Solution(s) to the data layout organization problem

In this subsection, we will present two possible solutions to the data layout organization problem as stated in subsection 7.5.8. First, we discuss the possible optimal solutions for the above problem formulation and secondly, we will present a pragmatic approach based on a heuristic approach which has been automated in a prototype tool.

7.5.9.1 Optimal solution. The tile size evaluation problem for multiple loop-nests is a classical integer linear programming problem [392]. Thus to obtain an optimal solution for the evaluation of tile sizes, ILP solver(s) need to be used. However one needs to take into account the advantages and disadvantages of using ILP solvers for larger problems that is for a general data layout organization problem with n arrays and m loop-nests, the ILP solver might need non-polynomial (NP) time, this remains to be investigated.

The tile size evaluation problem can also be solved as a specific case of the least squares problem with linear inequality constraints namely as a linear distance programming (LDP) problem whose solutions are characterized by Kuhn-Tucker theorem [310]. The general form of LDP problem is given below.

$$Minimize \ \| x \| \ subject \ to \ Gx \geq h \qquad (7.9)$$

Here x is our cost function and the constraint can be given as $Gx = C$, where G being the weights calculated from the effective sizes. The calculation of weights can be similar to array weights is discussed in the next sub-subsection.

---

[20]We refer to this problem as boundary effects since the impact of this mismatch is small but avoidable.

[21]The total number of variables are less than the total number of loop nests.

[22]In the worst case, one tile size for every loop nest in which the array is alive.

The tile size thus obtained for every variable is optimal since, the cost function (now) accounts for misses due to all other existing variables in different loop-nests. In practice, the issue of real concern or trade-off will be the amount of memory space overhead due to the mismatch of different tile sizes of different variables in different loop nests. We now discuss the heuristic solution which has been automated in a prototype tool.

7.5.9.2   Heuristic solution. We now discuss a pragmatic solution for the above problem, which requires less CPU time. This solution makes use of a heuristic approach, which is less complex and faster from the point of view of implementation in a tool. The main idea behind this heuristic was to first identify loop nests which have the most probability to cause conflict misses and then perform the data layout organization for all the arrays in that loop nest, based on their effective sizes and the number of memory accesses, and then find the loop nest with the next highest probability for causing conflict misses and so on until the data layout of all the arrays in the program are organized. The approach comprises the five steps explained below:

1. In the first step, we perform all the analysis. We evaluate the effective size of each array in each loop nest. Next, we also evaluate the number of accesses to every array in every loop nest.

2. In the second step, for every loop nest we evaluate the loop weights using the relation in equation 7.8.

3. Now, we visit the loop nest with highest loop weight. And we evaluate the individual array weights, where the array weight is the sum of reuse factors for the particular array in all the loop nests where it is alive times the effective size of the array in the considered loop nest.

4. In the fourth step, we obtain the tile size of all the arrays in the loop nest by proportionate allocation. The latter allocates larger tile sizes (in multiples of cache line sizes) to arrays with larger array weights and vice-versa. Once the tile size is obtained, we obtain the offset of the array in the cache through a global memory map used to keep track of all the array allocations.

5. We repeat the steps three and four for the loop nest with the next highest loop weight and so on, till all the arrays are covered. We perform code generation to fix the obtained data layout.

Note that in the above approach, we have solved both the tile size evaluation problem as well as the array merging problem in one step (step four). As mentioned earlier, this heuristic has been automated in a prototype C-to-C pre-compiler, which is based on the pseudo code shown in Fig. 7.48.

7.5.10   Experimental results and discussion

In this subsection, we will first present the experimental set-up used to evaluate the impact of the main memory data layout organization technique on different applications and then present a discussion of the obtained experimental results. Also, we will then briefly compare the data layout organization technique to the state-of-the-art.

7.5.10.1   Experimental set-up. The experimental setup comprises two parts, namely the prototype C-to-C pre-compiler and the cache simulator or processor to which this pre-compiler is coupled. The following two steps are carried out to obtain the experimental results :

1. We obtain the transformed C code from the prototype data layout transformation tool. Thus we now have the initial (non-transformed) and the data layout transformed codes.

2. The above two codes are compiled using the native compilers of different platforms and executed on the respective platforms. Two different platforms are used in this work:

    (a) The cache is simulated using the sim-cache simulator, part of the SimpleScalar tool set [74], to obtain the data cache statistics. To obtain the total number of cycles and instructions we have used the sim-outorder simulator for one case to illustrate the impact of the this approach on the execution time (using the simulator). In practice, the execution times of sim-cache and

```
begin
  for everyLoopnest do
    for everyVariable do
      EffectiveSize = f(loopindex, arrayindex) ;
      EvaluateNrofAccesses(array);
    end
    LoopWeight = Σ∀variables EffectiveSize ;
  end
  for HighestWeightLoopnest do
    for everyVariable do
      if (Variable Not Allocated)
        ArrayWeight = Σ_{i=1}^{loops} ReuseFactor_i × ES ;
        EvaluateTileSize(EffectiveSize, ArrayWeight) ;
        Offset = UpdateGlobalMemoryMap(TileSize) ;
      end
    end
    Repeat until No variables are left unallocated ;
  end
  for everyLoopnest do
    for everyVariable do
      ModifyCode(TileSize, Offset) ;
    end
  end
end
```

Figure 7.48.   Pseudo-code for the heuristic data layout organization technique.

sim-outorder vary by orders of magnitude for our drivers. Hence we have used the faster and functionally correct sim-cache simulator for measuring cache misses and transfers.

(b) Apart from the SimpleScalar architecture used for cache simulation we have also used real processors for observing the actual CPU performance. The processors used are PA-RISC 8000, Pentium-III, TriMedia TM1000 and MIPS R10000.

7.5.10.2 Results and discussion. We will now present a discussion of the experimental results obtained using the above experimental set-up.

Fig. 7.49 shows the miss rate and the number of off-chip accesses for the cavity detection and the QSDPCM algorithms. We observe in Fig. 7.49(a), (c) and (e) that the miss rate is consistently reduced by 65-70% on the average and for some cases it is reduced by up to 82%. This implies that we are able to remove a large majority of the conflict misses by the data layout organization technique. Note that even for a 2-way associative cache[23] we are able to reduce the miss rate by the same amount as that of a direct mapped cache as seen in Fig. 7.49(c). The reduction in the number of off-chip accesses also follows a similar pattern as observed in Fig. 7.49(b),(d) and (f), which also means that the data layout technique is able to reduce the write backs apart from the conflict misses (see definitions in subsection 7.5.3.1).

Fig. 7.50 shows the miss rate and the number of off-chip accesses for the SOR, the motion estimation and the 2D convolution algorithms. Here too we observe that the miss rate reduces consistently and so are the off-chip accesses, confirming again the large impact of the data layout organization technique. It is indeed interesting to note that for the motion estimation algorithm, the data layout organized algorithm performs much better when the cache size is increased from 256 bytes to 512

---

[23] A 2-way associative cache consumes more power than a direct mapped since the number of tag bits increase (more bit lines and related switching activity) and the increase in number of comparators.

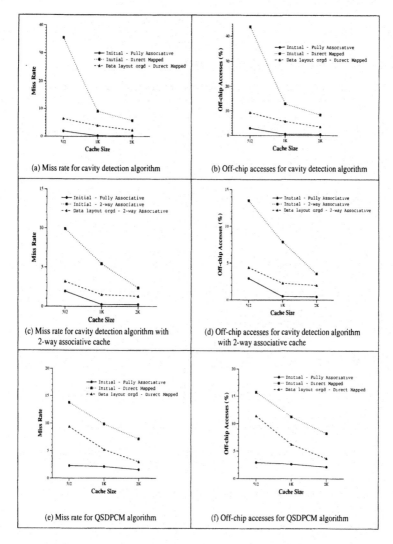

Figure 7.49.   Experimental results on cavity detection and QSDPCM algorithm.

bytes in comparison to the almost unmodified values for the initial motion estimation algorithm for the same change in the cache sizes.

Our initial goal was to achieve the performance (in terms of miss rate) of a fully associative cache using a direct mapped cache. This study was intended to show that by increasing the control complexity in the compiler, we can reach a performance close to the complex hardware control embedded in a fully associative cache (which is much more expensive in terms of power and area). We have indeed come very close to achieving this goal (within 18% of the ultimate target as seen in Fig. 7.49(a) & (c)). Note that the data layout organized case for a direct mapped cache (7% miss rate) performs better than the initial two way associative case (10% miss rate), which illustrates that our technique is able to outperform a 2-way associative cache without the hardware overhead.

We have earlier shown the impact on the data cache miss rates due to the data layout organization technique. The actual implementation of this technique involves modification of address values, which adds to the number of instructions. Table 7.2 shows the number of cycles as well as the number

	Initial code	Data layout organized
Number of cycles	290M	230M
Number of instructions	323M	391M

Table 7.2.   Simulated Number of cycles and number of instructions for cavity detection algorithm.

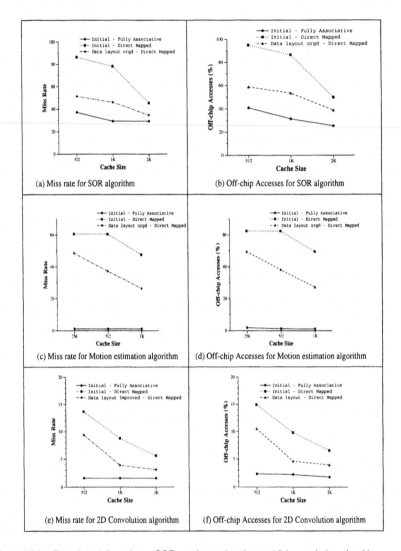

Figure 7.50.   Experimental results on SOR, motion estimation and 2d convolution algorithms.

of instructions for cavity detection algorithm on SimpleScalar machine simulated (sim-outorder) with a 512 byte direct mapped data cache, 2Kbyte 4-way instruction cache, single cycle access on a cache hit, a penalty of 18 cycles on a miss and no level two caches[24]. Note that the overhead in

---

[24]Most other parameters used were default sim-outorder values.

instructions is approximately 21%. We observe that we are able to gain in the total cycles due to the reduction in conflict misses for the data cache by 82% even though there is an increase in the total number of instructions.

It is worth noting that the SimpleScalar architecture has dedicated instructions for integer division and modulo operations, which can be performed within single cycle access. This though is not true for any of the existing (embedded) processors[25]. Hence we use existing processors to illustrate that we can remove all the overhead in cycles due to complex addressing and so overall we can reduce the number of cycles even further due to reduced cache misses. Thus, we need to perform source code level [201] [216] address optimizations to reduce the overhead in complex instruction especially for the integer division and modulo's introduced by the data layout technique. Note that until now we have not performed any address optimizations.

	Initial	DOECU (I)	DOECU (II)
Avg memory access time	0.482180	0.203100	0.187943
L1 Cache Line Reuse	423.219241	481.172092	471.098536
L2 Cache Line Reuse	4.960771	16.655451	23.198864
L1 Data Cache Hit Rate	0.997643	0.997926	0.997882
L2 Data Cache Hit Rate	0.832236	0.943360	0.958676
L1–L2 bandwidth (MB/s)	13.580039	4.828789	4.697513
Memory bandwidth (MB/s)	8.781437	1.017692	0.776886
Data transferred L1-L2 (MB)	6.94	4.02	3.70
Data transferred L2-Memory (MB)	4.48	0.84	0.61

Table 7.3.   Experimental Results for the cavity detection algorithm using the MIPS R10000 Processor.

Table 7.3, table 7.4 and table 7.5 show the obtained results for the different measures for all the three applications on a MIPS R10000 processor. Note that here we compare two different heuristic implementations in our prototype tool DOECU . The first heuristic (DOECU-I) performs a partly global optimization, whereas the second heuristic (DOECU-II) performs a completely global optimization. Note that table 7.5 has same result for both the heuristics since the motion estimation algorithm has only one (large) loop nest with a depth of six namely six nested loops with one body.

The main observations from the results are : main memory data layout optimized code has a larger spatial reuse of data both in the L1 and L2 cache. This increase in spatial reuse is due to the recursive allocation of simultaneously alive data for a particular cache size. This is observed from the L1 and L2 cache line reuse values. The L1 and L2 cache hit rates are consistently greater too, which indicates that the tile sizes evaluated by the tool were nearly optimal, since sub-optimal tile sizes will generate more self conflict cache misses.

	Initial	DOECU (I)	DOECU (II)
Avg memory access time	0.458275	0.293109	0.244632
L1 Cache Line Reuse	37.305497	72.854248	50.883783
L2 Cache Line Reuse	48.514644	253.450867	564.584270
L1 Data Cache Hit Rate	0.973894	0.986460	0.980726
L2 Data Cache Hit Rate	0.979804	0.996070	0.998232
L1–L2 bandwidth (MB/s)	115.431450	43.473854	49.821937
Memory bandwidth (MB/s)	10.130045	0.707163	0.315990
Data transferred L1-L2 (MB)	17.03	10.18	9.77
Data transferred L2-Memory (MB)	1.52	0.16	0.06

Table 7.4.   Experimental Results for the voice coder algorithm using the MIPS R10000 Processor.

[25]However another research architecture at HP labs, called PlayDoh [264], has similar instructions.

Since the spatial reuse of data is increased, the memory access time is reduced by an average factor 2 all the time. Similarly the bandwidth used between L1-L2 cache is reduced by a factor 0.7 to 2.5 and the bandwidth between L2 cache - main memory is reduced by factor 2-20. This indicates that though the initial algorithm had larger hit rates, the hardware was still performing many redundant data transfers between different levels of the memory hierarchy. These redundant transfers are removed by the modified data layout and heavily decrease the system bus loading. This has a large impact on the global system performance, since most (embedded) multimedia applications require to operate with peripheral devices connected using the off-chip bus. In addition also the system power/energy consumption goes down.

	Initial	DOECU (I/II)
Avg memory access time	0.782636	0.289850
L1 Cache Line Reuse	9132.917055	13106.610419
L2 Cache Line Reuse	13.500000	24.228571
L1 Data Cache Hit Rate	0.999891	0.999924
L2 Data Cache Hit Rate	0.931034	0.960362
L1–L2 bandwidth (MB/s)	0.991855	0.299435
Memory bandwidth (MB/s)	0.311270	0.113689
Data transferred L1-L2 (MB)	0.62	0.22
Data transferred L2-Memory (MB)	0.20	0.08

Table 7.5.   Experimental Results for the motion estimation algorithm using the MIPS R10000 Processor.

7.5.10.3 Comparison to state-of-the-art. Storage order optimizations [148] [93] are very helpful in reducing the capacity misses. Thus mostly conflict cache misses related to the sub-optimal data layout remain. Array padding has been proposed earlier to reduce the latter [340] [405] [407] [448]. These approaches are useful for reducing the (self-) conflict misses and to some extent the cross-conflict misses. Similarly, researchers have tried techniques like randomized placement of data and observed reduction of conflict misses [504] but their results do not indicate a significant reduction of conflict misses.

Besides [73], [260], [405] and [428] very little has been done to measure the impact of data layout(s) on the cache performance. Hence this study evaluates the extent to which we can reduce the total number of conflict misses by data layout organization. Note that earlier studies, in particular [340], [405] and [448], have shown that they are able to reduce the number of conflict misses by up to 50-60% on simple examples. However in our work, we demonstrate that we are able to reduce up to 82% of the total conflict misses for real-life applications.

7.5.11   Cache misses for spec92 versus multimedia applications

In this subsection we briefly comment on the disparities in cache miss distribution between the multimedia benchmarks that we have used and those of Spec92 benchmarks as given by [198].

We took the average of all the conflict and capacity misses of all our benchmarks and plotted them in Fig. 7.52. Similarly we took the figures from [198] and plotted them in Fig. 7.51. Note that in both the graphs we have used a direct mapped cache, hence we can safely compare the trends.

The main observation from these two figures is that in our multimedia benchmarks the total number of cache misses is dominated by conflict misses. In contrast in the Spec92 benchmarks the total number of cache misses are dominated by both the capacity and conflict misses equally. Thus we believe that conflict miss reduction will become very crucial in future multimedia applications and their implementation on programmable processors with hardware caches. Note that we have not found detailed figures of cache miss distribution for Spec2000 benchmarks yet and hence cannot make a comparison to those as yet.

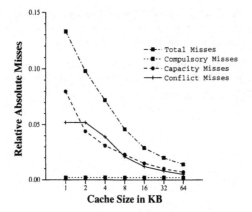

Figure 7.51.   Cache miss distribution for Spec92 benchmarks as observed by [198].

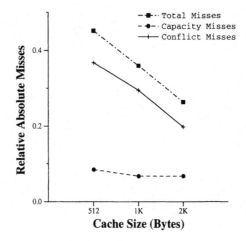

Figure 7.52.   Cache miss distribution for our multimedia benchmarks.

7.5.12 Summary

In this section we have presented a new data layout organization technique focusing on the conflict misses related to the main memory data layout. Issues related to the impact of this technique on conflict misses, the different steps involved in performing this optimization and the type of code generation have been discussed. We have also presented a complete problem formulation and solutions for the main memory data layout organization aspects related to conflict misses. Also we have presented a new cache miss estimation technique used to estimate the number of conflict misses. Issues related to automation of this technique have also been addressed.

In summary, the main conclusions from the initial experimental results are:

1. The above methodology always helps in reducing conflict misses to a (very) large extent. We reduce up to 25% of the conflict misses for applications where source-level transformations to improve execution order for locality of accesses are possible.

2. For applications where source-level transformations are not effective, this methodology is even more attractive, because then we reduce up to 80% (for example, SOR function) of the conflict misses. This is also confirmed by the experiments with our automated prototype tool on larger real-life applications like cavity detection and QSDPCM.

3. For embedded systems which are bandwidth constrained, this technique is able to reduce the off-chip accesses to a large extent. This is a significant design issue which makes this technique more useful than existing techniques focusing solely on performance.

4. We have also demonstrated the ability to automate this technique. Thus re-targeting this tool for different cache organizations is relatively easy. In our tool this can be done by the user by the command line options.

8 DEMONSTRATOR DESIGNS

In this chapter several complete applications will be explored to demonstrate the large impact which can be achieved by the DTSE approach. The main focus lies on target architecture styles where at least partly predefined memory organisations are present. But part of the organisation, especially on-chip, can also be customisable still. The demonstrators have been selected from different target application domains to substantiate our claims that data-dominant applications occur in a broad domain, of significant industrial relevance.

8.1 WIRELESS DEMONSTRATOR: DIGITAL AUDIO BROADCAST RECEIVER

This section describes the application of the DTSE (Data Transfer and Storage Exploration) methodology, on the critical modules of a Digital Audio Broadcasting (DAB) decoder. We explain the different optimization steps and compare the original Philips chip implementation with a DTSE optimized realisation in terms of power, area and speed results. We will show a power gain for the data storage and access related part which is one order of magnitude. The rest of the system is not negatively affected and even features a gain too due to the removal of redundant arithmetic operations during the DTSE stage. As a result, an impressive overall chip-level power gain of a factor 6.3 is achieved.

8.1.1 DAB decoder application

The simplified block diagram in Fig. 8.1 shows the different blocks in the DAB decoder. The decoder contains:

1. A 2048-point Fast Fourier Transformation (FFT)

2. Differential demodulation

3. Frequency de-interleaver

4. Time de-interleaver (Main Service Channel only, not for Fast Information Channel)

5. Error protection scheme (UEP/EEP)

6. Viterbi decoder

Other blocks are also present in the full DAB decoder but these are less critical in the overall design. Another important block is the synchro core which has been optimizated separately [69].

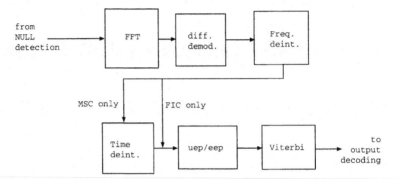

Figure 8.1.   Simplified DAB decoder block diagram.

Before optimizing the DAB decoder, a reference model (in C code) is built, containing the blocks described above. The functional reference does not fully reflect the Philips DAB decoder chip implementation, since current (ad hoc) optimizations can limit the freedom during the DTSE optimization steps. Both are compliant to the standard and are functionally equivalent. In addition, a Philips reference has been built which is a C-code description that fully reflects the data related implementation issues of the Philips DAB chip.

To verify the correctness of our reference models, we have used a 470 Mb baseband recording of a DAB signal, containing 1250 Mode I frames (2 minutes), excitating 7 out of the 64 possible sub-channels.

Not all DAB features and modes are implemented in the reference. In particular we have restricted ourselves to Mode I and a single channel audio stream. These do not affect the conclusions on the DTSE approach impact which is the main focus of this chapter.

8.1.2   DTSE preprocessing steps

8.1.2.1   Seperation in layers.   The reference code is preprocessed before entering the DTSE steps, as described in subsection 1.7.2. This version is called the pruned description, which has three layers described in Fig. 8.2. The first layer contains the test-bench, mode selection and the dynamic task control. The second layer contains the manifest control flow (especially loops but also several conditions that are statically analysable) and all the array-level data-flow. Finally, the third layer contains scalar data, data-path operations and some local control.

8.1.2.2   Separation of data storage, addressing and data-path.   The DTSE optimization will have a main focus on reducing the data related power. In order to show this gain, we make a clear separation in three categories (also shown in Fig. 8.3):

1   Data stored in the data memories.

2   Addressing of the array in the data memories *and* global control.

3   Data-path processing *and* local control of the data-path.

A major part of the power consumption takes place in the data/memory issues for data dominated applications like the DAB (see Section 8.1.4.4). Section 8.1.3 provides more detail on all types of data/memory related optimizations following the DTSE script.

Obviously, when not optimized properly, the memory addressing related part of these type of applications will be large as well. The ADOPT methodology [372, 369] reduces the power consumption related to the addressing. The focus in this section is however on the DTSE stage. Therefore we

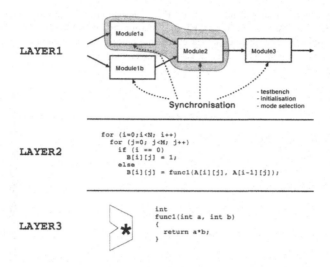

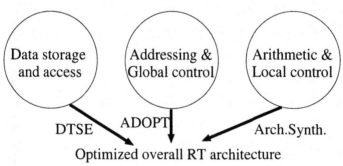

Figure 8.2.   Three different layers of pruning.

apply the ADOPT stage in a more ad hoc fashion. The ADOPT like optimization is needed to show that the data cost does not move to the addressing bubble.

The last bubble, related to data-path and local control, can be optimized with various optimization techniques (for instance using architecture synthesis approaches). The design experiment will however not touch this part directly, except for removing redundant data-path calculations. It will be shown in Section 8.1.4.3 that this has a very large positive impact however on the power consumption in the data-path. The power of that block is reduced significantly also by removing redundant calculations in the data-flow step of the DTSE approach.

Figure 8.3.   Separation in data, addressing and data-path allows focus and separate matched optimization techniques.

8.1.2.3   Data access analysis results.   After separation in layers (preprocessing), we can obtain the first array access count results (using the ATOMIUM tool set of IMEC), and use them to get a energy/area summary of the reference model. This allows us to identify and select signals with an important energy and/or area contribution for optimization. The Atomium array count tool has provided us with array access count numbers for a specific applied stimulus. Subchannel 3 is decoded for the first 10 frames (=960mS) of the stimulus CD.

Overall memory access related energy consumption results for the Philips chip reference of the DAB decoder are visualized in Fig. 8.4. Storage related area results are shown in Fig. 8.5. Note that every array is stored in a separate memory in the Philips design. From the energy figures in Fig. 8.4 it can be easily observed again that the FFT and Viterbi (esp. the metric's) related contributions are energy dominant. The time de-interleaver consumes also quite some energy and is the main contributor in area (see Fig. 8.5). That area usage is however unavoidable.

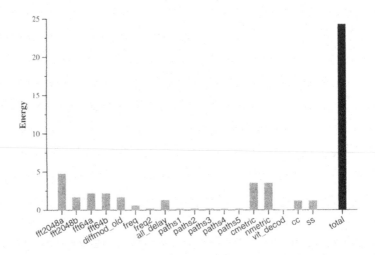

Figure 8.4.   Memory access energy estimates for the DAB decoder reference model (10 frames).

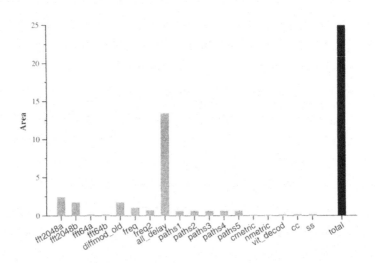

Figure 8.5.   Storage area estimates for the DAB decoder reference model (10 frames).

The energy number is *not* an absolute number and can be used as a relative measure only. This is because the memory model does not provide absolute energy numbers per memory access. Moreover, the energy estimation only includes the main stream data processing. For instance, the synchronisation core is not included in the total energy (see [69]). Also the FFT instruction ROM is not included as it is not data related memory. The area number ($36.750mm^2$) is an absolute number and is consistent with the estimate of Philips.

8.1.3 Optimizing reference model with DTSE methodology

The reference model is used as a starting point for the DTSE optimizations. The next subsections will handle the steps in the order implied by the DTSE script. As mentioned earlier, the preprocessing and pruning step is already applied. At the end of this section we will give a global overview of the achieved results.

Prior to the regular DTSE flow, several other important system-level exploration stages are needed. In particular, a task-level stage of DTSE steps is needed to deal with the dynamic concurrent task behaviour that is present in part of the DAB application. But that is not the focus of this chapter (see [94]). In addition, a static data type refinement stage is required to select optimized data type choices. Also this stage is left out here.

8.1.3.1 Global data flow transformations. As already indicated in subsection 1.7.3, two classes of data-flow transformations exist. The first class removes data-flow bottlenecks to make loop transformations possible. The second one, removes redundant storage and accesses. Four different subclasses of data-flow transformations of the second class are present, of which only (1) Advanced signal substitution including propagation of conditions and (2) Applying algebraic transformations (e.g. Modifying computation order in associative chains) have been extensively applied in the DAB decoder context. Because this book is not focussing on the data-flow trafo step, the details of this application will not be disscussed however. Only the final effect will be discussed further on in the result section 8.1.4.

8.1.3.2 Global loop transformations. The global loop transformation step optimizes the global regularity and locality of the complete application [124]. This step is steered by high-level array size (see chapter 4) and array reuse estimation. The optimization is achieved by globally transforming the loop nesting (see chapter 3). Besides the global merge, a more detailed loop nest transforming stage is performed on the most dominant loop nests.

In the original DAB many functional blocks require inter-function communication buffers (see Fig. 8.6). The functional blocks before the time de-interleaver are pipelined on the symbol level in three parts. In the first pipeline stage, the new incoming symbol is received in a 2048×16 buffer. The data is $\sim \frac{1}{4}$ of the time stable and unchanged. In this stable time, the FFT processor reads the data and processes the first radix-2 stage. The next pipeline stage processes the remaining stages of the FFT on the 2048×24 memory. The same stage performs the differential demodulation and the differential to metric conversion. The 4-bit metric output is stored in two 4-bit wide memory. The last pipeline stage performs the frequency de-interleaving and passes a single metric stream (including both real and imaginary metrics) to the time de-interleaver. A double buffer for the metrics array is introduced for half the buffer to meet the timing constraints. The time de-interleaver delay lines have been constructed such that each element was inserted before the element was extracted. This delay line increase of "1" has an increase of 15×3456 elements in practice. Finally, the Viterbi output is stored into a buffer to guarantee a continuous output bit stream. However, this buffer can easily be made smaller as it is too large and expensive.

The large FFT output buffer (size is 2048×24) is too large. This large buffer can be replaced by a single 3072×4 buffer by loop splitting and merge differently transformations. The double buffer between the frequency de-interleaver and time de-interleaver can be removed by merging the functionality. The frequency de-interleaver can be integrated in the time de-interleaving by substituting the pseudo random generator address in the read access. Moreover, the entire de-interleaving part is merged into the Viterbi decoder. When an additional metric is requested by the Viterbi decoder, a single metric is read from the differential demodulator input. When all the carriers are read, a new input symbol is processed. Finally, the PRBS is merged into the Viterbi-out by inversing the PRBS generation. In this way an additional read and write for PRBS processing is saved. The inverse PRBS calculation is slightly different but as complex as the normal PRBS calculation.

The entire code is merged into one big loop nest describing the complete DAB. The final required inter function buffers are shown in Fig. 8.7. A large reduction in buffer count and size is obtained in comparison with the reference of Fig. 8.6.

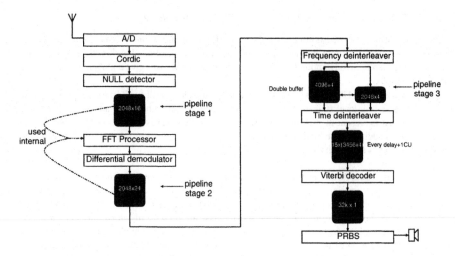

Figure 8.6.   The inter function buffering overhead in the Philips realisation.

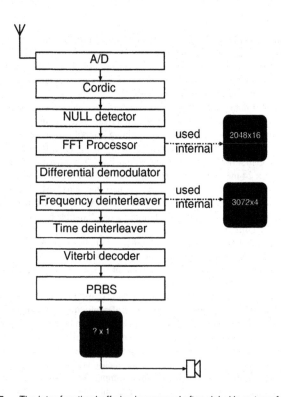

Figure 8.7.   The inter function buffering is removed after global loop transformations.

8.1.3.3 Data reuse decisions.   Only foreground level data reuse of scalar elements is present in the original design. So that DTSE step is not relevant here.

8.1.3.4   Platform independent optimization result.   The result in Table 8.1 summarizes the energy estimation for the source code version that is obtained after having applied all the platform-independent optimisation steps above.

Mem.	array	#words	$f_{acc.}$	Energy	Area	"cost"
						$\sqrt{S} \times BW \times f_{acc}$
SRAM1	fft2048	2048	1.298	0.855	2.198	5253
SRAM2	fft256	256	0.864	0.340	0.385	1236
SRAM3	fft16a	16	0.864	0.231	0.089	309
SRAM4	fft16b	16	0.864	0.231	0.089	309
SRAM5	diffmod_old	1536	0.498	0.239	1.343	1249
SRAM6	metr	3072	0.930	0.242	0.801	412
SRAM7	recover_metr	296	0.432	0.065	0.161	59
DRAM8	all_delay	98302	0.613	0.542	10.702	1537
DRAM9	small_delay	32768	0.357	0.191	5.773	516
SRAM10	paths	4096	0.000	0.000	0.587	0
SRAM14	cmetric	128	0.000	0.000	0.149	0
SRAM16	vit_decod	96	0.023	0.004	0.111	5
ROM1	cc	513	1.621	0.219	0.072	830
Total			**8.364**	**3.160**	**22.460**	**11715**

Table 8.1.   Data related energy estimation for platform-independent optimisation code (for 10 frames).

8.1.3.5   Storage Cycle Budget Distribution (SCBD).   In this subsection we construct a power optimized memory architecture implementation which meets all the timing constraints. The memory architecture must deliver just enough memory bandwidth. An over-dimensioned memory architecture will have too many ports and busses and hence more capacity than needed. Multiport memories consume more power than well designed single-port memories so the minimal memory architecture will be more power efficient also.

The power efficiency of the on-chip memories heavily depends on the applied voltage. So the well-known V_{dd}-scaling approach (see e.g. [100]) can be used also for on-chip memories (though not to the same extent as for the logic). Because of the fast increase of the cycle time below a certain voltage, we first determine the optimal voltage and derive from this the shortest abstract memory cycle length. For this purpose the Philips voltage derating table was used. The minimal energy is achieved for 2.0V. The maximum access frequency for this voltage is 30 M accesses for a DRAM and 59 M accesses for a SRAM.

As already indicated in chapter 6, the most time-consuming substep in the Storage Cycle Budget Distribution (SCBD) step is the automated Storage Bandwidth Optimization (SBO) substep [566, 64]. The SBO step should be used also to provide accurate feedback to the remaining SCBD substeps [524, 66, 63]. Our SBO tool provides accurate feedback about required number of cycles and localizes the bandwidth problems in the application.

The loop trafo for high-level inplace, I/O profile exploitation or creating freedom, and also the memory hierarchy layer assignment substeps will not be discussed in more detail here. Instead we will focus on an example of the important basic group structuring substep.

Basic Group structuring.   Basic group structuring has a large impact on power, area and memory bandwidth (see chapter 6). The principle of Basic Group Structuring is to store together in one array the elements of different arrays that are always read and written together. The goal is to use less memory instances but with a larger bit width which is less power consuming than a larger number of memories with a smaller bit width (on condition that they can use the same port and have the same access rate). Moreover, it is possible to adapt the bit width of arrays, so they can be placed together in the same memory resulting in a lower bit-waste.

This principle will be illustrated here only for the two arrays metr_re[1536] and metr_im[1536] in the Viterbi module. Some considerations have to be taken into account. Applying BG structuring involves a trade-off here because elements of those arrays are written together but not read together. Therefore, doing the same transformation on those arrays could lead to many read accesses to a memory wider than necessary. The different memory configurations are considered and the number of read/write related to the different cases. The Philips power model is used to determine the more appropriate solution. The different memory configurations are the following:

1. original: 2 memories of 1536 elements and a bit width of 4 bit, the total number of accesses is 3072 W and 3072 R.

2. packing: 1 memory of 1536 elements and a bit width of 8 bit, the total number of accesses is 1536 W and 3072 R.

3. 1 memory of 3072 elements and a bit width of 4 bit, the total number of accesses is 3072 W and 3072 R.

The results from the power model (with SRAM) are given in Table 8.2.

	1536 x 4 bit (x2)	1536 x 8 bit	3072 x 4 bit
power	0.861	0.928	1.040
area	1.110	0.845	0.801

Table 8.2.   Figures on power and area obtained with the Philips power model for different memory configuration.

The original configuration consumes lower power but has a large area. The third configuration has the opposite; a small area and a large power consumption. Moreover, the third solution has more accesses to the same memory which can form a potential bottleneck in parallel data accesses and therefore also system performance. The configuration with one memory of 1536 elements of 8 bit is the best solution for power, area and memory count. This solution is selected.

Storage Bandwidth Optimization (SBO).   The SBO step is an automated step. It provides a near optimal trade-off regarding the needed parallelism for a certain cycle budget. In combination with the automated Memory Allocation and Assignment step, it provides accurate power and area trade-offs for several cycle budget possibilities. It is up to the designer to choose the optimal operating points. It is of course also crucial to prune the set of alternatives to keep only the ones that allow to meet the overall timing constraints, given the available memory library options.

First the available time for the DAB computation is analysed. The easiest computation is made when a new input symbol comes in and all the data can be processed before the next symbol comes in. In this subsection we will focus on the most computation intensive mode I only. In mode I, a DAB frame contains 76 symbols and takes 96ms (T_F). The input symbol has a duration of ≈ 1.2461ms and contains 1536 carriers. Each carrier involves 2 metrics making a total of 3072 metrics per input symbol. Taking into account the lowest puncturing vector (index number 2) we find a rate of $\frac{8}{10}$ (this means 10 metrics are used at the Viterbi decoder input to produce 8 output bits). The maximum number of output bits generated during the execution time of the symbol is $3072 \times \frac{8}{10} = 2457.6 => 2464$ [1]. To summarize, this equals to a maximum of 308 output bytes per input symbol.

The worst case iteration count is passed to the SBO tool for finding the DAB application worst case cycle budget. This information is derived from the 3072 metrics per symbol (for the FFT and get metrics functionality) and 308 bytes (for the Viterbi functionality). [2]

The tightest Storage Bandwidth Optimization results, without self conflicts, have been selected for implementation (see Fig. 8.8). The overall cycle budget of 53099 cycles must be performed with timing constraint of 1.2461ms, causing an operating frequency of 43MHz. This satisfies the SRAM constraint but not the DRAM constraint of a maximum frequency of 30Mhz. But only the

---

[1] Rounded to the next multiple of 8 because the output is in bytes

[2] The worst case scenario for the Viterbi decoder is first a simple Viterbi decoding step. And at the end of the first traceback phase, a switch to the full decoder occurs because of an error.

delay lines buffer will be stored in DRAM. The accesses to the DRAM are modified in such a way that only 1 out of 4 cycles will access the delay buffer (thanks to additional basic group structuring transformations). As a result the DAB decoder can meet the cycle budget and run real-time on the most efficient voltage of 2.0V.

8.1.3.6   Memory Allocation and Assignment (MAA).

Pareto curves with cycle budget versus power.   In this subsection we present the interpretation of the Pareto curves with cycle budget versus power trade-offs for the DAB application. A trade-off has to be made between the performance gain of every conflict and the cost it incurs [66, 63]. On the one hand, every array in a separate memory is optimal for speed and seemingly also for power. But having many memories is very costly for area, interconnect and complexity and due to the routing overhead also for power in the end. Due to the presence of the real-time constraints and the complex control and data dependencies, a difficult trade-off has to be made. Therefore automatic tool support is crucial (see chapter 6).

A limited view is given in Fig. 8.8 where we only consider an allocation of 7 memories plus 2 ROMs. The rightmost point is the most relaxed cycle budget and the memory architecture can be freely defined in a cheap way. The tighter the cycle budget, the more memory architecture constraints pop up in the form of bandwidth requirements, and then energy-hungry multi-port memories are required. However, the power cost does not increase much when as long as no multi-port memories are used. This is reflected also in the point that was selected for our design (53099 cycles).

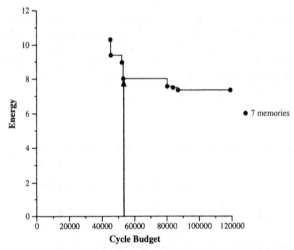

Figure 8.8.   Cycle budget versus energy Pareto curve for the complete DAB application with 7 memories.

A more extensive analysis with the tools has established that more than 13 (multi-port) memories does not improve the memory power consumption at all. When ignoring the interconnect and overall complexity of the physical design stage, the optimal memory count is 12 or 13. But the energy reduction compared to the case with 7 memories is negligible and therefore simpler and better.

Memory Allocation and Signal-to-Memory Assignment.   The optimal memory allocation is explored taking into account bandwidth and timing constraints as shown in Fig. 8.9. The useful range is defined from 6 memories up to 10 memories. Smaller amounts of memories are either infeasible within the bandwidth constraints or result in a too large penalty in both power and area. More than 10 memories is also not explored as it does not have a positive contribution to power any more.

Note that also during this MAA step, additional basic group structuring decisions have to be made.

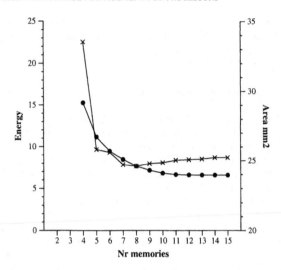

Figure 8.9. Memory allocation for the complete DAB application when bandwidth constraints are taken into account.

Transfer to bus assignment.   The optimized distributed memory architecture has been derived for this application to achieve real-time operation and a power-efficient data storage. To be able to perform this data-intensive process real-time, all important data-transfers in potentially different loops have been optimized globally. These data access can happen in parallel. A naive implementation will route every memory-port separately to the data-path but that is too costly. After bus-sharing exploration with our prototype tool, the solution of Fig. 8.10 is found. Here a memory configuration of 6 single-port memories and 2 ROMs has been selected.

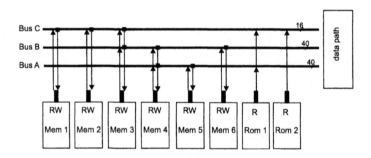

Figure 8.10.   Bus architecture: bus-sharing optimized.

8.1.4 Overall results

This section will summarize the power gains of the overall DTSE approach. In subsection 8.1.2.2 we have introduced the separation in three categories, data/memory, addressing/global control and data-path/local control. First we make a comparison for the data/memory related part between the reference implementation and the DTSE optimized version. Next we will make a comparison with respect to the two other sub-parts, namely addressing and data-path. The final subsection will bind the power results of all separate subparts and draw some conclusions.

8.1.4.1 Data/memory related power comparison between reference and optimized implementation. Memory power consumption is dependent on the size and access rate of every memory independently. Therefore we can estimate the overall power consumption accurately using a memory power model. The Atomium array count tool has been used to extract the memory access count (for the first 10 frames). These numbers are fed into the memory model provided by Philips. In this way we were able to extract relative numbers for the total energy consumption and area requirement for the Philips reference (subsection 8.1.2.3) and the DTSE optimized implementation (see Fig. 8.11 and Fig. 8.12).

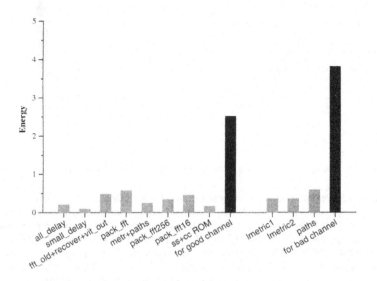

Figure 8.11.   Memory access energy estimates for the DAB decoder after DTSE optimisations (10 frames).

An order of magnitude is gained in energy by the DTSE optimizations. Actually a factor 11.7 is achieved for good channels. For bad channels, when the full fledged error correction is needed, then the very satisfactory factor 7.7 is remaining. Note, that this gain did not come at the cost of more area neither at the cost of a higher memory count. The area is reduced with 50% and the memory count is reduced from 16 to 9. The new implementation is able to meet all the timing constraints (see subsection 8.1.3.5).

8.1.4.2 Results and analysis for addressing and related control. The power estimation presented in this subsection concerns the power related to the computation in addressing. To perform this power estimation, the Orinoco environment [486] from the OFFIS research lab (Oldenburg - Germany) has been applied.

To use the Orinoco tool, it is first required to instrument the code. This instrumentation allows to collect data being computed and the operations being involved in these computations. The instrumented behavioural code is compiled and executed with a real and representative input stream. This creates an activity file that contains information about the computations present in the code. The activity file is then passed through the Orinoco tool which provides an estimation of the overall energy consumption due to the computations.

Only relative numbers will be given which compare the Philips and DTSE code and allow to get an overall power figure. Moreover, the purpose of the complementary energy estimation in this subsection is to check that bottlenecks have not been transferred from optimising the data transfer to the addressing computation.

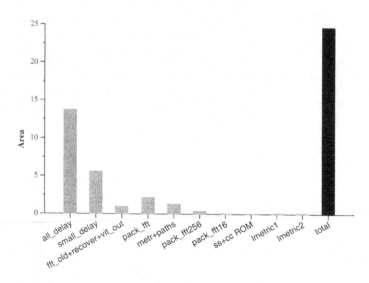

Figure 8.12.   Storage area estimates for the DAB decoder after DTSE optimisations (10 frames).

Tables 8.3 gives energy gain results for addressing. The normalised energy (1.00) corresponds to the total energy contribution of the reference version. The results show that not only no overhead on addressing has been introduced by the DTSE transformations but also that significant gains are achieved by combining the DTSE and Adopt [372] approach. Even when using the same functional Viterbi decoder (Viterbi 6 bit), a gain of a factor 2 is attained. The energy reduction obtained can also be explained by the fact that there are less memory accesses after DTSE, so less address computations. Furthermore, smaller and more narrow arrays are used, where, variables being involved in the address calculations have a smaller bit-width, which contributes to the energy reduction.

	Philips Reference	DTSE implementation Good channel	Bad channel
Addressing	1.00	0.37	0.47

Table 8.3.   Energy gains for addressing and related control.

8.1.4.3 Data path related power consumption and gains. The power estimation presented in this subsection concerns the computation related to data path. The energy in the data path has been reduced tremendously due to the removal of the many redundant calculations in the global data-flow transformation step in DTSE. In fact, a large amount of data was calculated which was not needed at all.

Tables 8.4 shows the energy gains. The normalised energy (1.00) corresponds to the total energy contribution of the reference version. Note, that this reference result is obtained by a detailed analysis with an accurate RT-level power estimator from Philips, called Petrol. A factor 3.6 in data path calculation reduction is achieved for a bad channel and a factor 5.9 for a good channel.

	Philips Reference	DTSE implementation Good channel	Bad channel
Data path	1.00	0.17	0.28

Table 8.4.   Energy gains for data path.

8.1.4.4   Final combined results for custom target architecture.   The power distribution over data, data-path and addressing in the Philips DabChic chip implementation shows a dominance in the data/memory related part (see left hand side of Fig. 8.13). Half of the power is devoted to storage (on-chip memory + large register files). About one third is spent in the code ROM. Both the actual data processing and memory addressing arithmetic consume about one tenth of the global power budget.

The high percentage of power consumption in the code ROM can however be significantly reduced using approaches described in e.g. [50, 270, 253, 330, 246] involving compression and the use of loop caches combined with compiler optimisations. At least a factor 3 should be obtainable then. In addition, the synchro code ROM power can be reduced with a factor 4, because it is clocked for every interval while it is used effectively one quarter of the time only. These relatively easy code ROM optimizations results in a global power reduction of 25% (see middle pie chart of Fig. 8.13). This increases the data memory dominance to two third of the overall power.

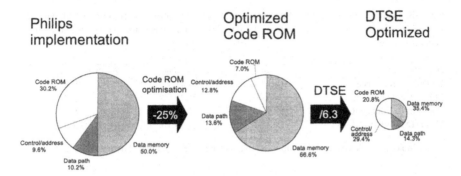

Figure 8.13.   Power consumed in different categories by the DAB application.

If we now combine the category percentage and the gain per category from the earlier discussed results in this section, in Table 8.5 an overall system-level energy gain factor of between 6.3 and 4.6 is obtained due to DTSE. The actual gain is indeed dependent on the signal to noise ratio of the received signal. A simpler channel decoder can be used when the reception is good. Both results are mentioned in the table below. All the percentages right of the vertical line in Table 8.5 are relative to full energy consumption of the code ROM optimized implementation. For good channels, the 66% of the data memory related power is reduced with a factor of 11.7 (see fig. 8.11) to 5.7%. The data-path and control/addressing are reduced respectively with a factor 5.9 (see Tab. 8.4) and 2.7 (see Tab. 8.3) to a percentage of respectively 2.3% and 4.7%. The code ROM for the FFT is reduced with a factor 4 because after the DTSE data-flow and loop trafo steps, it is actively used only a quarter of the cycles.

The final results are obtained by summing the optimized percentages of the individual categories. This results in 16% of the reference power which is an overall gain of a factor 6.3. A similar calculation can be made for a bad channel, which results in the factor 4.6. The gain due to DTSE and the new distribution over the categories are displayed at the right hand side of fig. 8.13.

8.1.4.5   DAB decoder running on Intel platform.   The DAB channel decoder including a MP-2 decoder can run real-time on a Linux machine powered with an Intel P-II 450MHz! Although the original target platform is not the Intel one, the three initial (platform independent) DTSE steps are good in general as we will show again on this performance demanding application. The introduced improvements made it feasible to run the DAB application in real-time (see second entry in Table 8.6).

The functional and Philips references are respectively a factor 9.8 and factor 2.9 too slow. In the DTSE optimized DAB application, the elapsed time is smaller than the length of the audio output (factor 0.77). This means that still 23% slack time is available to execute other processes. However, when further optimizing the application code with a customisable hardware platform in mind (like

Category	Code ROM Ref without DTSE	Good channel		Bad channel	
		Gain	Energy	Gain	Energy
Memories	66.6%	11.7×	5.7%	7.7×	8.6%
Data-path	13.6%	5.9×	2.3%	3.6×	3.8%
Control/addressing	12.8%	2.7×	4.7%	2.1×	6.1%
Code ROM (syncro)	2.1%		2.1%		2.1%
Code ROM (FFT)	4.9%	4.0×	1.2%	4.0×	1.2%
Total	100.0%		16.0%		21.8%
			≈**6.3**×		≈**4.6**×

Table 8.5. Overall energy results: the DTSE optimizations have brought a system-level gain of a factor 6.3 for good channels. All the percentages are relative to the full energy of the reference design which is the Philips design but then improved already with a more efficient code ROM implementation.

the ASIC platform that is discussed up to now) would severely worsen the performance of that code on a PC platform. The execution time of the final DTSE implementation exceeds by far the real audio play time.

Implementation	user time	elapse time	relative speed
Functional reference	1091.54	18:41.72	9.83
Philips reference	308.00	5:35.52	2.90
Platform independent	74.74	1:28.73	**0.77**
Full DTSE optimizations	282.13	4:56.16	2.60

Table 8.6.   Runtime of DAB application to produce 1:54 seconds of actual output data.

8.1.4.6 DAB decoder running on TriMedia TM1 platform. The DAB decoder code resulting after the platform-independent DTSE stage (which is memory platform specific but processor-independent) has been compiled also on a TriMedia TM-1300 platform running at 166MHz using its native compiler and enabling the most aggressive optimisation features. The power consumption due to the memory subsystem has been significantly reduced that way.

However, in this platform the DTSE stage on its own has not been sufficient to achieve real-time decoding. In order to solve that, we need to exploit also the sub-word level acceleration capabilities and specialised instruction set that the TriMedia processor offers. After profiling the application and analysing the critical path, we have focused on optimising the FFT function. However, ample opportunities for optimisations are also present in the Viterbi block. Optimising that block is part of our future work.

The test bench has been selected to generate a stream of 10 frames (about 1 second of audio) which is stored in an array inside the available off-chip memory banks of the TriMedia board. Several cost aspects have been taken into account while optimising the code, namely: the execution time, the memory related power consumption and the storage size of the arrays involved in the critical path of the application. Execution time has been measured by running the application on the board. However, to estimate the memory related power consumption, a TriMedia simulator has been used. With the simulator we have obtained the total number of accesses to the memory hierarchy and the number of hits and misses in the data cache. Given that previous experiments have shown that typically an access to main memory is at least 5 times more consuming than a cache access (using a conservative estimate), a relative estimation of the power consumption before and after the transformations has been performed.

Platform-independent DTSE transformations. The TM-1300 platform has a register file with 128 registers connected to a double port L1 data cache of 16KBytes followed by a single 32-bit port unified off-chip SDRAM memory. The application of DTSE platform independent code transformations have enabled the possibility to allocate two small but heavily accessed arrays of 16 elements of the FFT in the register file. Note that this extra amount of required registers is sufficiently small so the risk of creating register pressure and hence additional memory spilling operations is not present and it brings a significant gain in power reduction, and also a performance improvement.

Platform-dependent data-level DTSE and SIMD exploitation.   On the other hand, TriMedia's specialised SIMD instruction set allows to perform 2 complex operations in a single word with just one instruction. This fits very well in the FFT function where most of the data is accessed in pairs. This is because the need to process both a real and an imaginary part together using a specialised filter type operation (e.g., TriMedia's ifir16). However, to avoid extra data formatting operations prior to the execution of the SIMD instruction, a Basic Group matching transformation (platform-dependent DTSE stage at the data-level) step needs to be applied to format the data in memory in such way that the real and imaginary parts are packed together in one word of 32 bits.

The Basic Group matching transformation has been complemented with a manual selection of the SIMD instructions at the source level code. By doing this, a considerable speed-up of the FFT computation has been obtained. Indeed, about a factor two reduction in the number of cycles for the execution of the data processing present in the FFT butterfly is achieved. This translates in more than a 15% reduction in the execution time of the complete decoder (see Table 8.7).

Processor architecture-specific non-DTSE transformations.   After the application of DTSE and data-level parallelisation management transformations, high-level address code optimisations have been applied to remove redundant computations and to replace expensive operations by a cheaper arithmetic. These address related transformations have been complemented with data and control flow related optimisations, namely function inlining and unrolling of small loops that are processor architecture specific. For instance, calls to the butterfly4 function which are present in the critical path, have been inlined and many time consuming context saving operations have been avoided. In addition, by unrolling the most inner loops we break up the control dependencies between the different loop iterations and an increase in instruction level parallelism is obtained.

As a result, the DAB frames can be now decoded in real-time with a 45% gain in memory related power consumption.

Implementation	D-cache accesses (10^6)	D-cache misses (10^6)	elapse time(secs)	Total memory power (normalised)
Philips reference	336.5	4.7	9.20	1.00
Platf-indep DTSE	70.2	1.2	1.52	0.21
Data management	44.8	0.6	1.20	0.13
Proc.spec.	32.2	0.8	0.99	0.10

Table 8.7.   DAB results on the TriMedia obtained while producing 1 second of decoded data.

8.1.5 Summary

The Data Transfer and Storage Exploration (DTSE) methodology has been very successfully applied on the Digital Audio Broadcast (DAB) channel receiver application. For the DAB decoder realized on a fully customisable platform, an overall chip-level power gain of a factor 6.3 is obtained without sacrificing either performance or memory area. The DTSE methodology and tool support allows the designer to focus in every individual step on one aspect. In this way, it is possible to arrive at a good solution in a very effective way. The first part of the DTSE methodology is platform independent. This stage has delivered a transformed C code implementation which is capable of decoding a DAB signal faster than real-time on an Intel Pentium-II PC platform and also real-time on a Philips Trimedia TM1 platform operating at 100 MHz (with nearly a factor 2 in memory power savings due to our transformations). This application has much similarity with an OFDM-based wireless LAN receiver. So we expect similar gains in such wireless baseband modems.

8.2   GRAPHICS DEMONSTRATOR: MESA GRAPHICS LIBRARY OPTIMIZATION

Graphics is rapidly becoming an integral part of multimedia computing. The MPEG-4 standard is a good example of this coming together of the graphics and video compression worlds [482]. Rendering textures on different objects is one of the important graphics feature which is being incorporated. OpenGL is an emerging standard for developing graphics applications [469, 470]. Since OpenGL is a proprietary library of Silicon Graphics Inc, we use, in this experiment, a public domain version where source code is available, namely the Mesa graphics library (MGL) [366]. It provides a large number of functionalities needed for developing 2D/3D graphics applications.

In the sequel, we will show that significant reductions in both execution time and memory accesses can be obtained after our DTSE oriented program transformations.

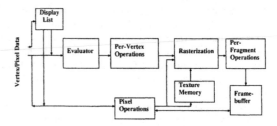

Figure 8.14.   OpenGL machine.

8.2.1   Mesa graphics library

MGL allows the programmer to specify graphical objects and to manipulate them. In the process, MGL also acts as a software interface to the graphics hardware. Fig. 8.14 shows the programming model (also called OpenGL machine) used to render objects. This model is also referred to as *graphics pipeline*. More details of this model are discussed in [470].

In the context of graphics accelerators used in personal computers (PC's), the architectural model is shown in Fig. 8.15. The current generation of PC graphics accelerators perform only the rasterization stage of the graphics pipeline. The earlier stages, namely the geometrical transformations, lighting and shading operations are performed in software on the PC and then these transformed vertices are handed over to the (hardware) accelerator for rendering. In recent years a trend to perform both the geometrical transformations as well as rendering on the programmable graphics processors is observed [309]. In this work, we will concentrate on a complete software implementation of the MGL.

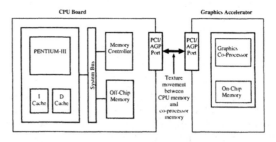

Figure 8.15.   PC and graphics accelerator interface and the transfer of texture between the two.

We will now briefly discuss the steps involved in the implementation of a texture mapping application using MGL on programmable (embedded) processors. The three main steps are listed below:

1. Specifying the co-ordinates of the object, namely the vertices and normals of the object and the co-ordinates for the texture.

2. Initializing the texture image or reading in texture from an existing source which can be a file or an image.

3. Transferring the texture from the intermediate storage to the actual object. This step involves first calculating the portion of the object where texture will be mapped[3] and then transferring the required number of pixels from the texture image to the (final) object.

The above process is shown in Fig. 8.16.

The test-vehicle used in this work is a 3D demonstrator program for texture mapping on objects. A 2D texture mapping function is used to obtain the final 3D texture mapped object (here a *taurus*[4]). The application comprises four main functions as discussed below and illustrated in Fig. 8.16.

1. **Initialization** - this function performs all the initial settings for rendering into the framebuffer like lighting, shading, clipping etc.

2. **Build figure** - this function computes the normals, vertices and texture co-ordinates for the object. This function also determines the computational complexity of the main "Draw Scene" function.

3. **Create texture** - this function defines the texture image or reads in from another source image. It also transfers the texture image to the internal data structure of the MGL.

4. **Draw scene** - this function performs the actual drawing of the object into the framebuffer and mapping texture onto the concerned object. We will show later on that this function takes up most of the total time in a software realization.

The number of sides and rings of the *taurus* are parameterized in the test-vehicle and hence can be varied. This has a large influence on the amount of computation required for calculating the vertices, normals and the texture co-ordinates. In this experiment we consider two cases, namely one with less computational complexity and the other with greater computational complexity. The less computational complexity case has a taurus with two rings and four sides whereas the greater computational complexity case has a taurus with eight rings and ten sides.

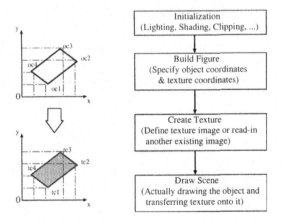

Figure 8.16.   The steps comprising the 3-D texture mapping test-vehicle used in this work.

We observe that the process of texture mapping involves the storage and transfer of the texture image as well as the object due to various conditions as specified by the application designer. The above issues have been our motivation to systematically explore the MGL for DTSE related optimizations, as described in this book and [93]. This has resulted in very promising results [297].

---

[3]This is because of the viewing angles and other effects like lighting, shading, clipping, etc.
[4]A taurus here refers to an object which comprises a ring of polygons.

Also studies on dynamic graphics workloads by [374], have concluded that even for performance, reduction in bandwidth usage due to data transfers on the system-bus is one of the key issues.

8.2.2 Experimental set-up

Two target platforms are used in this work to illustrate our DTSE principles and results:

1. **HP PA-RISC 8000 processor:** For this, we have used a HP workstation (HP-UX 10.20 E 9000/879). The HP "aCC" compiler was used with the +O (standard level two) option.

2. **TriMedia TM1000 processor:** We have used a TriMedia TM1000 simulator (tmsim v22.2.2 of tcs1.1y1016HP-UX) on the HP-UX platform as described above. The execution times, as given in Fig. 8.20, are obtained on a TM1000 board connected to a Windows 95 platform.

We will now also present the profile of the test-vehicle described above. The main intention is to identify the parts of the code which are critical to performance and lend themselves to possible optimizations. Fig. 8.17 and Fig. 8.18 shows the percentage of time spent in various routines of the test-vehicle for both PA-8000 and TM1000 processors respectively.

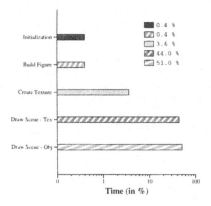

Figure 8.17.   Profile of the time spent in various routines in the graphics test-vehicle on HP PA-8000.

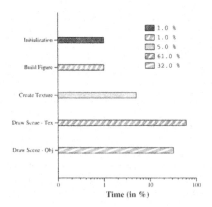

Figure 8.18.   Profile of the time spent in various routines in the graphics test-vehicle on a TriMedia TM1000 processor.

We observe that the texture mapping part in the function "Draw Scene" takes 44% of the total time for the HP PA-8000 processor and 61% of for the TM1000 processor. So that is our DTSE focus, together with the tightly coupled "Create Texture" function.

8.2.3   Experimental results for optimization strategies

In [297], several optimization strategies are described. We will only focus on the Low-Level and Hybrid-I strategies where mainly the underlying library is optimized without modifying the API to the application. These strategies involve two main stages namely first performing the enabling local optimizations and then the final global control flow transformation step in the DTSE approach [93] (see chapter 3). The details of the application to the MGL are described in [297].

We now present the experimental results using the set-up discussed in subsection 8.2.2. First we present the results of mapping the library and the test-vehicle on a HP PA-8000 processor and then on a Philips TM1000 processor. [5]

8.2.3.1   Experimental results on TM1000.   Fig. 8.19 shows a gain in total cycle count of approximately 44% (with a deviation of 3%), a reduction in the number of accesses to the instruction cache by 40% and lastly a reduction in the number of accesses to the data cache by a factor 2 [6]. Thus the effective gain in total cycles for the texture mapping function is almost a factor 3 compared to the initial case, considering the fact that we have optimized the part which consumed 66% of the total time. Similarly due to the increased data locality and reduced storage size, the data cache performance is greatly enhanced.

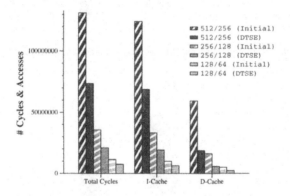

Figure 8.19.   Measured results for different image and texture sizes in the test-vehicle TM-1000 processor (based on simulator).

We observe that a major result of interest here is the additional large reduction in number of accesses to the instruction cache. The instruction cache performance is enhanced due to the fact that we have moved all the condition evaluations out of the iteration loops, which results in a lower number of instruction fetches for different paths out of a conditional node. This is especially true for the program memory, since if at compile-time we are not able to evaluate all the branches, then instructions for all the relevant branches will be stored in the program memory (if not in the instruction cache). Since we have done a compile-time analysis and have evaluated all the possible branches (almost 85% of all the branches inside the iteration loop). for the concerned functions we are now able to reduce a large number of the accesses to the instruction cache. This step was only due to the fact that we have used a DTSE oriented cost function[7], and not CPU performance alone.

---

[5]Note that all the results in this section are for the greater computational complexity case, which means the object (*taurus*) has eight rings and ten sides.

[6]"X/Y" on the x-axis stands for an image size of $X \times X$ pixels and a texture size of $Y \times Y$ pixels.

[7]In this specific case, we have used the number of memory accesses and the temporary storage size as the cost function.

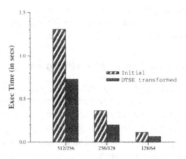

Figure 8.20.   Execution time for different image and texture sizes on TM-1000 processor (on TriMedia board).

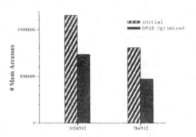

Figure 8.21.   Number of memory accesses for two different image and texture sizes in the test-vehicle.

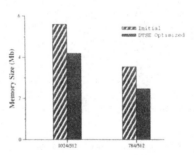

Figure 8.22.   Memory size for two different image and texture sizes in the test-vehicle.

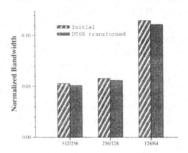

Figure 8.23.   System bus loading for different image and texture sizes in the test-vehicle on TM-1000 processors.

Fig. 8.20 shows a reduction in the execution time by approximately 47%. Similarly Fig. 8.21 shows a reduction in the number of memory accesses by 44% and Fig. 8.22 shows a reduction in memory size by approximately 25-30% respectively. Note that this positive effect applies both for the I-cache and the D-cache misses. This confirms our claim that the DTSE approach, if applied well, does not introduce an I-cache penalty but instead usually shows a significant gain. In addition, Fig. 8.23 shows that we are able to reduce the system bus loading, which is as important for the global system performance at the board level. Also note that, although we have optimized only part of the library relating to the texture mapping, the other parts which were not optimized are not effected (either positively or negatively) in a significant way. We have observed a change of (maximum) 2-4% in the execution time for other non-optimized functions, which does not impact the overall application design. This shows that we can perform our DTSE approach indeed in a decoupled way.

8.2.3.2 Experimental results on HP PA-RISC. Fig. 8.24 shows a consistent gain of 28% in total execution time (with a deviation of 1.5%) for all the different image and texture sizes. Thus if we consider the actual impact of our optimizations, then we have a gain in execution time by a factor 2.4. Note that also here we have optimized only part of the whole application, namely the part that concerns the texture mapping and accounts for approximately 48% of the total time.

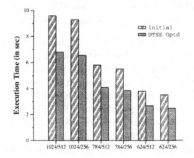

Figure 8.24.   Execution time for different image and texture sizes in the test-vehicle measured on HP PA-8000 processor.

8.2.4 Summary

In this section we have presented a study comprising analysis and DTSE optimization strategies for texture mapping applications using MGL. The following are the main contributions:

1. We show that applications involving graphics lend themselves to DTSE related optimizations and there is a large potential scope for further optimizations. This has triggered further research that is currently going on to apply also the (many) other DTSE steps in our global methodology [209].

2. Even though these applications require library support with a fixed API, we can optimize the library as well as the application for meeting our objectives. This indeed involves more design time but when good tool (compiler) support is provided the involved design time can be significantly reduced.

3. On the average, we are able to reduce the total execution time of the test-vehicle by 28% for the HP PA-8000 processor and by 44% for the TM1000 processor. Additionally we also have reduced the energy due to memory accesses and bus loading.

8.3 VIDEO DEMONSTRATOR: MPEG4 MOTION ESTIMATION

8.3.1 Platform independent program transformation steps

Very promising results have been obtained for the dominant video motion estimation module in the new multi-media compression standard MPEG4. The recent MPEG-4 standard [482] is a key to multi-media applications. The exploration below has been performed on the MoMuSys Video VM Version 7.0. It involves complex data-dominant algorithms. An embedded instruction-set processor realization of such a (de)coder has to be power efficient in order to reduce the size of the chip packages (where it is embedded) or the battery (if used in a mobile application). We have shown that it is feasible to exploit the dominance of data transfer and storage related costs, to achieve large savings in the system power of a crucial part of MPEG-4, namely the video motion estimation (see Fig. 8.25), without sacrificing on the performance or on the system latency [67, 70].

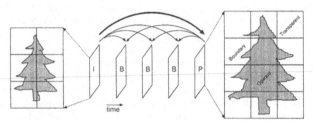

Figure 8.25.   Illustration of MPEG-4 video object plane (VOP) sequence for content based object coding. The arrows represent all motion estimation steps for one group of VOPs.

Profiling experiments have shown that 77% of the original accesses take place for the luminance pixels in (full- and half-pel) motion estimation, which we have focused on[8]. The primary design goal has been to reduce memory transfers between large frame memories and processor data-paths. In this more global demonstrator we have incorporated the entire video motion estimation code, but we have focused only on the most promising code transformation steps in our DTSE method-ology. Different also from the previous experiments in the other chapters, we have now applied a two-stage DTSE approach instead of our standard single stage. First, we have applied the DTSE transformations at the task-level abstraction where only a (relatively small) subset of the M-D sig-nals has to be considered, which allows to reduce the design complexity in the global DTSE stage, and hence saving design time. By analyzing these task-level loop structures and large signals, the data locality of the algorithm implementation can be improved compared to the original sequence of motion estimation steps [482] (see also Fig. 8.25). However, not all motion estimation steps can be merged (infeasible due to data dependencies present in the actual code) and a thorough exploration is required of the huge search space, for which we have used our systematic loop transformation methodology.

This task-level transformed algorithm code result is now passed to the data reuse decision step [154]. In our systematic multi-step methodology several issues can be decoupled, allowing the po-tential reuse to be (more extensively) explored without letting the design complexity explode. At the task-level abstraction, only the main signals shared or communicated between the interacting tasks are considered. Moreover, only the first levels of data reuse are explored because the lower levels involve too small signal copies which have to be decided on during the more detailed processor-level DTSE stage. This leads to the results marked as task-level DTSE in Fig. 8.26. The left hand side steps belong to the task-level stage where only the main VOP signal access was considered. A reduc-tion by a factor 65 in background luminance pixel reads has been obtained this way. Experiments have however confirmed that after this substantial saving on the VOP access, significant power is also related to the "second order" signals which are only visible when the abstraction level is ex-panded to the processor-level where the instruction/operation details become explicit again. Similar steps, applying our more conventional processor-level DTSE methodology (explained earlier), have been performed for each of the distinct tasks in the motion estimation code. We will not provide the details here. The right-hand side steps in Fig. 8.26 show that our processor-level DTSE code

---

[8]The cost of a data transfer is a function of the memory size, memory type, and especially the access frequency F_{real}. An accurate power model, derived from a VTI data sheet, has been used for the exploration.

transformations allow to reduce also these contributions to a fraction of the original contribution. A substantial gain in global memory related average access power of a factor 13 is reached. This has been obtained and verified by simulation for the first 96 frames of the "hal" sequence in CIF resolution.

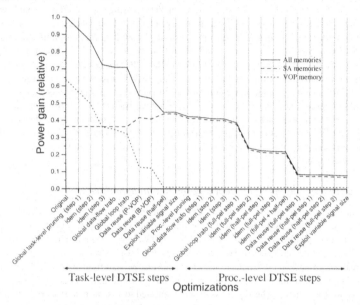

Figure 8.26.   Total relative power for VOP memory and search area memories during application of task-level and processor-level DTSE stages.

The overall advantages of this new two-stage approach are that:

1. The task-level DTSE stage can have a truly global scope because only the most important inter-task signals (and only a few crucial intra-task ones) have to be considered. The processor-level DTSE stage can focus in full detail on each task separately without having to consider the task interaction any longer. These two facts result in significant savings on design time as experienced in our exploration of the MPEG4 motion estimation routines.

2. In many applications, the concurrent tasks exhibit asynchronous process control interactions which prohibit the expansion of the full code into a single statically analyzable piece of code. In these cases, the detailed processor-level analysis and optimisations do not make sense beyond the boundaries of such tasks. In contrast, our experiments have shown that the coarser view of the task-level DTSE stage still allows to perform relevant optimizations across the task boundaries.

The exploration space to optimize data storage and transfers is very large, and our experiments clearly show the importance of a formalized methodology to traverse it. It also demonstrates again that the DTSE optimizations have to be performed aggressively before the multi-media algorithms are realized on instruction-set processor targets.

8.3.2 Cache and address optimisation steps

The ever increasing gap between processor and memory speeds has motivated the design of embedded systems with deeper cache hierarchies. To avoid excessive miss rates, instead of using bigger cache memories and more complex cache controllers [424], program transformations have been proposed to reduce the amount of capacity and conflict misses [149, 295, 404, 285]. This is achieved however by complicating the memory index arithmetic code which results in performance degradation when executing the code on programmable processors with limited address capabilities. However, when these are complemented by high-level address code transformations, the overhead introduced can be largely eliminated at compile time. Here, the clear benefits of the combined approach is illustrated on a MPEG4 Motion Estimation kernel, using popular programmable processor architectures and showing important gains in energy (a factor 2 less) with a relatively small penalty in execution time (8-25%) instead of factors overhead without the address optimisation stage. The results lead to a systematic Pareto optimal trade-off (supported by tools) between memory power and CPU cycles which has up to now not been feasible for the targeted systems.

The clear benefits of the combined approach to aggressively improve cache utilisation by source code transformations which introduce complex addressing which is then reduced again by a high-level address optimisation stage, has to our knowledge not been studied in the literature. We will provide a systematic approach to achieve that which leads to Pareto optimal speed power trade-off curves. That approach is illustrated on a real-life application driver of industrial relevance, using three popular programmable processor architectures, showing important gains in cycle count and energy consumption.

8.3.3 Experimental setup for cache optimisation experiments

Due to flexibility requirements of the implementation an instruction-set processor architecture is preferred over a fully custom one. The architectures which we target in this paper, consist of multi-media (extended) processors (such as TriMedia's TM1000, Intel's Pentium-III MMX) but also more general RISC architectures such as Hewlett-Packard's PA800 architecture.

For realistic embedded implementations we assume that we start from their "core" versions which have a relatively small local (L1) on-chip cache memory to reduce the energy per L1 cache access (which is our focus here). These processors are connected to a (distributed) shared off-chip memory (hierarchy). Therefore, meeting real-time constraints in such architectures, requires an optimal mapping of the application onto the target platform.

For our experimental setup we have used the SimpleScalar simulator tool set [74] for simulating cache performance for varying cache sizes. The cache is simulated using the faster and functionally correct sim-cache simulator for measuring cache performance. Existing processors have been used to measure the performance values. The code versions have been compiled and run on the different platforms using the native compilers and enabling their most aggressive speed oriented optimisation features.

Our exploration approach is illustrated on a MPEG4 Full Search Motion Estimation kernel for CIF video frames.

When this algorithm is mapped onto an embedded system, the system bus load and power requirements are quite high. At our laboratory, much effort has been already spent on optimising the drivers at the source-level code, mainly by applying the platform independent transformation stage of the DTSE methodology [151]. Here we will focus only on the data cache related speed-power trade-offs during the platform specific mapping stage.

8.3.4 Memory data-layout optimisation for D-caches

For the D-cache, optimisations are available which are oriented to minimise both capacity and conflict misses (see chapter 7). Capacity misses can be greatly reduced by exploiting the limited life-time of the data (inplace oriented optimisations [149]) and the conflict misses by applying data-layout oriented transformations [285]. Using prototype tools supporting these techniques [289], we obtain the transformed C code. Thus, we now have the initial (reference) and the data layout transformed codes.

After cache simulations we have observed that our driver is dominated by misses in the D-cache (the miss rate for I-cache is about three orders of magnitude smaller than for the D-cache). Therefore, we have focused on optimising the D-cache miss rate by applying main memory data-layout oriented program code transformations. Simulations have been performed using different cache controller types: a 4-way associative cache for the non-optimised version and a direct mapped for both

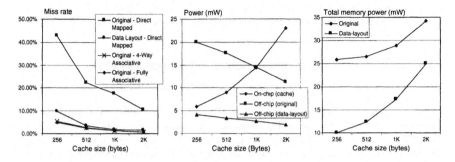

Figure 8.27.   D-cache miss-rate evolution for the MPEG4 Motion Estimation driver for different cache sizes and controller types, including off-chip/on-chip power breakdown and total power evolution for different cache sizes (for a direct mapped cache controller).

optimised and non-optimised versions. Also, different capacity sizes (with the number of cache lines ranging from 256 till 2024) have been evaluated. Fig. 8.27 shows the miss rate after simulations for both controller types.

For the non-optimised code, the miss rate obtained using a direct mapped controller is rather large when we compare it to the one obtained when using the 4-way associative controller. However, when using the cache optimised code, the miss rate behaves equally well for both controller types. This is clearly an important result because it enables the use of the simpler direct mapped cache controllers for embedded applications, instead of the more costly, and power hungry, set-associative controller types.

8.3.5   High-level optimisation of the index/address computation

After cache oriented code transformations, the new addressing code is dominated by index expressions of piece-wise linear nature and it becomes a bottleneck for the auto-increment nature of most commercial Address Calculation Unit (ACU) architectures. Therefore, it should be removed before going to the more pointer level specific address optimisation techniques like those found in conventional [8] or state-of-the-art compilers [381]. Moreover, conventional automated address optimisation approaches [318, 324, 197, 492, 518] do not allow this for address code containing modulo/division operations.

Indeed, a considerable number of costly modulo operations are executed and traditional compilers completely fail to efficiently eliminate them. Hence, the execution time increases considerably: a factor between 2-20 depending on the application and the target processor architecture (see Section 8.3.6).

To remove this large bottleneck we have developed efficient address optimisation techniques. These are based on the use of program transformations which are largely independent of the targeted instruction-set architecture [372, 216], but which need also to be complemented by more architecture specific ones [201]. Using these techniques, the addressing code is globally optimised, resulting in factors overall improvement in execution cycles when compared to their original data-organised versions.

High-level address transformations also have a positive effect on memory accesses and in L1-misses both for I-caches and D-caches. For the D-cache, we have observed that the amount of data memory accesses as well the number of data misses decreases slightly ($\simeq < 10\%$). However, this improvement in data memory power is very limited when compared by the one achieved using the memory data-layout oriented transformations (see Section 8.3.6). For the I-cache, a larger reduction in instruction memory fetches ($>40\%$) is obtained (due to the more efficient addressing code) while the reduction in instruction fetch misses stays also limited ($\simeq$<10%). Still, the miss-rate for the I-cache is about three orders of magnitude smaller than for the D-cache and so the off-chip memory related power.

8.3.6 Trade-off between performance, power and main-memory size

We now discuss how the above technique can be applied to achieve important system level trade-offs.

We have started with an initial algorithm which has been optimised from a DTSE perspective. In the first step, we obtain miss rates for different cache sizes when using a cost efficient direct mapped cache. This is shown in Fig. 8.27 for the Motion Estimation kernel. Once we have obtained the miss rates for different cache sizes, we compute the total (off-chip plus on-chip) required power. We observe that a cache size with 256 bytes consumes the least total power. For smaller cache sizes, the total consumed power would be dominated by the accesses (due to misses) to the off-chip memory. For larger sizes the accesses to the on-chip memory dominates the total power though.

In a next step, we choose the power optimal cache size of 256 bytes to decide on a trade-off between main-memory size (the overhead in memory space in the data layout optimisation process as compared to the initial algorithm) and the reduction in miss rate as shown in Table 8.8. Thus depending on the design constraints, the designer can now either choose a lower power solution with some overhead in main-memory size and vice-versa. For the Motion Estimation kernel, the points selected are responsible for a 30% overhead in main-memory size (labelled "Light") and a larger factor 3 overhead (labelled "Aggressive").

In the final stage, we perform address optimisations (also supported by prototype tools [201]) to remove the overhead in addressing operations introduced in the data layout optimisation process.

Fig. 8.28 shows that we are able to largely remove this overhead in addressing. This is visible by comparing the projected values on the horizontal axis of the curve before the high-level address optimisation phase (right-hand side) and after it (left-hand side). Using the "Light" data-layout version, we are able to reduce memory related power by almost a factor 2. This is achieved by trading-off a 30% in main-memory size with a 14-40% in execution time (depending on the target processor), instead of the 80-280% CPU cycle overhead without the address post-processing stage (see Fig. 8.28). Note that the data related memory power does not significantly change with the address code transformations stage but mainly with the memory data-layout stage as motivated in Section 8.3.5.

Table 8.8.   Main-memory size/power/speed trade-offs for the Motion Estimation.

Data-layout trade-off	Memory-size Overhead	Avrg.Power Gain	Avrg.Speed Degradation
Light	30 %	45 %	14-40 %
Aggressive	3X	60 %	18-60 %

Note that the approach we have used in this work is mostly power centric, i.e. we first optimise for power (by selection of the optimal cache size) then for main-memory size (by deciding the amount of data-layout) and lastly for performance (by optimising the addressing code). But the above technique can be used for a performance centric approach too. In this case, the designer should first choose the optimal cache size for data miss rate (e.g. by selecting the cache size from which the decrease in number of misses starts to saturate) and then select the amount of data-layout corresponding to the required trade-off between main-memory size and memory power. Following this approach, a cache size of 1K bytes for the Motion Estimation would lead to better system performance (less data misses) at the expenses of some overhead in memory power (see total memory power evolution in Fig. 8.27). We have performed a full SimpleScalar [74] machine simulation of a memory hierarchy with 1 cycle latency for cache hit and 20 cycle latency for a cache miss. We have observed a 15% gain in the total number of cycles together with a gain in total memory power of about 80%. Still, Pareto curves like those shown in Fig. 8.28 can also be used in this scenario for finer tunning of the desired operation point.

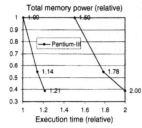
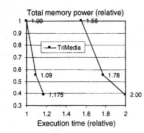
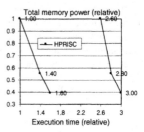

Figure 8.28.   Power and performance trade-offs for the Motion Estimation kernel for different processors and memory data-layouts. Lines at the right represent trade-offs before the address optimisation phase and at the left after it.

8.4   MEDICAL IMAGING DEMONSTRATOR: CAVITY DETECTOR

First we will introduce the cavity detector test-vehicle and we then we will explore the different DTSE steps. A more detailed tutorial on this will become available from the autumn of 2002 in a companion book to this one (also available from Kluwer).

8.4.1   Cavity detector functionality

The cavity detection algorithm extracts edges from medical images, such that they can more easily be interpreted by physicians [57]. The initial algorithm consists of four functions, each of which has an image frame as input and one as output (Fig. 8.29). The LabelRoots module is however not data-dominated and is not considered here.

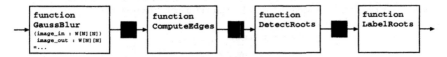

Figure 8.29.   Initial cavity detection algorithm broken up in 4 main modules organized in an image pipeline.

For each of the modules, their C definition and a brief explanation of their functionality is provided[9]:

```
void GaussBlur (uchar image_in[N][M], uchar image_out[N][M])
  { Perform horizontal horizontal and vertical gauss blurring on each pixel
    by applying a filter mask.}

void ComputeEdges (uchar image_in[N][M], uchar image_out[N][M])
  { Replace every pixel with the maximum difference with its neighbours to
    highlight the presence of edges.}

void Reverse (uchar image_in[N][M], uchar image_out[N][M])
  { Search for the maximum value that occurs in the image: maxval.
    Replace every pixel by this maximum value minus the initial value.
  }

void DetectRoots (uchar image_in[N][M], uchar image_out[N][M])
  { Reverse (image_in, image_tmp);
    image_out[x][y] is true if no neighbours are bigger than image_tmp[x][y]
  }

void main ()
  { GaussBlur(in_image, g_image);
    ComputeEdges(g_image, c_image);
    DetectRoots (c_image, out_image);
  }
```

---

[9]Note that *Reverse* is called within *DetectRoots*

When this initial algorithm is mapped to an embedded system, the system bus load and power requirements are (too) high. The main reason is that each of the functions reads an image from shared background memory, and writes the result back to this memory. After applying our DTSE methodology each module will be share the memory space and the number of accesses to the large arrays is significantly reduced.

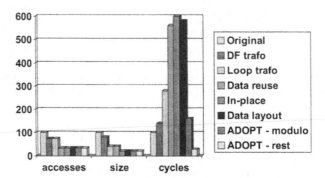

Figure 8.30.   Global results for all DTSE and address optimisation steps applied to the cavity detector, estimated on a programmable processor platform with a single on-chip and off-chip memory. The accesses and memory size are for the off-chip DRAM.

8.4.2   DTSE transformations applied to cavity detector

The effect of all the DTSE transformations is illustrated in the separate bars of the cavity detector results overview in fig. 8.30. More details will be provided in the companion book.

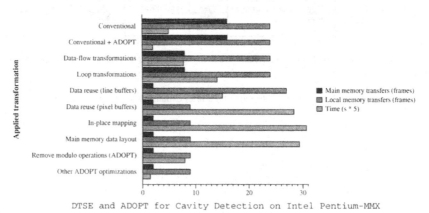

Figure 8.31.   Application of DTSE (first 6 bars after initial) and ADOPT (2 last bars) steps on the cavity detection application. Note the heavy reduction in main and local memory accesses/system bus loading and the significant improvement in pure speed.

8.4.3   Global cost-speed trade-offs

The main goal of this subsection is to illustrate how using the data layout techniques discussed above, combined with a subsequent address optimisation stage, we can perform a system-wide trade-off between speed, power and memory size in the design of the cavity detection algorithm, at the level of data layout decisions.

8.4.3.1 Impact on Cache Miss Rate. First, we will briefly discuss the impact of data layout organization on the cavity detection algorithm for different cache size and associativities. Fig. 8.32(a) shows the miss rate for the cavity detection algorithm for different cache sizes and associativities. The main conclusion regarding the miss rate is, with our technique we are able to perform as well as a 4-way associative cache.

For every cache size we are able to outperform the initial direct mapped and the initial 2-way associative case. Indeed we are able to come quite close in terms of miss rate to that of a 4-way associative cache. Here we will discuss the aspects of using this technique in the context of designing embedded systems with variable cache sizes and the corresponding trade-offs. This is becoming very important in the context of embedded processor platforms that exhibit parameterized memory organizations [238]. Moreover also memory architectures which allow a power down of non-used cache space are becoming available [221] enabling this exploration even at run time.

8.4.3.2 Trade-off between power, size and speed. We now discuss how the above technique can be applied to achieve important system-level trade-off's. Note that we have started with an initial algorithm (labeled as "initial") which was not optimized at all. Next, we have performed local loop transformations like loop fusion (labeled as "local trfd"). After this we performed global transformations, both execution as well as storage order (labeled as "global trfd"). Then we begin the step of applying data layout optimization and the exploration for cache size.

In the first step, we obtain the cache miss rates for different cache sizes using the experimental set-up as described earlier. This is shown in Fig. 8.32(a). Note the difference in miss rates for the initial algorithm and the data layout optimized algorithm. Also observe that with our technique we are able to perform as well as a 4-way associative cache. Nevertheless, a 4-way associative cache consumes much more power than a direct mapped one, hence we will use the direct mapped cache. Once we have obtained the miss rates for different cache sizes, we compute the required power for both the algorithms as shown in figure 8.32(b) and (c). We observe that for a cache size with 256 bytes the power consumption is the least. Also observe the individual power components namely the on-chip and off-chip power consumptions for the initial and data layout optimized case, which shows that we are indeed able to reduce the dominance of off-chip power consumption to a large extent. In a next step we choose the power optimal cache size of 256 bytes. We now decide on a trade-off between main memory size[10] and the reduction in miss rate as shown in Fig. 8.32(e). Thus depending on the design constraints, the designer can now either choose a lower power solution with some overhead in size and vice-versa. We have decided to choose the two versions with 2% (labeled DL1) and 2X (labeled DL2) overhead and observe the trade-offs.

In the final stage, we perform address optimizations [201], so as to remove the overhead in addressing operations introduced in the data layout optimization process. The address optimization stage is referred to as ADOPT and details of this methodology can be found in [201, 216, 372]. The results in table 8.9 show that we are not only able to remove the overhead in addressing but obtain additional performance gains on various platforms as compared to the initial algorithm. We observe that both the data layout cases after performing address optimization are performing far better than the initial and locally transformed cases. Thus the main trade-off exists between the globally transformed and the two data layout optimized versions. In the current case, we can achieve a reduction in cache related power by almost 45% (difference between 47.5% and 26.5%) by trading-off approximately 10% of the performance. Similarly we can perform a more larger trade-off with the second data layout case for highly power constrained systems.

Note that for the MIPS processor, our data layout case (with 2% overhead) with more overhead in addressing and area performs even better (in execution time) than the globally transformed case. This shows that the data layout (as well as address optimization) technique is highly platform dependent and automated tools are very much required[11], which will allow designers to trade-off and simulate for different alternatives.

---

[10]The overhead in memory space in the data layout optimization process as compared to the initial algorithm.
[11]In the above case study, we have applied the loop transformations manually, and the data layout and address optimizations steps with our prototype tools.

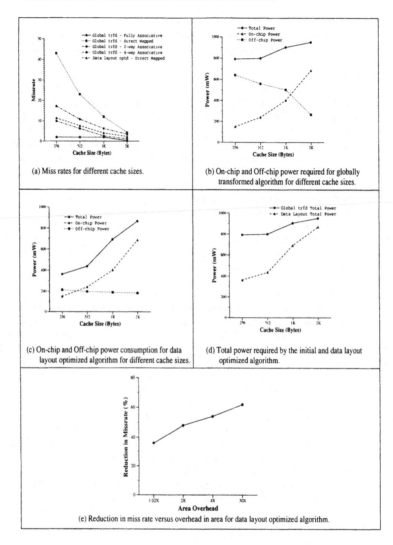

Figure 8.32. Power and Area trade-offs for the cavity detection algorithm for different cache sizes and associativities.

Note that the approach we have used in this work is mostly power centric namely we first optimize for power then size and lastly for performance. But the above technique can be used for a performance centric approach too. In that case, the designer should first choose the cache size with least number of misses and then optimize for power by trading-off performance (and/or size) with power.

8.4.4 Experimental results on several platforms

For a single-processor context, very promising results on the combination of energy reduction and performance improvement in both cycle count and system bus load have been obtained for this cavity detection application running on a Pentium P2 with MMX support. The results are illustrated in Fig. 8.31. Note the reduction of the main memory/L2 cache data transfers (expressed in average

	PA-8000	Pentium III	TriMedia TM1000	MIPS R10k	Area Overhead	Power
Initial	0.77	3.86	9.4	2.90	0	100 %
Local Trf	0.72	3.55	8.6	2.21	0	90 %
Global Trfd + Adopt	0.23	0.96	3.4	1.19	0	47.5 %
Global Trfd + DL1 + Adopt	0.26	1.04	3.7	0.97	2 %	26.5 %
Global Trfd + DL2 + Adopt	0.43	1.23	4.2	1.52	2X	24 %

Table 8.9.   Execution times (in seconds), required area and power consumption for cavity detection algorithm on different processors and the corresponding trade-offs.

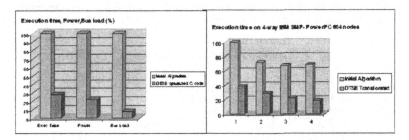

Figure 8.33.   Illustration of platform-independent execution time, power and bus load improvements for cavity detection algorithm on TriMedia TM1 (left) and 4-way parallel IBM machine with PowerPC nodes (right).

number of transfers per image pixel) with a relative factor of 7.5 after the application of the different stages in our Acropolis DTSE approach that were also described in section 1.7 (first 6 steps in bar-chart after initial). This heavily reduces both energy and system bus load. Similar reductions have been obtained for the L1 cache transfers which are extremely difficult to measure accurately. Therefore we have estimated the reduction of local memory accesses (without a hardware cache controller) and we have observed a gain with a factor 3. After the DTSE stage, the bottleneck in terms of cycle count lies in the address arithmetic. However, after a subsequent application of our ADOPT methodology [372] where the address overhead is removed (last 2 steps in bar-chart of Fig. 8.31), we notice that due to the prior removal of the data transfer bottleneck the only remaining cycles are the minimally required ones for the data-paths and local control on the Pentium MMX processor. As a result also the pure cycle count (expressed in relative units) is reduced with a factor 3 compared to the conventional approach in the end!

In addition, we have demonstrated on this cavity detection algorithm that the top-level steps of our overall approach can be seen as platform-independent (see also section 1.7). Indeed, similar improvements on performance and bus load/power figures, starting from exactly the same DTSE-optimized C source code and using the native compilers in their most optimized mode, have been obtained on widely different processor platforms (see Fig. 8.33). This is in particular so for a HP-PA8000 RISC, a TriMedia TM1 VLIW, and even on a 4-way parallel IBM machine in combination with a state-of-the-part parallelizing compiler, namely the IBM C compiler (xlc) for AIX 4.2. This compiler uses a language-independent optimizer, the Toronto Portable Optimizer (TPO), for global program optimizations and parallelization, and finally, an optimizing back-end for target-specific optimizations. TPO even includes experimental inter-procedurally analysis apart from classical data flow optimizations like constant propagation, copy propagation, dead code elimination, and loop-invariant code motion [381]. It also performs loop transformations like fusion, unimodular transformations, tiling, and distribution to improve data locality and parallelism detection [402]. Finally, TPO supports parallelization of loops based on both data dependence analysis and user directives. It supports various static and dynamic scheduling policies for nested parallelization of loops [234].

This experiment clearly shows that our high-level DTSE optimizations can be performed independent of the target architecture, and are fully complementary to even the recent parallelism-oriented compiler research (at the data or instruction-level).

This desirable platform-independence is not true any more for the lower-level DTSE stages though. Here the memory organization parameters like cache size/line size, number of ports and main memory bank organization have to be incorporated to obtain really good results. So for this purpose, we need a retargetable DTSE pre-compiler to speed up the exploration of different memory architectures and different versions of the application code, by providing fast feedback [524]. With the introduction of our compiler technology, which has been made available in a down-loadable demonstrator version from the web for educational illustration purposes (look for Acropolis project at http://www.imec.be/acropolis) at the end of 1999, the design time can indeed be heavily reduced in this stage.

Promising results have been obtained for instance on the parallel TI C80 processor, leading to a significant reduction in area and power by combining global data-flow [95, 352] and loop transformations [149] [12]:

algo	parallelism	Main mem	Frame trfs	LRAM trfs
Initial	data	2 (+1)	36	72
	task	8 (+1)	36	72
Locally	data	1 (+1)	8	44
Optimized	task	4 (+1)	8	44
Globally	data	0	0	14
optimized	task	0	0	38

It should be stressed that the new solutions still allow to achieve the required amount of speed-up by exploiting the parallel processors, but with different parallelisation choices as in the original cases.

8.5   IMAGE CODING DEMONSTRATOR: QUAD-TREE STRUCTURED DPCM ALGORITHM

For the QSDPCM application, we have obtained very promising results on a parallel IBM machine in combination with a state-of-the-part parallelizing compiler, namely the IBM TPO compiler (see also section 8.4) for AIX 4.2 [299, 286].

Most of the submodules of the QSDPCM algorithm operate in cycles with an iteration period of one frame. For the first step of the algorithm (motion estimation in the 4×4 subsampled frames) the previous frame must already have been reconstructed and 4×4 subsampled. The only submodules that are not in the critical cycle (path) are both the 4×4 and 2×2 subsampling of the current (input) frame. It is worth noting that the algorithm spans 20 pages of complex C code. Hence it is practically impossible to present all the detailed dependencies and profile of the algorithm.

Table 8.10, gives the number of memories required and the cor responding number of accesses for the QSDPCM algorithm. Note that we have two memories, one for the frame signals (206Kb) and another for the non-frame signals (282Kb for original case). A significant reduction is achieved in the total number of off-chip accesses for both the memories. Hence the total power consumption, for the global DTSE-transformed case is heavily reduced compared to the original. Also observe the reduction in the size of memory, from 282K bytes for the original algorithm to 1334 bytes for the globally transformed algorithm.

---

[12]Main memory= storage size in terms of frames of 512×512. LRAM transfers (trfs) refer to the number of transfers to local (on-chip) SRAMs per frame pixel. The extra frame transfers (+1) indicated in column 3 refer to the additional accesses when also external I/O is incorporated.

Version	Memory Size	# Accesses
Initial	206K	5020K
	282K	8610K
Globally	206K	4900K
transformed	1334	8548K

Table 8.10.   Amount of memory and the corresponding total number of accesses to each of these memories for the QSDPCM algorithm for different optimization levels. The memory size is in bytes.

Function	Original	TPO-Orig.	DTSE Opt.	TPO-DTSE Opt
QSDPCM	12.95	8.70	6.96	5.55

Table 8.11.   Performance of sequential QSDPCM algorithm for various cases using a Power PC 604e. The execution times are in seconds.

Version	Frame Size	P=1	P=2	P=3	P=4
Original	288 × 528	5.89	3.11	2.26	2.00
	576 × 1056	22.83	12.32	9.09	7.62
DTSE	288 × 528	3.48	1.75	1.18	0.98
transformed	576 × 1056	13.45	7.28	4.79	3.47

Table 8.12.   Performance of parallelized QSDPCM algorithm for various cases using a Power PC 604 based 4-way SMP machine. Here P represents the number of processors and the execution time is in seconds.

Table 8.11 gives the execution times for different optimizations for the QSDPCM algorithm on a single processor. A quite significant gain in total execution time is present for the DTSE case and a further gain due to the TPO related transformations. Similarly, table 8.12 shows that the DTSE-transformed parallelized code has clearly better execution times compared to that of the original parallelized code. Note that the execution time for the original case reduces by a factor three with four processors, whereas for the DTSE transformed case there is even a reduction in execution time by an almost ideal factor 3.9. This once again shows that DTSE is complementary and is even augmenting the effect of the performance and parallelization related optimizations.

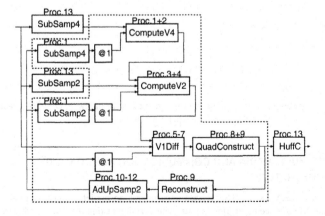

Figure 8.34.   Principal operation of Quad-tree Structured DPCM algorithm and proposed partitioning.

For the entire QSDPCM application (see Fig. 8.34), the following table gives the size of the frame memories and the number of transfers needed for the processing of 1 frame. The results are provided for different mapping alternatives related to a software/hardware partitioning experiment over several processors (13 in total) [129, 130].

Algorithm	Par.proc.mapping	Frame size (Kwords)	Transfers (K)
Initial	trivial	742	2245
Initial	optimal	325	2245
SLMM optimized	optimal	166	1238

By combining this with a good selection of hybrids between data and task-level parallelism exploitation [128, 132] and a distributed memory organisation [40, 483], we obtain the following relative area and power figures for different memory libraries (note the very large saving factors):

Version	Partitioning	Area	Power
Initial	Pure data	1	1
	Pure task	0.92	1.33
	Modified task	0.53	0.64
	Hybrid 1	0.45	0.51
	Hybrid 2	0.52	0.63
Globally transformed	Pure task	0.0041	0.0080
	Modified task	0.0022	0.0040
	Hybrid 1	0.0030	0.0050
	Hybrid 2	0.0024	0.0045
Globally transformed	Modified task	0.0022	0.0002
(distr. mem.)	Hybrid 1	0.0030	0.0003
	Hybrid 2	0.0024	0.0002

Interesting results have also been obtained by applying the cache optimisation methods of chapter 7 on the QSDPCM test-vehicle. Results for hardware caching on a single processor are available in Fig. 8.35.

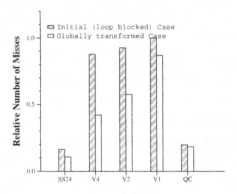

Figure 8.35. Number of misses (power and speed impact) for hardware controlled caching on QS-DPCM video coder: SLMM optimisation results for different modules.

8.6 OTHER MULTI-MEDIA AND COMMUNICATION APPLICATION DEMONSTRATORS

Since end 1994, many other DTSE related experiments on predefined programmable processors (including cores) have been performed by us, with very interesting results [88, 141, 143, 142, 149, 208, 378, 388]. Experiments have shown for instance that our methodology allows to heavily reduce the power of the dominant storage components in a GSM voice coder application (a factor 2 or more) and this without a penalty in processor instruction count. For this purpose, we have performed aggressive transformations on the data-flow and the loop structure, we have introduced a custom memory hierarchy (see Fig. 8.36) and we have performed careful in-place mapping of array signals on top of one another in the memory space.

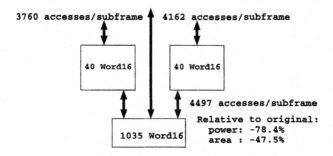

Figure 8.36.   Power optimized GSM memory architecture.

When mapping video coding test-vehicles on a programmable embedded ARM processor, at the same time, also significant power reductions can be obtained still [378, 388, 298].

Also for the well-known FFT algorithm we have shown that we can obtain additional gains in memory power and system bus access and this has resulted in a parametrizable soft IP module [68]. A similar parametrizable code has been obtained for a turbo decoder, applying the full DTSE script [467, 465, 466, 337, 338, 473]. In that case nearly a factor 20 has been gained in total chip power for a heavily customized memory organisation. The original design had more than 95% of its power located in the storage related units. For wavelet coder modules [303, 302, 462, 139] also a large factor has been gained in power for a customized memory organisation. Moreover, on programmable processors, up to a factor 9 in D-cache miss reduction (and an even higher factor for the L2-cache) is achieved compared to a conventional row-column wavelet transform [24].

Other experiments on single-processor based realisations for data-dominated submodules, oriented to instruction count improvement and power or external memory size reduction, have been performed too. These have shown that also the cycle budget on (programmable) application-specific instruction-set processors (ASIPs) is positively affected by our methodology, especially due to the reduced data transfer time to/from the storage units and the more efficient address generation (which is dealt with by our high-level address optimisation methodology and the partial support within our ADOPT environment [372, 201]). That methodology has been extended to programmable processor contexts [216], including modulo addressing reduction [201].

Since 1996 also work focused on data transfer and storage exploration in a multi-processor context has become a full-fledged activity. We have demonstrated the usefulness of this novel system-level DTSE approach based on several realistic test-vehicles, including the cavity detector for edge detection in medical image processing [128, 131, 299] (see above), the quad-tree based image coding (QSDPCM) application [129, 132, 130, 299, 352] (see above), and a demanding 3D image reconstruction algorithm [514, 513, 511]. All these experiments have confirmed that our novel techniques (almost) do not interfere with traditional parallelization techniques and are in that sense complementary to them. In order to fully exploit the potential they should however be performed upfront in a source-to-source precompilation stage.

9 CONCLUSIONS AND FUTURE WORK

Power and cost efficient design, while meeting performance constraints on a (partly) programmable platform, has become a very important issue for a broad class of embedded applications. Many of the applications for low-cost, low-power design turn out to be data-dominated. Experiments have shown that for such data-dominated applications, a very large part of the power consumption is due to the data storage and data transfer, also on programmable platforms. Also the area cost (memory footprint) is for a large part dominated by the memories. However, until recently, outside our group (together with the universities directly co-opeation with us) and U.C.Irvine, little research has been done on systematically optimizing the storage and transfer of background data. Since the late 80's our research group is convinced that the memory organisation should be optimized as a first step in the system design/compiler methodology for this type of applications. This has been formalized in the DTSE methodology which is presented here for the first time with its overall trajectory for platforms with partly predefined memory architectures.

9.1 CONTRIBUTIONS

The current starting point of this DTSE methodology for partly predefined platform architectures [96, 128], is a system specification with accesses on multi-dimensional (M-D) signals which can be statically ordered. The output is a transformed source code specification, potentially combined with a (partial) netlist of memories which is the input for the final platform architecture design/linking stage when partly customisable memory realizations are envisioned. The transformed source code is input for the software compilation stage in the case of instruction-set processors. The address code or address calculation unit (ACU) generators are produced by a separate methodology (ADOPT) [371, 372]. This DTSE methodology is partly supported with prototype tools developed in our ACROPOLIS research project. Reuse is also made of the tools in our more mature ATOMIUM research project [87, 93, 561].

The major tasks in the methodology and the main research contributions on supporting (pre)compiler techniques and tools are briefly discussed now (see also figure 1.17). More details about our past and current research (including PS files of selected papers) are available at our web site: http://www.imec.be/acropolis/

275

- The global loop and control-flow transformations (*loop trafo*) task aims at improving the data access locality for M-D signals and at removing the intermediate buffers introduced due to mismatches in production and consumption ordering. In chapter 3, we have made contributions to a global loop transformation steering methodology and automatable techniques, using a Polyhedral Dependency Graph (PDG) model. The PDG model that we employ has previously been developed at IMEC for performing global loop transformations for DTSE in the ATOMIUM context [541, 182, 93]. However, the overall transformation steering approach was not complete. In this work, we have made several contributions to this methodology:

    - We have designed more accurate cost functions for the first phase (placement phase) of the method [123].

    - We have outlined a constraint-based exploration method for this placement phase [123].

    - We have shown how a parallelization and partitioning (estimation) step can be effectively integrated into the methodology [124].

- In relation to this loop transformation step, also a method has been developed for combining parallelization and (HW/SW) partitioning with the DTSE methodology. In [128] we have first mentioned that for parallel target architectures, DTSE has to be combined with the parallelization stage to achieve power and area efficient implementations. In [129, 130] we have demonstrated a method for integrating a (HW/SW) partitioning step into the DTSE methodology. In [351, 347] we have shown how this partitioning can be performed within performance constraints. Furthermore, we have presented the trade-offs between task and data parallelism (and hybrid cases) in a DTSE context in [132]. More detailed interactions between data parallel compilation and DTSE have been presented in [299, 286].

- To provide feedback to the system designer or to the loop transformation and data reuse decision steps, *array-oriented memory size estimation* techniques are required. Lower and upper bounds for a partially defined loop organisation and order that are needed to really steer the many possible loop transformations for a given application, have been described in chapter 4 [274, 275]. They include both size estimation technique for individual dependencies [273] and one for inter-signal dependencies with overlapping life-times [276]. In addition, guidelines for loop transformation steering have been developed [272].

- In the *data reuse decision* task, we decide on the exploitation of the memory hierarchy [154, 562]. Important considerations here are the distribution of the data (copies) over the hierarchy levels as these determine the access frequency and the size of each resulting memory. Obviously, the most frequently accessed memories should be the smallest ones. We have proposed formalized techniques to steer this partitioning of the background transfers, which is driven by high-level estimates on bandwidth and sizes. An effective exploration technique and a prototype tool to automate this has been proposed in chapter 5 [512]. This systematic exploration step is not present in conventional (parallelizing) compilers.

    At this stage of the methodology, the transformed behavioural description can already be used as input for compilation to any type of processor (it is "platform-independent" code still) or for more efficient simulation, or it can be further optimized in the next platform-dependent steps.

- Several additional steps are involved to define the detailed memory organisation. Given the cycle budget, the goal is to allocate memory units and ports (including their types) from a memory library and to assign the data to the best suited memory units. In a first step, we determine the bandwidth requirements and the balancing of the available cycle budget over the different memory accesses in the algorithmic specification (*storage cycle budget distribution* or SCBD) [562, 564, 566]. We also achieve a balanced distribution of the globally available cycle budget over the loop nests such that the maximally required band-width is reduced (full storage *cycle budget distribution* step) [566, 66, 64]. This ensures that fewer multi-port memories are required to meet the timing constraints and also fewer memories overall. Very fast techniques have been developed to make this step feasible in an interactive way with different variants [394, 398] including more adaptive [395] and space-time extensions [396, 397] of the underlying access ordering technique. See chapter 6.

- The necessary number of powered-up memory/bank units and their type (e.g. single- or dual-port) are determined during *memory allocation* [35], matching the access conflicts in the balanced flow-graph. Even for fully predefined memory allocations, freedom remains also in the assignment of data to the different memory alternatives compatible with the real-time access constraints defined in the previous SCBD step, and this especially in the banks of e.g. the external SDRAM. The latter assignment has major impact on power consumption. This *signal-to-memory assignment* [483, 524] is not really incorporated in the book. Some information is presented in section 1.7. The combination of these tools allows us to derive real Pareto curves of the background memory related cost (e.g. power) versus the cycle budget [63]. See chapter 6. This systematic system-wide trade-off exploration step is not present in conventional compilers.

- Finally, the *memory data layout organisation issues* are addressed. The *in-place optimisation* step exploits the fact that for data with (partially) exclusive life-times, the memory space can be (partially) shared. This in-place mapping approach enables a better exploitation of the caches and it allows to remove nearly all capacity misses, as illustrated in our experiments (see chapter 7). We have also developed advanced main memory layout organisation techniques (which go beyond padding) and which allow to remove also most of the present conflict misses due to the limited cache associativity [285].

In addition, several other memory organization and data storage management related contributions have been made by us which are not detailed in this book. They are briefly described below (with appropriate publication pointers):

- *Set data type optimization:* In the context of ADT refinement for power, we have developed a model for set data types. Using this model and a power cost function we have developed a methodology for determining an optimized set data structure without exhaustively scanning the search space [566]. This is also combined with a *virtual management exploration* step in our MATISSE methodology and tool environment.

- *Extension to parallel processor targets:* All the above remains true also for a parallel processor context [351, 347, 348], whether it is a data-/task-parallel hybrid combination [132] or a heterogeneous parallel processor [130]. It is however possible to break up the DTSE problems in several separate stages that are dealt with sequentially in the complete system design trajectory. In that case they are separated by concurrency or parallelisation management stages and they are, linked through constraint propagation. This is supported by unified system design flow work that is partly described for data-dominated multi-media and telecom applications in [81, 83].

  For complex concurrent applications involving multiple threads of control and a large amount of concurrent processes, we are splitting up the DTSE stage in a two-layer approach, where the problem is partly solved globally first at the more abstract task level, and then on a processor-scope level in a second more detailed stage. Experiments on e.g. an MPEG-4 demonstrator application have shown very promising results [94, 70]. This also allows to couple the DTSE approach to a system-level abstraction stage dealing with a set of dynamic concurrent "grey-box" tasks [500]. That flow includes a data management stage at the concurrent task level [342, 82] where both DTSE and dynamic memory oriented transformations [343] are applied.

  For the (regular) data-level parallelism stage it has been shown also that the methodology has a strong enabling effect when integrated in more conventional array processor mapping methods [461, 460]. This includes the data management related to the (un)packing of the subwords for effectively exploiting the sub-word parallel instructions in modern multimedia processors [355]. So also for this purpose data-level DTSE is a crucial preprocessing step.

- *Address optimization:* This stage is not part of the DTSE methodology itself, but is incorporated into another methodology developed at IMEC, namely ADOPT (ADdress OPTimization) [372].

The DTSE steps described above will significantly increase the addressing complexity of the applications, e.g. by introducing more complex loop bounds, index expressions, conditions, ... When this computational overhead is neglected, the optimized system will, although being more power efficient, suffer a severe performance degradation.

Most of the additional complexity introduced by DTSE, can however be removed again by source code optimizations such as constant propagation, code hoisting, modulo addressing reduction, strength reduction and others [216, 201]. Moreover, the fact that it originates from DTSE optimizations makes it easily recognizable for the ADOPT tool. For custom processors, address generation hardware can be synthesized [371]. The final result is that for most applications not only the power consumption is reduced, but also the performance is increased.

- *Application of DTSE to many real-life driver and demonstrators:* Very promising results on real-life multi-media and telecom network applications have been obtained for all of the DTSE tasks, as indicated in chapter 8.

9.2   FUTURE WORK

It is clear from the above that much progress has been made in the past decade but many relevant research issues still need to be solved in order to obtain a full design support for the complex DTSE task. Some of the most relevant objectives are:

- a detailed methodology for system-level data-flow transformations, which is the least explored step of the entire DTSE script.

- a further refined methodology and especially design tool support for preprocessing/pruning, the global loop and reindexing transformations, and for the data-reuse decision steps.

- even better high-level estimators of the storage related power and area costs in early steps of the methodology.

- a further refined methodology and especially mature design tool support for SDRAM bank assignment and SDRAM mode selection, and for cache locking and capacity miss removal steps.

- further extensions of the concepts and methodologies are required to go to parallel target architectures, both at the data- and task-level.

- extensions towards other application domains (see e.g. [566] for network component applications).

- support for system-level design validation using formal methods to void as much as possible the very costly resimulation after the complex code transformations (see e.g. [457, 119, 120]).

References

[1] W.Abu-Sufah, D.Kuck, D.Lawrie, "On the performance enhancement of paging systems through program analysis and transformations", *IEEE Trans. on Computers*, Vol.C-30, No.5, pp.341-355, May 1981.

[2] Multi-media compilation project (Acropolis) at IMEC
http://www.imec.be/acropolis/

[3] OAT/SEMP club (ed. F.Catthoor), "The ADOPT address equation optimisation and generation environment", IMEC, Internal Report, April 1996.

[4] S.Adve, D.Burger, R.Eigenmann, A.Rawsthorne, M.Smith, C.Gebotys, M.Kandemir, D.Lilja, A.Choudhary, J.Fang, P-C.Yew, "The interaction of architecture and compilation technology for high-performance processor design", *IEEE Computer Magazine*, Vol.30, No.12, pp.51-58, Dec. 1997.

[5] A.Agarwal, D.Krantz, V.Nataranjan, "Automatic partitioning of parallel loops and data arrays for distributed shared-memory multiprocessors", *IEEE Trans. on Parallel and Distributed Systems*, Vol.6, No.9, pp.943-962, Sep. 1995.

[6] R.Agrawal, H.Jagadish, "Partitioning techniques for large-grained parallelism", *IEEE Trans. on Computers*, Vol.37, No.12, pp.1627-1634, Dec. 1988.

[7] I.Ahmad, C.Y.R.Chen, "Post-processor for data path synthesis using multiport memories", *Proc. IEEE Intnl. Conf. Comp. Aided Design*, Santa Clara CA, pp.276-279, Nov. 1991.

[8] A.V.Aho, R.Sethi, J.D.Ulmann, "Compilers: principles, techniques and tools", Addison-Wesley, 1986.

[9] A.V.Aho, R.Sethi, J.D.Ullman, "Compilers : Principles, techniques and tools", *Addison Wesley Publishing Co.*, 1993.

[10] G.Albera, I.Bahar, "Power/performance advantages of victum buffer in high-performance processors", *IEEE Alessandro Volta Memorial Intnl. Wsh. on Low Power Design (VOLTA)*, Como, Italy, pp.43-51, March 1999.

[11] D.Alekseevskij, E.Vinberg, A.Solodovnikov, "Geometry of spaces of constant curvature", *Encyclopaedia of Mathematical Sciences*, Vol.29 on "Geometry II" (ed. E.Vinberg), pp.1-138, Springer-Verlag, 1993.

[12] R.Allen, K.Kennedy, "Automatic translation of Fortran programs to vector form", *ACM Trans. on Programming Languages and Systems*, Vol.9, No.4, pp.491-542, Oct. 1987.

[13] R.Allen, K.Kennedy, "Vector register allocation," *IEEE Trans. on Computers*, Vol.41, No.10, pp.1290-1316, Oct. 1992.

[14] J.Allen, D.Schimmel, "Issues in the design of high-performance SIMD architectures", *IEEE Trans. on Parallel and Distributed Systems*, Vol.7, No.8, pp.818-829, Aug. 1996.

[15] M.Al-Mouhamed, "Analysis of macro-dataflow dynamic scheduling on nonuniform memory access architectures", *IEEE Trans. on Parallel and Distributed Systems*, Vol.4, No.8, pp.875-888, Aug. 1993.

[16] M.Al-Mouhamed, S.Seiden, "A Heuristic Storage for Minimizing Access Time of Arbitrary Data Paterns", *IEEE Trans. on Parallel and Distributed Systems*, Vol.8, No.4, pp.441-447, Apr. 1997.

[17] E.Altman, G.Gao, "Optimal software pipelining through enumeration of schedules", *Proc. EuroPar Conf.*, Lyon, France, Aug. 1996. "Lecture notes in computer science" series, Springer Verlag, pp.833-840, 1996.

[18] S.Amarasinghe, J.Anderson, M.Lam, and C.Tseng, "The SUIF compiler for scalable parallel machines", in *Proc. of the 7th SIAM Conf. on Parallel Proc. for Scientific Computing*, 1995.

[19] C.Amza, A.Cox, S.Dwarkadas, P.Keleher, H.Lu, R.Rajmony, W.Yu, W.Zwaenepoel, "Treadmarks: shared memory computing on network of workstations", *IEEE Computer Magazine*, Vol.29, No.2, pp.18-28, Feb. 1996.

[20] C.Ancourt, F.Coelho, F.Irigoin and R.Keryell, "A linear algebra framework for static HPF code distribution", *Proc. Wsh. on Compilers for parallel computers*, Delft, Dec. 1993.

[21] C.Ancourt, F.Irigoin and Y.Yang, "Minimal data dependence abstractions for loop transformations", *Intnl. J. of Parallel Programming*, Vol.23, No.4, pp.359-388, 1995.

[22] C.Ancourt, D.Barthou, C.Guettier, F.Irigoin, B.Jeannet, J.Jourdan, J.Mattioli, "Automatic data mapping of signal processing applications", *Proc. Intnl. Conf. on Applic.-Spec. Array Processors*, Zurich, Switzerland, pp.350-362, July 1997.

[23] J.Anderson, S.Amarasinghe, M.Lam, "Data and computation transformations for multiprocessors", in *5th ACM SIGPLAN Symp. on Principles and Practice of Parallel Programming*, pp.39-50, Aug. 1995.

[24] Y.Andreopoulos, P.Schelkens, G.Lafruit, K.Masselos, J.Cornelis, "Analysis of wavelet transform implementations for image and texture coding applications in programmable platforms", *Proc. IEEE Wsh. on Signal Processing Systems (SIPS)*, Antwerp, Belgium, IEEE Press, Sep. 2001.

[25] A.Antola, A.Avai, L.Breveglieri, "Modular design methodologies for image processing architectures", *IEEE Trans. on VLSI Systems*, Vol.1, No.4, pp.408-414, Dec. 1993.

[26] G.Araujo, S.Malik, M.T.Lee, "Using register-transfer paths in code generation for heterogeneous memory-register architectures", *Proc. 33rd ACM/IEEE Design Automation Conf.*, Las Vegas NV, pp.597-600, June 1996.

[27] J.Armer, J-M.Bard, B.Canfield, D.Charlot, S.Freeman, A.Graf, R.Kessler, G.Lamouroux, W.Mayweather, M.Patti, P.Paul, A.Pirson, F.Rominger, D.Teichner, "A chip set for MPEG-2 video encoding", *Proc. IEEE Custom Integrated Circuits Conf.*, Santa Clara CA, pp.401-404, May 1995.

[28] O.Arregi, C.Rodriguez, A.Ibarra, "Evaluation of the optimal strategy for managing the register file", *Microprocessing and microprogramming*, No.30, pp.143-150, 1990.

[29] The Atomium club, "Information models for the new ATOMIUM basic software kernels", IMEC Internal Report (restricted document), May 1998.

[30] Atomium web site http://www.imec.be/atomium/.

[31] D.Bailey, "RISC Microprocessors and scientific computing", Technical Report, RNR-93-004, NASA Ames Research Center, March 1993.

[32] S.Bakshi and D.Gajski, "A memory selection algorithm for high-performance pipelines", *Proc. 6th ACM/IEEE Europ. Design and Test Conf.*, Brighton, U.K. pp.124-129, Feb. 1995.

[33] M.Balakrishnan, A.K.Majumdar, D.K.Banerji, J.G.Linders, J.C.Majithia, "Allocation of multiport memories in data path synthesis" *IEEE Trans. on Comp.-aided Design*, Vol.CAD-7, No.4, pp.536-540, April 1988.

[34] F.Balasa, F.Catthoor, H.De Man, "Exact Evaluation of Memory Area for Multi-dimensional Processing Systems", *Proc. IEEE Intnl. Conf. Comp. Aided Design*, Santa Clara CA, pp.669-672, Nov. 1993.

[35] F.Balasa, F.Catthoor, H.De Man, "Dataflow-driven Memory Allocation for Multi-dimensional Signal Processing Systems", *Proc. IEEE Intnl. Conf. Comp. Aided Design*, San Jose CA, pp.31-34 Nov. 1994.

[36] F.Balasa, F.Franssen, F.Catthoor, H.De Man, "Transformation of Nested Loops with Modulo Indexing to Affine Recurrences", in special issue of *Parallel Processing Letters* on "Parallelization techniques for uniform algorithms", C.Lengauer, P.Quinton, Y.Robert, L.Thiele (eds.), World Scientific Pub., pp.271-280, 1994.

[37] K.Balmer, N.Ing-Simmons, P.Moyse, I.Robertson, J.Keay, M.Hammes, E.Oakland, R.Simpson, G.Barr, D.Roskell, "A single chip multi-media video processor" *Proc. IEEE Custom Integrated Circuits Conf.*, San Diego CA, pp.91-94, May 1994.

[38] F.Balasa, "Background memory allocation for multi-dimensional signal processing", *Doctoral dissertation*, ESAT/EE Dept., K.U.Leuven, Belgium, Nov. 1995.

[39] F.Balasa, F.Catthoor, H.De Man, "Background Memory Area Estimation for Multi-dimensional Signal Processing Systems", *IEEE Trans. on VLSI Systems*, Vol.3, No.2, pp.157-172, June 1995.

[40] F.Balasa, F.Catthoor, H.De Man, "Practical Solutions for Counting Scalars and Dependences in ATOMIUM – a Memory Management System for Multi-dimensional Signal Processing", *IEEE Trans. on Comp.-aided Design*, Vol.CAD-16, No.2, pp.133-145, Feb. 1997.

[41] U.Banerjee, "An introduction to a formal theory of dependency analysis", *J. of Supercomputing*, Vol.2, No.2, pp.133-149, Oct. 1988.

[42] U.Banerjee, "Loop Transformations for Restructuring Compilers: the Foundations", Kluwer, Boston, 1993.

[43] U.Banerjee, R.Eigenmann, A.Nicolau, D.Padua, "Automatic program parallelisation", *Proc. of the IEEE*, invited paper, Vol.81, No.2, pp.211-243, Feb. 1993.

[44] U.Banerjee, "Loop Parallelization", Kluwer, Boston, 1994.

[45] P.Banerjee, J.Chandy, M.Gupta, E.Hodges, J.Holm, A.Lain, D.Palermo, S.Ramaswamy, E.Su, "The Paradigm compiler for distributed-memory multicomputers", *IEEE Computer Magazine*, Vol.28, No.10, pp.37-47, Oct. 1995.

[46] S.Bartolini, C.A.Prete, "A software strategy to improve cache performance", *IEEE TC on Computer Architecture Newsletter*, pp.35-40, Jan. 2001.

[47] M.Barreteau, P.Feautrier, "Efficient mapping of interdependent scans", *Proc. EuroPar Conf.*, Lyon, France, Aug. 1996. "Lecture notes in computer science" series, Springer Verlag, pp.463-466, 1996.

[48] S.J.Beaty, "List Scheduling: Alone, with Foresight, and with Lookahead", *Conf. on Massively Parallel Computing Systems*, May 1994.

[49] L.A.Belady, "A study of replacement algorithms for a virtual-storage computer", *IBM Systems J.*, Vol.5, No.6, pp.78-101, 1966.

[50] N.Bellas, I.Hajj, C.Polychronopoulos, G.Stamoulis, "Architectural and compiler support for energy reduction in the memory hierarchy of high-performance microprocessors", *Proc. IEEE Intnl. Symp. on Low Power Design*, Monterey CA, pp.70-75, Aug. 1998.

[51] L.Benini, A.Macii, M.Poncino, "A recursive algorithm for low-power memory partitioning", *Proc. IEEE Intnl. Symp. on Low Power Design*, Rapallo, Italy, pp.78-83, Aug. 2000.

[52] L.Benini, G.De Micheli, "System-level power optimization techniques and tools", *ACM Trans. on Design Automation for Embedded Systems (TODAES)*, Vol.5, No.2, pp.115-192, April 2000.

[53] O.Bentz, J.Rabaey, D.Lidsky, "A Dynamic Design Estimation And Exploration Environment", *Proc. 34th ACM/IEEE Design Automation Conf.*, pp.190-195, June 1997.

[54] D.Bernstein, M.Rodeh, I.Gertner, "On the complexity of scheduling problems for parallel/pipelined machines", *IEEE Trans. on Computers*, Vol.C-38, No.9, pp.1308-1313, Sep. 1989.

[55] S.Bhattacharyya, P.Murthy, E.Lee, "Optimal parenthesization of lexical orderings for DSP block diagrams", *IEEE Wsh. on VLSI signal processing*, Osaka, Japan, Oct. 1995. Also in *VLSI Signal Processing VIII*, I.Kuroda, T.Nishitani, (eds.), IEEE Press, New York, pp.177-186, 1995.

[56] V.Bhaskaran and K.Konstantinides, "Image and Video Compression Standards : Algorithms and Architectures", Kluwer Acad. Publ., Boston, 1995.

[57] M.Bister, Y.Taeymans, J.Cornelis, "Automatic Segmentation of Cardiac MR Images", *Computers in Cardiology*, IEEE Computer Society Press, pp.215-218, 1989.

[58] W.Blume, R.Eigenmann, J.Hoeflinger, D.Padua, P.Petersen, L.Rauchwerger, P.Tu, "Automatic detection of parallelism", *IEEE Parallel and Distributed Technology*, Vol.2, No.3, pp.37-47, Fall 1994.

[59] F.Bodin, W.Jalby, C.Eisenbeis, D.Windheiser, "A quantitative algorithm for data locality optimization", *Proc. Intnl. Wsh. on Code Generation*, pp.119-145, 1991.

[60] F.Bodin, W.Jalby, D.Windheiser, C.Eisenbeis, "A quantitative algorithm for data locality optimization", Technical Report IRISA/INRIA, Rennes, France, 1992.

[61] J.Bormans, K.Denolf, S.Wuytack, L.Nachtergaele and I.Bolsens, "Integrating System-Level Low Power Methodologies into a Real-Life Design Flow", *Proc. IEEE Wsh. on Power and Timing Modeling, Optimization and Simulation (PATMOS)*, Kos, Greece, pp.19-28, Oct. 1999.

[62] D.Bradlee, S.Eggers, R.Henry, "Integrated register allocation and instruction scheduling for RISCs", *Proc. 4th Intnl. Conf. on Architectural Support for Prog. Lang. and Operating Systems (ASPLOS)*, Santa Clara CA, pp.122-131, April 1991.

[63] E.Brockmeyer, A.Vandecappelle, F.Catthoor, "Systematic Cycle budget versus System Power Trade-off: a New Perspective on System Exploration of Real-time Data-dominated Applications", *Proc. IEEE Intnl. Symp. on Low Power Design*, Rapallo, Italy, pp.137-142, Aug. 2000.

[64] E.Brockmeyer, S.Wuytack, A.Vandecappelle, F.Catthoor, "Low power storage cycle budget tool support for hierarchical graphs", *Proc. 13th ACM/IEEE Intnl. Symp. on System-Level Synthesis* (ISSS), Madrid, Spain, pp.20-22, Sep. 2000.

[65] E.Brockmeyer, C.Ghez, W.Baetens, F.Catthoor, "Low-power processor-level data transfer and storage exploration", chapter in "Unified low-power design flow for data-dominated multi-media and telecom applications", F.Catthoor (ed.), Kluwer Acad. Publ., Boston, pp.25-63, 2000.

[66] E.Brockmeyer, S.Wuytack, A.Vandecappelle, F.Catthoor, "Low power storage for hierarchical graphs", *Proc. 3rd ACM/IEEE Design and Test in Europe Conf.*, Paris, France, User-Forum pp.249-254, April 2000.

[67] E.Brockmeyer, F.Catthoor, J.Bormans, H.De Man, "Code transformations for reduced data transfer and storage in low power realisation of MPEG-4 full-pel motion estimation", *Proc. IEEE Intnl. Conf. on Image Proc.*, Chicago, IL, pp.III.985-989, Oct. 1998.

[68] E.Brockmeyer, J.D'Eer, F.Catthoor, H.De Man, "Parametrizable behavioral IP module for a data-localized low-power FFT", *Proc. IEEE Wsh. on Signal Processing Systems (SIPS)*, Taipeh, Taiwan, IEEE Press, pp.635-644, Oct. 1999.

[69] E.Brockmeyer, J.D'Eer, F.Catthoor, N.Busa, P.Lippens, J.Huisken, "Code transformations for reduced data transfer and storage in low power realization of DAB synchro core", *Proc. IEEE Wsh. on Power and Timing Modeling, Optimization and Simulation (PATMOS)*, Kos, Greece, pp.51-60, Oct. 1999.

[70] E.Brockmeyer, L.Nachtergaele, F.Catthoor, J.Bormans, H.De Man, "Low power memory storage and transfer organization for the MPEG-4 full pel motion estimation on a multi media processor", *IEEE Trans. on Multi-Media*, Vol.1, No.2, pp.202-216, June 1999.

[71] M.Bromley, S.Heller, T.McNerney, G.Steele, "Fortran at then gigaflops: the Connection Machine convolution compiler", *Proc. ACM Intnl. Conf. on Programming Language Design and Implementation*, pp.145-156, June 1991.

[72] J.Bu, E.Deprettere, P.Dewilde, "A design methodology for fixed-size systolic arrays", in *Application-specific array processors* (S.Y.Kung, E.Swartzlander, J.Fortes eds.), IEEE computer society Press, pp.591-602, 1990.

[73] D.C.Burger, J.R.Goodman and A.Kagi, "The declining effectiveness of dynamic caching for general purpose multiprocessor", *Technical Report*, University of Wisconsin, Madison, 1261,1995.

[74] D.Burger, T.Austin, "The Simplescalar Toolset", version 2.0, online document available at http://ww.cs.wisc.edu/ mscalar/simplescalar.html.

[75] G.Cabillic, I.Puaut, "Dealing with heterogeneity in Stardust: an environment for parallel programming on networks of heterogeneous workstations", *Proc. EuroPar Conf.*, Lyon, France, Aug. 1996. "Lecture notes in computer science" series, Springer Verlag, pp.114-119, 1996.

[76] ..., "Cadence Signal Processing and Design Verification – SPW Home Page", *http://www.cadence.com/eda_solutions/sld_spdv_l3_index.html*

[77] D.Callahan, S.Carr, K.Kennedy, "Improving register allocation for subscripted variables", *Proc. of Programming Languages, Design and Implementation (PLDI)*, pp.53-65, 1990.

[78] S.Carr, K.McKinley, C.Tseng, "Compiler optimizations for improving data locality", *Proc. 6th ACM Intnl. Conf. on Architectural Support for Programming Languages and Operating Systems*, pp.252-262, Aug. 1994.

[79] F.Catthoor, N.Dutt, C.Kozyrakis, "How to solve the current memory access and data transfer bottlenecks: at the processor architecture or at the compiler level?", Hot topic session, *Proc. 3rd ACM/IEEE Design and Test in Europe Conf.*, Paris, France, pp.426-433, March 2000.

[80] F.Catthoor, "Multiple to single assignment convertion: some ideas on a systematic approach", IMEC, Internal Report, Feb. 2000.

[81] F.Catthoor (ed.), "Unified low-power design flow for data-dominated multi-media and telecom applications", ISBN 0-7923-7947-0, Kluwer Acad. Publ., Boston, 2000.

[82] F.Catthoor, "Task concurrency management and data transfer-storage exploration for embedded dynamic IT systems", contribution to panel on "System Level Design: Solid Advances or Survival Strategies?", at *Proc. 13th ACM/IEEE Intnl. Symp. on System-Level Synthesis*, Madrid, Spain, pp.VI, Sep. 2000.

[83] F.Catthoor, E.Brockmeyer, "Unified meta-flow summary for low-power data-dominated applications", chapter in "Unified low-power design flow for data-dominated multi-media and telecom applications", F.Catthoor (ed.), Kluwer Acad. Publ., Boston, pp.7-23, 2000.

[84] F.Catthoor, E.Brockmeyer, K.Danckaert, C.Kulkarni, L.Nachtergaele, A.Vandecappelle, "Custom memory organisation and data transfer: architecture issues and exploration methods", chapter in "VLSI section of Electrical and Electronics Engineering Handbook" (ed. M.Bayoumi), Academic Press, 2001.

[85] F.Catthoor, K.Danckaert, S.Wuytack, N.Dutt, "Code transformations for data transfer and storage exploration preprocessing in multimedia processors", *IEEE Design and Test of Computers*, Vol.18, No.2, pp.70-82, June 2001.

[86] F.Catthoor, K.Danckaert, C.Kulkarni, T.Omnes, "Data transfer and storage architecture issues and exploration in multimedia processors", book chapter in "Programmable Digital Signal Processors: Architecture, Programming, and Applications" (ed. Y.H.Yu), Marcel Dekker, Inc., New York, 2001.

[87] F.Catthoor, F.Franssen, S.Wuytack, L.Nachtergaele, H.De Man, "Global communication and memory optimizing transformations for low power signal processing systems", *IEEE Wsh. on VLSI signal processing*, La Jolla CA, Oct. 1994. Also in *VLSI Signal Processing VII*, J.Rabaey, P.Chau, J.Eldon (eds.), IEEE Press, New York, pp.178-187, 1994.

[88] F.Catthoor, M.Moonen (eds.), "Parallel Programmable Architectures and Compilation for Multi-dimensional Processing", special issue in *Microprocessors and Microprogramming*, Elsevier, Amsterdam, pp.333-439, Oct. 1995.

[89] F.Catthoor, M.Janssen, L.Nachtergaele, H.De Man, "System-level data-flow transformations for power reduction in image and video processing", *Proc. Intnl. Conf. on Electronic Circuits and Systems* (ICECS), Rhodos, Greece, pp.1025-1028, Oct. 1996.

[90] F.Catthoor, S.Wuytack, E.De Greef, F.Franssen, L.Nachtergaele, H.De Man, "System-level transformations for low power data transfer and storage", in paper collection on "Low power CMOS design" (eds. A.Chandrakasan, R.Brodersen), IEEE Press, pp.609-618, 1998.

[91] F.Catthoor, "Energy-delay efficient data storage and transfer architectures: circuit technology versus design methodology solutions", invited paper, *Proc. 1st ACM/IEEE Design and Test in Europe Conf.*, Paris, France, pp.709-714, Feb. 1998.

[92] F.Catthoor, "Power-efficient data storage and transfer methodologies: current solutions and remaining problems", invited paper, *Proc. IEEE CS Annual Wsh. on VLSI*, Orlando, FL, pp.53-59, April 1998.

[93] F.Catthoor, S.Wuytack, E.De Greef, F.Balasa, L.Nachtergaele, A.Vandecappelle, "Custom Memory Management Methodology – Exploration of Memory Organisation for Embedded Multimedia System Design", ISBN 0-7923-8288-9, Kluwer Acad. Publ., Boston, 1998.

[94] F.Catthoor, D.Verkest, E.Brockmeyer, "Proposal for unified system design meta flow in task-level and instruction-level design technology research for multi-media applications", *Proc. 11th ACM/IEEE Intnl. Symp. on System-Level Synthesis* (ISSS), Hsinchu, Taiwan, pp.89-95, Dec. 1998.

[95] F.Catthoor, M.Janssen, L.Nachtergaele, H.De Man, "System-level data-flow transformation exploration and power-area trade-offs demonstrated on video codecs", special issue on "Systematic trade-off analysis in signal processing systems design" (eds. M.Ibrahim, W.Wolf) in *J. of VLSI Signal Processing*, Vol.18, No.1, Kluwer, Boston, pp.39-50, 1998.

[96] F.Catthoor, S.Wuytack, E.De Greef, F.Balasa, P.Slock, "System exploration for custom low power data storage and transfer", chapter in "Digital Signal Processing for Multimedia Systems" (eds. K.Parhi, T.Nishitani), Marcel Dekker, Inc., pp.773-814, New York, 1999.

[97] F.Catthoor, "Energy-delay efficient data storage and transfer architectures and methodologies: current solutions and remaining problems", special issue on "IEEE CS Annual Wsh. on VLSI" (eds. A.Smailagic, R.Brodersen, H.De Man) in *J. of VLSI Signal Processing*, Vol.21, No.3, Kluwer, Boston, pp.219-232, July 1999.

[98] F.Catthoor, "Data transfer and storage optimizing code transformations for embedded multi-media applications", in tutorial on "Embedded memories in system design - from technology to systems architecture", *36th ACM/IEEE Design Automation Conf.*, New Orleans LA, June 1999.

[99] G.Chaitin, M.Auslander, A.Chandra, J.Cocke, M.Hopkins, P.Markstein, "Register allocation and spilling via graph coloring", *Computer languages*, No.6, pp.47-57, 1981.

[100] A.Chandrakasan, S.Scheng, R.W.Brodersen, "Low power CMOS digital design", *IEEE J. of Solid-state Circ.*, Vol.SC-27, No.4, pp.473-483, April 1992.

[101] A.Chandrakasan, V.Gutnik, T.Xanthopoulos, "Data driven signal processing: an approach for energy efficient computing", *Proc. IEEE Intnl. Symp. on Low Power Design*, Monterey CA, pp.347-352, Aug. 1996.

[102] B.Chapman, P.Mehrotra, H.Zima, "Programming in Vienna Fortran", *Scientific Programming*, Vol.1, pp.31-50, 1992.

[103] B.Chapman, H.Zima, P.Mehrotra, "Extending HPF for advanced data-parallel applications", *IEEE Parallel and Distributed Technology*, Vol.2, No.3, pp.59-71, Fall 1994.

[104] S.Chen, A.Postula, "Synthesis of custom interleaved memory systems", *IEEE Trans. on VLSI Systems*, Vol.8, No.1, pp.74-83, Feb. 2000.

[105] Y-Y.Chen, Y-C.Hsu, C-T.King, "MULTIPAR: behavioral partition for synthesizing multiprocessor architectures", *IEEE Trans. on VLSI Systems*, Vol.2, No.1, pp.21–32, March 1994.

[106] T-S.Chen, J-P.Sheu, "Communication-free data allocation techniques for parallelizing compilers on multicomputers", *IEEE Trans. on Parallel and Distributed Systems*, Vol.5, No.9, pp.924-938, Sep. 1994.

[107] F.Chen, S.Tongsima, E.H.Sha, "Loop scheduling algorithm for timing and memory operation minimization with register constraint", *Proc. IEEE Wsh. on Signal Processing Systems (SIPS)*, Boston MA, IEEE Press, pp.579-588, Oct. 1998.

[108] F.Chen, E.H.Sha, "Loop scheduling and partitions for hiding memory latencies", *Proc. 12th ACM/IEEE Intnl. Symp. on System-Level Synthesis* (ISSS), San Jose CA, pp.64-70, Dec. 1999.

[109] M.J.Chen and E.A.Lee, "Design and implementation of a multidimensional synchronous dataflow environment", *28th Asilomar Conf. on Signals, Systems and Computers*, Vol.1, pp.519-524, Oct.-Nov. 1994.

[110] T.Chilimbi, M.Hill, J.Larus, "Making pointer-based data structures cache conscious", *IEEE Computer Magazine*, Vol.33, No.12, pp.67-74, Dec. 2000.

[111] W.Chin, J.Darlington, Y.Guo, "Parallelizing conditional recurrences", *Proc. EuroPar Conf.*, Lyon, France, Aug. 1996. "Lecture notes in computer science" series, Springer Verlag, pp.579-586, 1996.

[112] M.Cierniak, W.Li, "Unifying Data and Control Transformations for Distributed Shared-Memory Machines", *Proc. of the SIGPLAN'95 Conf. on Programming Language Design and Implementation*, La Jolla, pp.205-217, Feb. 1995.

[113] A.Cohen, V.Lefebre, "Storage mapping optimization for parallel programs", *Proc. EuroPar Conf.*, Toulouse, France, pp.375-382, Sep. 1999.

[114] J.F.Collard, "Space-time transformations of while-loops using speculative execution", *Proc. IEEE Scalable High-Performance Computing Conf.*, SHPCC'94, Knoxville TN, pp.429-436, May 1994.

[115] J.F.Collard, D.Barthou, P.Feautrier, "Fuzzy data flow analysis", *Proc. ACM Principles and Practice of Parallel Programming, PPoPP'95*, Santa Barbara CA, July 1995.

[116] C.Cook, C.Pancake, R.Walpole, 'Are expectations for parallelism too high? A survey of potential parallel users", *Proc. Supercomputing*, Washington DC, Nov. 1994.

[117] J.Covino et al., "A 2ns zero-wait state 32kB semi-associate L1 cache", *Proc. IEEE Intnl. Solid-State Circ. Conf.*, San Francisco CA, pp.154-155, Feb. 1996.

[118] B.Creusillet, F.Irigoin, "Interprocedural array region analysis", Technical report CRI, Ecole des Mines de Paris, 1995.

[119] M.Cupak, F.Catthoor, "Efficient functional validation of system-level loop transformations for multi-media applications", *Proc. Electronic Circuits and Systems Conf.*, Bratislava, Slovakia, pp.39-43, Sep. 1997.

[120] M.Cupak, F.Catthoor, H.De Man, "Verification of loop transformations for complex data dominated applications", *Proc. Intnl. High Level Design Validation and Test Wsh.*, La Jolla, CA, pp.72-79, Nov. 1998.

[121] M.Cupak, C.Kulkarni, F.Catthoor, H.De Man, "Functional Validation of System-level Loop Transformations for Power Efficient Caching" *Proc. Wsh. on System Design Automation*, Dresden, Germany, March, 1998.

[122] R.Cypher, J.Sanz, "SIMD architectures and algorithms for image processing and computer vision", *IEEE Trans. on Acoustics, Speech and Signal Processing*, Vol.ASSP-37, No.12, pp.2158-2174, Dec. 1989.

[123] K.Danckaert, F.Catthoor, H.De Man, "A preprocessing step for global loop transformations for data transfer and storage optimization", *Proc. Intnl. Conf. on Compilers, Arch. and Synth. for Emb. Sys.*, San Jose CA, pp.34-40, Nov. 2000.

[124] K.Danckaert, F.Catthoor, H.De Man, "A loop transformation approach for combined parallelization and data transfer and storage optimization", *Proc. ACM Conf. on Par. and Dist. Proc. Techniques and Applications*, PDPTA'00, pp.2591-2597, Las Vegas NV, June 2000.

[125] K.Danckaert, C.Kulkarni, F.Catthoor, H.De Man, V.Tiwari, "A systematic approach to reduce the system bus load and power in multi-media algorithms", accepted for Special Issue on "Low Power System Design" of *VLSI Design J.*, Gordon and Beach Sc. Publ., 2001.

[126] K.Danckaert, C.Kulkarni, F.Catthoor, H.De Man, V.Tiwari, "A systematic approach for system bus load reduction applied to medical imaging", *Proc. IEEE Intnl. Conf. on VLSI Design*, Bangalore, India, pp.48-53, Jan. 2001.

[127] K.Danckaert, "Loop transformations for data transfer and storage reduction on multiprocessor systems", *Doctoral dissertation*, ESAT/EE Dept., K.U.Leuven, Belgium, May. 2001.

[128] K.Danckaert, F.Catthoor, H.De Man, "System-level memory management for weakly parallel image processing", *Proc. EuroPar Conf.*, Lyon, France, Aug. 1996. "Lecture notes in computer science" series, Vol.1124, Springer Verlag, pp.217-225, 1996.

[129] K.Danckaert, F.Catthoor, H.De Man, "System level memory optimization for hardware-software co-design", *Proc. IEEE Intnl. Wsh. on Hardware/Software Co-design*, Braunschweig, Germany, pp.55-59, March 1997.

[130] K.Danckaert, K.Masselos, F.Catthoor, H.De Man, C.Goutis, "Strategy for power efficient design of parallel systems", *IEEE Trans. on VLSI Systems*, Vol.7, No.2, pp.258-265, June 1999.

[131] K.Danckaert, F.Catthoor, H.De Man, "Platform independent data transfer and storage exploration illustrated on a parallel cavity detection algorithm", *Proc. ACM Conf. on Par. and Dist. Proc. Techniques and Applications*, PDPTA'99, Vol.III, pp.1669-1675, Las Vegas NV, June 1999.

[132] K.Danckaert, K.Masselos, F.Catthoor, H.De Man, "Strategy for power efficient combined task and data parallelism exploration illustrated on a QSDPCM video codec", *J. of Systems Architecture*, Elsevier, Amsterdam, Vol.45, No.10, pp.791-808, 1999.

[133] A.Darte, Y.Robert, "Scheduling Uniform Loop Nests", Internal report 92-10, ENSL/IMAG, Lyon, France, Feb. 1992.

[134] A.Darte, Y.Robert, "Affine-by-statement scheduling of uniform loop nests over parametric domains", Technical report 92-16, LIP, Ecole Normale Sup. de Lyon, April 1992.

[135] A.Darte, Y.Robert, "Mapping uniform loop nests onto distributed memory architectures", Internal report 93-03, ENSL/IMAG, Jan. 1993.

[136] A.Darte, T.Risset, Y.Robert, "Loop nest scheduling and transformations", in *Environments and Tools for Parallel Scientific Computing*, J.J.Dongarra et al. (eds.), Advances in Parallel Computing 6, North Holland, Amsterdam, pp.309-332, 1993.

[137] A.Darte, Y.Robert, "Affine-by-statement scheduling of uniform and affine loop nests over parametric domains", Internal report, ENSL/IMAG, Lyon, France, Sep. 1994.

[138] R.Davoli, L-A.Giachini, O.Baboglu, A.Amoroso, L.Alvisi, "Parallel computing in networks of workstations with Paralex", *IEEE Trans. on Parallel and Distributed Systems*, Vol.7, No.4, pp.371-384, April 1996.

[139] F.Decroos, P.Schelkens, G.Lafruit, J.Cornelis, F.Catthoor, "An Integer Wavelet Transform Implemented on a Parallel TI TMS320C40 Platform" *DSP Symp. FEST'99*, Houston, TX, (only web proc.), Aug. 1999.

[140] E.De Greef, "Pruning in a High-level Memory Management Context" IMEC, Internal Report, Dec. 1993.

[141] E.De Greef, F.Catthoor, H.De Man, "A memory-efficient, programmable multi-processor architecture for real-time motion estimation type algorithms", *Intnl. Wsh. on Algorithms and Parallel VLSI Architectures*, Leuven, Belgium, Aug. 1994. Also in "Algorithms and Parallel VLSI Architectures III" (eds. M.Moonen, F.Catthoor), Elsevier, pp.191-202, 1995.

[142] E.De Greef, F.Catthoor, H.De Man, "Memory organization for video algorithms on programmable signal processors", *Proc. IEEE Intnl. Conf. on Computer Design*, Austin TX, pp.552-557, Oct. 1995.

[143] E.De Greef, F.Catthoor, H.De Man, "Mapping real-time motion estimation type algorithms to memory-efficient, programmable multi-processor architectures", *Microprocessors and Microprogramming*, special issue on "Parallel Programmable Architectures and Compilation for Multi-dimensional Processing" (eds. F.Catthoor, M.Moonen), Elsevier, pp.409-423, Oct. 1995.

[144] E.De Greef, F.Catthoor, H.De Man, "Reducing storage size for static control programs mapped onto parallel architectures", presented at *Dagstuhl Seminar on Loop Parallelisation*, Schloss Dagstuhl, Germany, April 1996.

[145] E.De Greef, F.Catthoor, H.De Man, "Array Placement for Storage Size Reduction in Embedded Multimedia Systems", *Proc. Intnl. Conf. on Applic.-Spec. Array Processors*, Zurich, Switzerland, pp.66-75, July 1997.

[146] E.De Greef, F.Catthoor, H.De Man, "Memory Size Reduction through Storage Order Optimization for Embedded Parallel Multimedia Applications", special issue on "Parallel Processing and Multi-media" (ed. A.Krikelis), in *Parallel Computing* Elsevier, Vol.23, No.12, Dec. 1997.

[147] E.De Greef, F.Catthoor, H.De Man, "Memory Size Reduction through Storage Order Optimization for Embedded Parallel Multimedia Applications", *Intnl. Parallel Proc. Symp.(IPPS)* in Proc. Wsh. on "Parallel Processing and Multimedia", Geneva, Switzerland, pp.84-98, April 1997.

[148] E.De Greef, "Storage size reduction for multimedia applications", *Doctoral dissertation*, ESAT/EE Dept., K.U.Leuven, Belgium, Jan. 1998.

[149] E.De Greef, F.Catthoor, H.De Man, "Program transformation strategies for reduced power and memory size in pseudo-regular multimedia applications", *IEEE Trans. on Circuits and Systems for Video Technology*, Vol.8, No.6, pp.719-733, Oct. 1998.

[150] R.Deklerck, J.Cornelis, M.Bister, "Segmentation of Medical Images", *Image and Vision Computing J.*, Vol.11, Nr.8, pp.486-503, Oct. 1993.

[151] K.Denolf, P.Vos, J.Bormans and I.Bolsens, "Cost-efficient C-Level Design of an MPEG-4 Video Decoder", *Proc. IEEE Wsh. on Power and Timing Modeling, Optimization and Simulation (PATMOS)*, Goettingen, Germany, Oct. 2000.

[152] C.Dezan, H.Le Verge, P.Quinton, and Y.Saouter, "The Alpha du CENTAUR experiment", in *Algorithms and parallel VLSI architectures II*, P.Quinton and Y.Robert (eds.), Elsevier, Amsterdam, pp.325-334, 1992.

[153] C.Diderich, M.Gengler, "Solving the constant-degree parallelism alignment problem", *Proc. EuroPar Conf.*, Lyon, France, Aug. 1996. "Lecture notes in computer science" series, Springer Verlag, pp.451-454, 1996.

[154] J.P.Diguet, S.Wuytack, F.Catthoor, H.De Man, "Formalized methodology for data reuse exploration in hierarchical memory mappings", *Proc. IEEE Intnl. Symp. on Low Power Design*, Monterey CA, pp.30-35, Aug. 1997.

[155] C.Ding, K.Kennedy, "The memory bandwidth bottleneck and its amelioration by a compiler", *Proc. Intnl. Parallel and Distr. Proc. Symp.(IPDPS)* in Cancun, Mexico, pp.181-189, May 2000.

[156] M.Dion, Y.Robert, "Mapping affine loop nests: new results", *Intnl. Conf. on High-performance computing and networking*, Milan, Italy, pp.184-189, May 1995.

[157] M.Dion, T.Risset, Y.Robert, "Resource-constrained scheduling of partitioned algorithms on processor arrays", *Integration, the VLSI J.*, Elsevier, Amsterdam, No.20, pp.139-159, 1996.

[158] R.Doalla, B.Fraguela, E.Zapata, "Set associative cache behaviour optimization", *Proc. EuroPar Conf.*, Toulouse, France, pp.229-238, Sep. 1999.

[159] J.Donovan, "Systems programming", *McGraw Hill Incoporated*, Ninth Edition, New York, 1995.

[160] U.Eckhardt, R.Merker, "Scheduling in co-partitioned array architectures", *Proc. Intnl. Conf. on Applic.-Spec. Array Processors*, Zurich, Switzerland, pp.219-228, July 1997.

[161] U.Eckhardt, R.Merker, "Hierarchical algorithm partitioning at system level for an improved utilisation of memory structures", *IEEE Trans. on Comp.-aided Design*, Vol.CAD-18, No.1, pp.14-24, Jan. 1999.

[162] C.Eisenbeis, W.Jalby, D.Windheiser, F.Bodin, "A Strategy for Array Management in Local Memory", *Proc. of the 4th Wsh. on Languages and Compilers for Parallel Computing*, Aug. 1991.

[163] C.Eisenbeis, D.Windheiser, "A new class of algorithms for software pipelining with resource constraints", *Technical Report*, INRIA, Le Chesnay, France, 1992.

[164] P.Ellervee, M.Miranda, F.Catthoor, A.Hemani, "High-level Memory Mapping Exploration for Dynamically Allocated Data Structures", *Proc. 37th ACM/IEEE Design Automation Conf.*, Los Angeles CA, pp.556-559, June 2000.

[165] P.Ellervee, M.Miranda, F.Catthoor, A.Hemani, "System-level data format exploration for dynamically allocated data structures", accepted for *IEEE Trans. on Computer-aided design*, Vol.20, No., 2001.

[166] P.Ellervee, M.Miranda, F.Catthoor, A.Hemani, "Exploiting data transfer locality in memory mapping", *Proc. 25th EuroMicro Conf.*, Milan, Italy, pp.14-21, Sep. 1999.

[167] H.El-Rewini, H.Ali, T.Lewis, "Task scheduling in multiprocessing systems", *IEEE Computer Magazine*, Vol.28, No.12, pp.27-37, Dec. 1995.

[168] D.Eppstein, Z.Galil and G.Italiano, "Dynamic graph algorithms (Chapter 8)", *CRC Handbook of Algorithms and Theory of Computation*, 1999.

[169] J.Z.Fang, M.Lu, "An iteration partition approach for cache or local memory thrashing on parallel processing", *IEEE Trans. on Computers*, Vol.C-42, No.5, pp.529-546, May 1993.

[170] A.Faruque, D.Fong, "Performance analysis through memory of a proposed parallel architecture for the efficient use of memory in image processing applications", *Proc. SPIE'91, Visual communications and image processing*, Boston MA, pp.865-877, Oct. 1991.

[171] P.Feautrier, "Dataflow analysis of array and scalar references", *Intnl. J. of Parallel Programming*, Vol.20, No.1, pp.23-53, 1991.

[172] P.Feautrier, "Some efficient solutions to the affine scheduling problems", *Intnl. J. of Parallel Programming*, Vol.21, No.5, pp.389-420, 1992.

[173] P.Feautrier, "Toward automatic partitioning of arrays for distributed memory computers", *Proc. 7th ACM Intnl. Conf. on Supercomputers*, pp.175-184, Tokyo, Japan, 1993.

[174] P.Feautrier, "Compiling for massively parallel architectures: a perspective", *Intnl. Wsh. on Algorithms and Parallel VLSI Architectures*, Leuven, Belgium, Aug. 1994. Also in "Algorithms and Parallel VLSI Architectures III" (eds. M.Moonen, F.Catthoor), Elsevier, pp.259-270, 1995.

[175] P.Feautrier, "Automatic parallelization in the polytope model", to appear in *Lecture Notes in Computer Science*.

[176] A.Fernandez, J.Llaberia, J.Navarro, M.Valero-Garcia, "Transformation of systolic algorithms for interleaving partitions", *Proc. Intnl. Conf. on Applic.-Spec. Array Processors*, Barcelona, Spain, pp.56-70, Sep. 1991.

[177] C.Ferdinand, F.Martin, R.Wilhelm, "Cache behaviour prediction by abstract interpretation", *Science of Computer Programming, Special issue based on SAS'96*, 1998.

[178] S.Fitzgerald, R.Oldehoeft, "Update-in-place analysis for true multidimensional arrays", *Scientific Programming*, J.Wiley & Sons Inc, Vol.5, pp.147-160, 1996.

[179] I.Foster, "High-performance distributed computing: the I-WAY experiment and beyond", *Proc. EuroPar Conf.*, Lyon, France, Aug. 1996. "Lecture notes in computer science" series, Springer Verlag, pp.3-10, 1996.

[180] T.Franzetti, "Survey of Scheduling Algorithms for the Design of Embedded Multimedia Systems", *M.S. thesis*, Dep. of Comp. Sc. and Applied Math., Univ. Toulouse, June 2000.

[181] F.Franssen, M.van Swaaij, F.Catthoor, H.De Man, "Modelling piece-wise linear and data dependent signal indexing for multi-dimensional signal processing", *Proc. 6th Intnl. Wsh. on High-Level Synthesis*, Laguna Beach CA, Nov. 1992.

[182] F.Franssen, F.Balasa, M.van Swaaij, F.Catthoor, H.De Man, "Modeling Multi-Dimensional Data and Control flow", *IEEE Trans. on VLSI systems*, Vol.1, No.3, pp.319-327, Sep. 1993.

[183] F.Franssen, L.Nachtergaele, H.Samsom, F.Catthoor, H.De Man, "Control flow optimization for fast system simulation and storage minimization", *Proc. 5th ACM/IEEE Europ. Design and Test Conf.*, Paris, France, pp.20-24, Feb. 1994.

[184] F.Franssen, "MASAI's future", Internal report, IMEC, Leuven, Belgium, 1995.

[185] A.Fraboulet, G.Huard, A.Mignotte, "Loop alignment for memory access optimisation", *Proc. 12th ACM/IEEE Intnl. Symp. on System-Level Synthesis* (ISSS), San Jose CA, pp.71-70, Dec. 1999.

[186] A.Fraboulet, "Optimisation de la memoire et de la consommation des systemes multimedia embarques", *Doctoral Dissertation*, Univ. INSA de Lyon, Nov. 2001.

[187] T.Fujii, N.Ohta, "A load balancing technique for video signal processing on a multicomputer type DSP", *Proc. IEEE Intnl. Conf. on Acoustics, Speech and Signal Processing*, New York, pp.1981-1984, April 1988.

[188] B.Furht, "Parallel computing: glory and collapse", *IEEE Computer Magazine*, Vol.27, No.11, pp.74-75, Nov. 1994.

[189] D.Gajski, F.Vahid, S.Narayan, J.Gong, "Specification and design of embedded systems", Prentice Hall, Englewood Cliffs NJ, 1994.

[190] D.Gajski, F.Vahid, S.Narayan, J.Gong, "System-level exploration with SpecSyn", *35th ACM/IEEE Design Automation Conf.*, pp.812-817, June 1998.

[191] D.Gajski, F.Vahid, S.Narayan, J.Gong, "SpecSyn: an environment supporting the specify-explore-refine paradigm for hardware/software system design", *IEEE Trans. on VLSI Systems*, Vol.6, No.1, pp.84-100, March 1998.

[192] D.Gannon, W.Jalby, K.Gallivan, "Strategies for cache and local memory management by global program transformations", *J. of Parallel and Distributed Computing*, Vol.5, pp.568-586, 1988.

[193] G.Gao, C.Eisenbeis, J.Wang, "Introduction to Instruction Level Parallelism Wsh.ʳ, *Proc. EuroPar Conf.*, Lyon, France, Aug. 1996. "Lecture notes in computer science" series, Springer Verlag, pp.745-746, 1996.

[194] C.H.Gebotys, "Utilizing memory bandwidth in DSP embedded processors", *38th ACM/IEEE Design Automation Conf.*, Las Vegas NV, pp.347-351, June 2001.

[195] C.H.Gebotys, M.I.Elmasry, "Simultaneous scheduling and allocation for cost constrained optimal architectural synthesis", *Proc. of the 28th ACM/IEEE Design Automation Conf.*, pp.2-7, San Jose CA, Nov. 1991.

[196] C.H.Gebotys, "Low energy memory component design for cost-sensitive high-performance embedded systems", *Proc. IEEE Custom Integrated Circuits Conf.*, San Diego CA, pp.397-400, May 1996.

[197] C.H.Gebotys, "DSP address optimisation using a minimum cost circulation technique", *Proc. IEEE Intnl. Conf. Comp. Aided Design*, San Jose CA, pp.100-104, Nov. 1997.

[198] J.D.Gee, M.D.Hill, D.N.Pnevmatikatos, A.J.Smith, "Cache performance of Spec92 benchmark suite", *IEEE Micro Magazine*, pp.17-27, Aug. 1993.

[199] R.Gerber, S.Hong, M.Saksena, "Guaranteeing real-time requirements with resource-based calibration of periodic processes", *IEEE Trans. on Software Engineering*, Vol.SE-21, No.7, pp.579-592, July 1995.

[200] A.Gerasoulis and J.Jiao and T.Yang, "A multistage approach to scheduling task graphs", *Series in Discrete Mathematics and Theoretical Computer Science*, Vol.22, pp.81-103, 1995.

[201] C.Ghez, M.Miranda, A.Vandecappelle, F.Catthoor, D.Verkest, "Systematic high-level address code transformations for piece-wise linear indexing: illustration on a medical imaging algorithm", *Proc. IEEE Wsh. on Signal Processing Systems (SIPS)*, Lafayette LA, IEEE Press, pp.623-632, Oct. 2000.

[202] S.Ghosh, M.Martonosi, S.Malik, "Cache miss equations: an analytical representation of cache misses", *IEEE TC on Computer Architecture Newsletter*, special issue on "Interaction between Compilers and Computer Architectures", pp.52-54, June 1997.

[203] S.Ghosh, M.Martonosi, S.Malik, "Cache miss equations: a compiler framework for analyzing and tuning memory behavior", *ACM Transactions on Programming Languages and Systems*, Vol.21, No.4, pp.702-746, July 1999.

[204] R.Gonzales, M.Horowitz, "Energy dissipation in general-purpose microprocessors", *IEEE J. of Solid-state Circ.*, Vol.SC-31, No.9, pp.1277-1283, Sep. 1996.

[205] J.Goodman, W.Hsu, "Code scheduling and register allocation in large basic blocks", *Proc. 2th ACM Intnl. Conf. on Supercomputers*, pp.442-452, 1988.

[206] G.Goossens, J.Rabaey, J.Vandewalle, H.De Man, "An efficient microcode-compiler for custom DSP-processors", *Proc. IEEE Intnl. Conf. Comp. Aided Design*, Santa Clara CA, pp.24-27, Nov. 1987.

[207] G.Goossens, J.Rabaey, J.Vandewalle, H.De Man, "An efficient microcode compiler for application-specific DSP processors", *IEEE Trans. on Comp.-aided Design*, Vol.9, No.9, pp.925-937, Sep. 1990.

[208] G.Goossens, D.Lanneer, M.Pauwels, F.Depuydt, K.Schoofs, A.Kifli, M.Cornero, P.Petroni, F.Catthoor, H.De Man, "Integration of medium-throughput signal processing algorithms on flexible instruction-set architectures", *J. of VLSI signal processing*, special issue on "Design environments for DSP" (eds.I.Verbauwhede, J.Rabaey), No.9, Kluwer, Boston, pp.49-65, Jan. 1995.

[209] C.Grenier, "Optimizing mesa graphics library", IMEC Internal Report, Nov. 2000.

[210] M.Griebl, C.Lengauer, S.Wetzel, "Code Generation in the Polytope Model", *Proc. Intnl. Conf. on Parallel Architectures and Compilation Techniques*, pp.106-111, Paris, France, 1998.

[211] P.Grun, N.Dutt, and A.Nicolau, "MIST: an algorithm for memory miss traffic management", *Proc. IEEE Intnl. Conf. on Comp. Aided Design*, Santa Clara CA, pp.431-437, Nov. 2000.

[212] P.Grun, N.Dutt, and A.Nicolau, "Memory aware compilation through accerate timing extraction", *Proc. 37th ACM/IEEE Design Automation Conf.*, Los Angeles CA, pp.316-321, June 2000.

[213] P.Grun, F.Balasa, and N.Dutt, "Memory Size Estimation for Multimedia Applications", *Proc. ACM/IEEE Wsh. on Hardware/Software Co-Design (Codes)*, Seattle WA, pp.145-149, March 1998.

[214] G.Gudjonsson, W.Winsborough, "Compile-time memory reuse in logic programming languages through update-in-place", *ACM Trans. on Programming Languages and Systems*, Vol.21, No.3, pp.430-501, May 1999.

[215] N.Guil, E.Zapata, "A parallel pipelined Hough transform", *Proc. EuroPar Conf.*, Lyon, France, Aug. 1996. "Lecture notes in computer science" series, Springer Verlag, pp.131-138, 1996.

[216] S.Gupta, M.Miranda, F.Catthoor, R.Gupta, "Analysis of high-level address code transformations for programmable processors", *Proc. 3rd ACM/IEEE Design and Test in Europe Conf.*, Paris, France, pp.9-13, April 2000.

[217] P.Gupta, C-T.Chen, J.DeSouza-Batista, A.Parker, "Experience with image compression chip design using Unified System Construction tools", *31st ACM/IEEE Design Automation Conf.*, San Diego, CA, pp.250-256, June 1994.

[218] A.Gupta, "Fast and effective algorithms for graph partitioning and sparse-matrix ordering", *IBM J. of Research and Development*, Vol.41, No.1/2, 1997.

[219] T.Halfhill, J.Montgomery, "Chip fashion: multi-media chips", *Byte Magazine*, pp.171-178, Nov.1995.

[220] M.Hall, J.Anderson, S.Amarasinghe, B.Murphy, S.Liao, E.Bugnion, M.Lam, "Maximizing multiprocessor performance with the SUIF compiler", *IEEE Computer Magazine*, Vol.30, No.12, pp.84-89, Dec. 1996.

[221] F.Hamzaoglu, Y.Ye, A.Keshavarzi, K.Zhang, S.Narendra, S.Borkar, V.De, M.Stan, "Dual-VT SRAM Cells with Full-Swing Single-Ended Bit Line Sensing for High-Performance On-Chip Cache in 0.13 um Technology Generation", *IEEE Intnl. Symp. on Low Power Electronics and Devices*, Rapallo, Italy, July 2000.

[222] F.Harmsze, A.Timmer, J.van Meerbergen, "Memory arbitration and cache management in stream-based systems", *Proc. 3rd ACM/IEEE Design and Test in Europe Conf.*, Paris, France, pp.257-262, April 2000.

[223] W.Hardt, W.Rosenstiel, "Prototyping of tightly coupled hardware/software systems", *Design Automation for Embedded Systems*, Vol.2, Kluwer Acad. Publ., Boston, pp.283-317, 1997.

[224] L.S.Haynes (ed.), *IEEE Computer*, special issue on "Highly Parallel Computing", Vol.15, No.1, Jan. 1982.

[225] S.Heemstra de Groot, O.Herrman, "Evaluation of some multiprocessor scheduling techniques of atomic operations for recursive DSP filters", *Proc. Europ. Conf. on Circ. Theory and Design*, ECCTD, Brighton, U.K., pp.400-404, Sep. 1989.

[226] M.Hill, "Aspects of Cache Memory and Instruction Buffer Performance", *PhD thesis*, U.C.Berkeley, Nov. 1987.

[227] P.N.Hilfinger, J.Rabaey, D.Genin, C.Scheers, H.De Man, "DSP specification using the Silage language", *Proc. Intnl. Conf. on Acoustics, Speech and Signal Processing*, Albuquerque, NM, pp.1057-1060, April 1990.

[228] M.Hill, "Dinero III cache simulator", Online document available via http://www.cs.wisc.edu/ markhill, 1989.

[229] P.Hoang, J.Rabaey, "Program partitioning for a reconfigurable multiprocessor system", presented at *IEEE Wsh. on VLSI signal processing*, San Diego CA, Nov. 1990.

[230] P.Hoang, J.Rabaey, "Scheduling of DSP programs onto multiprocessors for maximum throughput", *IEEE Trans. on Signal Processing*, Vol.SP-41, No.6, pp.2225-2235, June 1993.

[231] D.Hochbaum, "Approximation Algorithms for NP-hard problems", *PWS Publishing Company*, ISBN 0-534-94968-1, 1997.

[232] High Performance Fortran Forum, "High Performance Fortran Language Specification", Version 1.0, Technical Report CRPC-TR92225, Rice Univ., May 1993.

[233] C-H.Huang, P.Sadayappan, "Communication-free hyperplane partitioning of nested loops", in "Languages and compilers for parallel computing" (U.Banerjee et al., eds.), 1991.

[234] S.Hummel and E.Schonberg, "Low-overhead scheduling of nested parallelism", *IBM J. of Research and Development*, 1991.

[235] Y-T.Hwang, Y-H.Hu, "A unified partitioning and scheduling scheme for mapping multi-stagwe regular iterative algorithms onto processor arrays", *J. of VLSI Signal Processing*, No.11, Kluwer, Boston, pp.133-150, 1995.

[236] K.Hwang, Z.Xu, "Scalable parallel computers for real-time signal processing", *IEEE Signal Processing Magazine*, No.4, pp.50-66, July 1996.

[237] "IBM 16Mb single data rate synchronous DRAM", IBM Corporation, 1998. Online document available at http://www.chips.ibm.com/techlib/products/memory/datasheets/.

[238] C.Ussery, "Configurable VLIW is well suited for SoC Design", *Electronic Engineering Times*, Aug. 14, 2000. Online document available via http://www.improvsys.com/Press/ArticleLinks.html

[239] F.Irigoin, R.Triolet, "Dependency approximation and global parallel code generation for nested loops", In *Parallel and Distributed Algorithms* (eds. M.Cosnard et al.), pp.297-308, Elsevier, 1989.

[240] M.J.Irwin, M.Kandemir, N.Vijaykrishnan, A.Sivasubramaniam, "A holistic approach to system level energy optimisation", *Proc. IEEE Wsh. on Power and Timing Modeling, Optimization and Simulation (PATMOS)*, Goettingen, Germany, pp.88-107, Oct. 2000.

[241] T.Ben Ismail, K.O'Brien, A.A.Jerraya, "Interactive system-level partitioning with Partif", *Proc. 5th ACM/IEEE Europ. Design and Test Conf.*, Paris, Paris, France, pp.464-468, Feb. 1994.

[242] P.Italiano, P.Cattaneo, U.Nanni, G.Pasqualone, C.Zaroliagis and D.Alberts, "LEDA extension package for dynamic graph algorithms", Technical Report, Dep. of Comp.Sc., Univ. of Halle, Germany, March 1998.

[243] K.Itoh, K.Sasaki, Y.Nakagome, "Trends in low-power RAM circuit technologies", special issue on "Low power electronics" of the *Proc. of the IEEE*, Vol.83, No.4, pp.524-543, April 1995.

[244] K.Itoh, "Low Voltage Memory Design", in tutorial on "Low voltage technologies and circuits", *IEEE Intnl. Symp. on Low Power Design*, Monterey CA, Aug. 1997.

[245] B.Jacob, P.chen, S.Silverman, T.Mudge, "An analytical model for designing memory hierarchies", *IEEE Trans. on Computers*, Vol.C-45, No.10, pp.1180-1193, Oct. 1996.

[246] M.Jayapala, F.Barat, P.Op de Beeck, F.Catthoor, G.De Coninck, "Low Energy Clustered Instruction Fetch and Split Loop Cache Architecture for Long Instruction Word Processors", *Wsh. on Compilers and Operating Systems for Low Power (COLP'01)* in conjunction with *Intnl. Conf. on Parallel Arch. and Compilation Techniques (PACT)*, Barcelona, Spain, Sep. 2001.

[247] J.M.Janssen, F.Catthoor, H.De Man, "A Specification Invariant Technique for Operation Cost Minimisation in Flow-graphs". *Proc. 7th ACM/IEEE Intnl. Symp. on High-Level Synthesis*, Niagara-on-the-Lake, Canada, pp.146-151, May 1994.

[248] A.Jantsch, P.Ellervee, J.Oberg, A.Hemani, H.Tenhunen, "Hardware/software partitioning and minimizing memory interface traffic", *Proc. ACM EuroDAC'94*, pp.226-231, Sep. 1994.

[249] P.K.Jha and N.Dutt, Library mapping for memories, *Proc. European Design Automation Conf.*, Paris, France, pp.288-292, March 1997.

[250] M.Jimenez, J.Llaberia, A.Fernandez, E.Morancho, "A unified transformation technique for multi-level blocking" *Proc. EuroPar Conf.*, Lyon, France, Aug. 1996. "Lecture notes in computer science" series, Springer Verlag, pp.402-405, 1996.

[251] T.Johnson, W.Hwu, "Run-time adaptive cache hierarchy management via reference analysis", *Proc. ACM Intnl. Symp. on Computer Arch.*, pp.315-326, June 1997.

[252] L.John, R.Radhakrishnan, "c_ICE: A compiler-based instruction cache exclusion scheme", *IEEE TC on Computer Architecture Newsletter*, special issue on "Interaction between Compilers and Computer Architectures", pp.61-63, June 1997.

[253] N.Jouppi, "Improving direct-mapped cache performance by the addition of a small fully-associative cache and prefetch buffers", *Proc. ACM Intnl. Symp. on Computer Arch.*, pp.364-373, May 1990.

[254] M.Kampe, F.Dahlgren, "Exploration of spatial locality on emerging applications and the consequences for cache performance", *Proc. Intnl. Parallel and Distr. Proc. Symp.(IPDPS)* in Cancun, Mexico, pp.163-170, May 2000.

[255] M.Kamble, K.Ghose, "Analytical Energy Dissipation Models for Low Power Caches", *Proc. IEEE Intnl. Symp. on Low Power Design*, Monterey CA, pp.143-148, Aug. 1997.

[256] M.Kandemir, N.Vijaykrishnan, M.J.Irwin, W.Ye, "Influence of compiler optimisations on system power", *Proc. 37th ACM/IEEE Design Automation Conf.*, Los Angeles CA, pp.304-307, June 2000.

[257] M.Kandemir, J.Ramanujam, M.J.Irwin, N.Vijaykrishnan, I.Kadyif, A.Parikh, "Dynamic management of scratch-pad memory space", *38th ACM/IEEE Design Automation Conf.*, Las Vegas NV, pp.690-695, June 2001.

[258] M.Kandemir, U.Sezer, V.Delaluz, "Improving memory energy using access pattern classification", *Proc. IEEE Intnl. Conf. Comp. Aided Design*, San Jose CA, pp.201-206, Nov. 2001.

[259] M.Kandemir, A.Choudhary, J.Ramanujam, R.Bordawekar, "Optimizing out-of-core computations in uniprocessors", *IEEE TC on Computer Architecture Newsletter*, special issue on "Interaction between Compilers and Computer Architectures", pp.25-27, June 1997.

[260] M.Kandemir, J.Ramanujam, A.Choudhary, "Improving cache locality by a combination of loop and data transformations", *IEEE Trans. on Computers*, Vol.48, No.2, pp.159-167, Feb. 1999.

[261] J.Kang, A.van der Werf, P.Lippens, "Mapping array communication onto FIFO communication - towards an implementation", *Proc. 13th ACM/IEEE Intnl. Symp. on System-Level Synthesis* (ISSS), Madrid, Spain, pp.207-213, Sep. 2000.

[262] D.Karchmer and J.Rose, "Definition and solution of the memory packing problem for field-programmable systems", *Proc. IEEE Intnl. Conf. Comp. Aided Design*, San Jose CA, pp.20-26, Nov. 1994.

[263] G.Karypis, V.Kumar, "hMETIS, Multilevel k-way hypergraph partitioning", Technical Report, Univ. of Minnesota, Dep. of CS, Minneapolis MN, 1998.

[264] V.Kathail, M.Schlansker, B.R.Rau, "HPL PlayDoh architecture specification : Version 1.1", *Technical Report HPL-93-80(R.1)*, Compiler and Architecture Research, Hewlett Packard Labs, Feb. 2000.

[265] W.Kelly, W.Pugh, "Generating schedules and code within a unified reordering transformation framework", Technical Report UMIACS-TR-92-126, CS-TR-2995, Institute for Advanced Computer Studies Dept. of Computer Science, Univ. of Maryland, College Park, MD 20742, 1992.

[266] W.Kelly, W.Pugh, E.Rosser, "Code generation for multiple mappings", Technical report CS-TR-3317, Dept. of CS, Univ. of Maryland, College Park MD, USA, 1994.

[267] W.Kelly, W.Pugh, "A framework for unifying reordering transformations", Technical report CS-TR-3193, Dept. of CS, Univ. of Maryland, College Park MD, USA, April 1993.

[268] B.Kim, T.Barnwell III, "Resource allocation and code generation for pointer based pipelined DSP multiprocessors", *Proc. IEEE Intnl. Symp. on Circuits and Systems*, New Orleans, pp., May 1990.

[269] C.Kim et al, "A 64 Mbit, 640 MB/s bidirectional data-strobed, double data rate SDRAM with a 40 mW DLL for a 256 MB memory system", *IEEE J. of Solid-state Circ.*, Vol.SC-33, pp.1703-1710, Nov. 1998.

[270] J.Kin, M.Gupta, W.Mangione-Smith, "The filter cache: an energy efficient memory structure", *Proc. 30th Intnl. Symp. on Microarchitecture*, pp.184-193, Dec. 1997.

[271] T.Kirihata et al, "A 220 mm2, four- and eight-bank, 256 Mb SDRAM with single-sided stitched WL architecture", *IEEE J. of Solid-state Circ.*, Vol.SC-33, pp.1711-1719, Nov. 1998.

[272] P.G.Kjeldsberg, F.Catthoor, E.J.Aas, "Application of high-level memory size estimation for guidance of loop transformations in multimedia design", *Proc. Nordic Signal Proc. Symp.*, Norrkvping, Sweden, pp.371-374, June 2000.

[273] P.G.Kjeldsberg, F.Catthoor, E.J.Aas, "Automated data dependency size estimation with a partially fixed execution ordering", *Proc. IEEE Intnl. Conf. on Comp. Aided Design*, Santa Clara CA, pp.44-50, Nov. 2000.

[274] P.G.Kjeldsberg, F.Catthoor, E.J.Aas, "Storage requirement estimation for data-intensive applications with partially fixed execution ordering", *Proc. ACM/IEEE Wsh. on Hardware/Software Co-Design (Codes)*, San Diego CA, pp.56-60, May 2000.

[275] P.G.Kjeldsberg, "Storage requirement estimation and optimisation for data-intensive applications", *Doctoral dissertation*, Norwegian Univ. of Science and Technology, Trondheim, Norway, March 2001.

[276] P.G.Kjeldsberg, F.Catthoor, E.J.Aas, "Detection of partially simultaneously alive signals in storage requirement estimation for data-intensive applications", *38th ACM/IEEE Design Automation Conf.*, Las Vegas NV, pp.365-370, June 2001.

[277] U.Ko, P.Balsara, A.Nanda, "Energy optimization of multi-level processor cache architectures", *Proc. IEEE Intnl. Wsh. on Low Power Design*, Laguna Beach CA, pp.45-50, April 1995.

[278] I.Kodukula, N.Ahmed, K.Pingali, "Data-centric multi-level blocking", *Proc. ACM Intnl. Conf. on Programming Language Design and Implementation*, pp.346-357, 1997.

[279] D.Kolson, A.Nicolau, N.Dutt, "Minimization of memory traffic in high-level synthesis", *Proc. 31st ACM/IEEE Design Automation Conf.*, San Diego, CA, pp.149-154, June 1994.

[280] D.Kolson, A.Nicolau, N.Dutt, "Elimination of redundant memory traffic in high-level synthesis", *IEEE Trans. on Comp.-aided Design*, Vol.15, No.11, pp.1354-1363, Nov. 1996.

[281] T.Komarek, P.Pirsch, "Array Architectures for Block Matching Algorithms", *IEEE Trans. on Circuits and Systems*, Vol.36, No.10, Oct. 1989.

[282] K.Konstantinides, R.Kaneshiro, J.Tani, "Task allocation and scheduling models for multi-processor digital signal processing", *IEEE Trans. on Acoustics, Speech and Signal Processing*, Vol.ASSP-38, No.12, pp.2151-2161, Dec. 1990.

[283] A.Krikelis, "Wsh. on parallel processing and multimedia: a new research forum", *IEEE Concurrency Magazine*, pp.5-7, April-June 1997.

[284] D.Kuck, "Parallel Processing of Ordinary Programs", *Advances in Computers*, Vol.15, pp.119-179, 1976.

[285] C.Kulkarni, F.Catthoor, H.De Man, "Advanced data layout organization for multi-media applications", *Intnl. Parallel and Distr. Proc. Symp.(IPDPS)* in Proc. Wsh. on "Parallel and Distrib. Computing in Image Proc., Video Proc., and Multimedia (PDIVM'2000)", Cancun, Mexico, May 2000.

[286] C.Kulkarni, K.Danckaert, F.Catthoor, M.Gupta, "Interaction Between Parallel Compilation And Data Transfer and Storage Cost Minimization for Multimedia Applications", accepted for *Intnl. J. of Computer Research*, Special Issue on "Industrial Applications of Parallel Computing", 2001.

[287] C.Kulkarni, F.Catthoor, H.De Man, "Cache optimization and global code transformations for low power embedded multimedia applications", accepted to *IEEE Trans. on Parallel and Distributed Systems*, Vol.10, No., pp., 2001.

[288] C.Kulkarni, "Cache optimization for multimedia applications", *Doctoral dissertation*, ESAT/EE Dept., K.U.Leuven, Belgium, Feb. 2001.

[289] C.Kulkarni, M.Miranda, C.Ghez, F.Catthoor, H.De Man, "Cache Conscious Data Layout Organization For Embedded Multimedia Applications", *Proc. 4th ACM/IEEE Design and Test in Europe Conf.*, Munich, Germany, March 2001.

[290] D.Kulkarni, M.Stumm, "Loop and Data Transformations: A Tutorial", Technical Report CSRI-337, Computer Systems Research Inst., Univ. of Toronto, pp.1-53, June 1993.

[291] D.Kulkarni, M.Stumm, "Linear loop transformations in optimizing compilers for parallel machines", Technical report, Comp. Systems Res. Inst. Univ. of Toronto, Canada, Oct. 1994.

[292] D.Kulkarni, M.Stumm, R.Unrau, "Implementing flexible computation rules with subexpression-level loop transformations", Technical report, Comp. Systems Res. Inst. Univ. of Toronto, Canada, 1995.

[293] D.Kulkarni, M.Stumm, "Linear loop transformations in optimizing compilers for parallel machines", *The Australian Computer J.*, pp.41-50, May 1995.

[294] C.Kulkarni, "Caching strategies for voice coder application on embedded processors", *M.S. thesis*, Dep. of Electrical Engineering, K.U.Leuven, June 1997.

[295] C.Kulkarni, F.Catthoor, H.De Man, "Hardware cache optimization for parallel multimedia applications", *Proc. EuroPar Conf.*, Southampton, U.K.,pp.923-931, Sep. 1998.

[296] C.Kulkarni, F.Catthoor, H.De Man, "Cache transformations for low power caching in embedded multimedia processors", *Proc. Intnl. Parallel Proc. Symp.(IPPS)*, Orlando FL, pp.292-297, April 1998.

[297] C.Kulkarni, F.Catthoor, "Analysis and Optimization of Texture Mapping Applications Using OpenGL", IMEC Internal Report, April 1999.

[298] C.Kulkarni, D.Moolenaar, L.Nachtergaele, F.Catthoor, H.De Man, "System-level energy-delay exploration for multi-media applications on embedded cores with hardware caches", *J. of VLSI Signal Processing*, special issue on SIPS'97, No.19, Kluwer, Boston, pp.45-58, 1999.

[299] C.Kulkarni, K.Danckaert, F.Catthoor and M.Gupta, "Interaction Between Data Parallel Compilation and Data Transfer and Storage Cost Minimization for Multimedia Applications", *Proc. EuroPar Conf.*, Toulouse, France, pp.668-676, Sep. 1999.

[300] C.Kulkarni, F.Catthoor, H.De Man, "Optimizing Graphics Applications: A Data Transfer and Storage Exploration Perspective", *Proc. 1st Wsh. on Media Proc. and DSPs*, in IEEE/ACM Intnl. Symp. on Microarchitecture, MICRO-32, Haifa, Israel, Nov. 1999.

[301] F.J.Kurdahi, A.C.Parker, "REAL: a program for register allocation", *Proc. 24th ACM/IEEE Design Automation Conf.*, Miami FL, pp.210-215, June 1987.

[302] G.Lafruit, B.Vanhoof, L.Nachtergaele, F.Catthoor, J.Bormans, "The local wavelet transform: a memory-efficient, high speed architecture for a Region-Oriented Zero Tree Coder", *Integrated Computer-Aided Engineering*, Vol.7, No.2, pp.89-103, March 2000.

[303] G.Lafruit, F.Catthoor, J.Cornelis, H.De Man, "An efficient VLSI architecture for the 2-D wavelet transform with novel image scan", *IEEE Trans. on VLSI Systems*, Vol.7, No.1, pp.56-68, March 1999.

[304] A.Lain, D.Chakrabarti, P.Banerjee, "Compiler and run-time support for exploiting regularity within irregular applications", *IEEE Trans. on Parallel and Distributed Systems*, Vol.11, No.2, pp.119-135, Feb. 2000.

[305] L.Lamport, "The parallel execution of DO loops", *Communications of the ACM*, Vol.17, No.2, pp.83-93, Feb. 1974.

[306] M.Lam, E.Rothberg and M.Wolf, " The cache performance and optimizations of blocked algorithms", *Proc. 4th Intnl. Conf. on Architectural Support for Prog. Lang. and Operating Systems (ASPLOS)*, Santa Clara CA, pp.63-74, April 1991.

[307] D.Lanneer, M.Cornero, G.Goossens, H.De Man, "Data routing: a paradigm for efficient data-path synthesis and code generation", *Proc. 7th ACM/IEEE Intnl. Symp. on High-Level Synthesis*, Niagara-on-the-Lake, Canada, May 1994.

[308] P.Landman, "Low power architectural design methodologies", *Doctoral Dissertation*, U.C.Berkeley, Aug. 1994.

[309] G.Lawton, "The wild world of 3D graphics chips", *IEEE Computer Magazine*, pp.12-16, Sep. 2000.

[310] C.L.Lawson, R.J.Hanson, "Solving least squares problems", *Classics in Applied mathematics*, SIAM, Philadelphia, 1995.

[311] G.Lawton, "Storage technology takes the center stage", *IEEE Computer Magazine*, Vol.32, No.11, pp.10-13, Nov. 1999.

[312] E.Lee, D.Messerschmitt, "Pipeline interleaved programmable DSP's: synchronous data-flow programming", *IEEE Trans. on Acoustics, Speech and Signal Processing*, Vol.35, No.9, pp.1334-1346, Sep. 1987.

[313] E.Lee, D.Messerschmitt, "Synchronous data flow", *Proc. of the IEEE*, Vol.75, No.9, pp.1235-1245, Sep. 1987.

[314] V.Lefebvre, P.Feautrier, "Optimizing storage size for static control programs in automatic parallelizers", *Proc. EuroPar Conf.*, Passau, Germany, Aug. 1997. "Lecture notes in computer science" series, Springer Verlag, Vol.1300, 1997.

[315] C.E.Leiserson, J.B.Saxe, "Optimizing synchronous circuitry by retiming", *Proc. Third Caltech Conf. of VLSI*, R.Bryant (ed.), Comp. Science Press, 1983.

[316] C.Lengauer. "Loop parallelization in the polytope model", *Proc. of the Fourth Intnl. Conf. on Concurrency Theory (CONCUR93)*, Hildesheim, Germany, Aug. 1993.

[317] S-T.Leung, J.Zahorjan, "Restructuring arrays for efficient parallel loop execution", Technical Report, Dep. of CSE, Univ. of Washington, Feb. 1994.

[318] R.Leupers, P.Marwedel, "Algorithms for address assignment in DSP code generation", *Proc. IEEE Intnl. Conf. Comp. Aided Design*, San Jose CA, pp.109-112, Nov. 1996.

[319] W.Li, K.Pingali. "Access normalization: loop restructuring for NUMA compilers", *Proc. 5th Intnl. Conf. on Architectural Support for Prog. Lang. and Operating Systems (ASPLOS)*, April 1992.

[320] W.Li, K.Pingali. "A singular loop transformation framework based on non-singular matrices", *Proc. 5th Annual Wsh. on Languages and Compilers for Parallelism*, New Haven CN, Aug. 1992.

[321] Y.Li, W.Wolf, "Hardware-software co-synthesis with memory hierarchies", *Proc. IEEE Intnl. Conf. on Comp. Aided Design*, Santa Clara CA, pp.430-436, Nov. 1998.

[322] Y.T.Li, S.Malik, "Performance analysis of real-time embedded software", *Kluwer Academic Publishers*, Boston, MA, 1999.

[323] C.Liem, T.May, P.Paulin, "Register assignment through resource classification for ASIP microcode generation", *Proc. IEEE Intnl. Conf. Comp. Aided Design*, San Jose CA, pp.397-402, Nov. 1994.

[324] C.Liem, P.Paulin, A.Jerraya, "Address calculation for retargetable code generation and exploration of instruction-set architectures", *Proc. 33rd ACM/IEEE Design Automation Conf.*, Las Vegas NV, pp.597-600, June 1996.

[325] D.Lilja, "The impact of parallel loop scheduling strategies on prefetching in a shared memory multi-processor", *IEEE Trans. on Parallel and Distributed Systems*, Vol.5, No.6, pp.573-584, June 1994.

[326] H-B.Lim, P-C.Yew, "Efficient integration of compiler-directed cache coherence and data prefetching", *Proc. Intnl. Parallel and Distr. Proc. Symp.(IPDPS)* in Cancun, Mexico, pp.331-339, May 2000.

[327] B.Kernighan, S.Lin, "An effective heuristic procedure for partitioning graphs", *The Bell System Technical J.*, pp.291-308, Feb. 1970.

[328] P.Lippens, J.van Meerbergen, W.Verhaegh, A.van der Werf, "Allocation of multiport memories for hierarchical data streams", *Proc. IEEE Intnl. Conf. Comp. Aided Design*, Santa Clara CA, Nov. 1993.

[329] L.Liu, "Issues in multi-level cache design", *Proc. IEEE Intnl. Conf. on Computer Design*, Cambridge MA, pp.46-52, Oct. 1994.

[330] N.Liveris, N.D.Zervas, C.E.Goutis, "A Code Transformation-based Methodology for Improving I-Cache Performance", accepted for *Proc. Intnl. Conf. on Electronic Circuits and Systems*, Malta, pp., Sep. 2001.

[331] R.Lo, S.Chan, J.Dehnert, R.Towle, "Aggregrate operation movement: a min-cut approach to global code motion", *Proc. EuroPar Conf.*, Lyon, France, Aug. 1996. "Lecture notes in computer science" series, Springer Verlag, pp.801-814, 1996.

[332] D.B.Loveman, "Program improvement by source-to-source transformation", *J. of the ACM*, Vol.24, No.1, pp.121-145, 1977.

[333] W.Löwe, J.Eisenbiegler, W.Zimmermann, "Optimization of parallel programs on machines with expensive communication", *Proc. EuroPar Conf.*, Lyon, France, Aug. 1996. "Lecture notes in computer science" series, Springer Verlag, pp.602-610, 1996.

[334] C-K.Luk, T.Mowry, "Automatic compiler-inserted prefetching for pointer-based applications", *IEEE Trans. on Computers*, Vol.48, No.2, pp.134-141, Feb. 1999.

[335] J.Ma, E.Deprettere, K.Parhi, "Pipelined CORDIC based QRD-RLS adaptive filtering using matrix lookahead", *Proc. IEEE Wsh. on Signal Processing Systems (SIPS)*, Leicester, UK, pp.131-140, Nov. 1997.

[336] M.Mace, "Memory storage patterns in parallel processing", Kluwer Acad. Publ., Boston, 1987.

[337] F.Maessen, L.van der Perre, F.Willems, B.Gyselinckx, M.Engels, F.Catthoor, "Memory power reduction for the high-speed implementation of turbo coders", *Proc. Symp. on Communications and Vehicular Technology (VTC'00)*, Leuven, Belgium, Oct. 2000.

[338] F.Maessen, A.Giulietti, B.Bougard, V.Derudder, L.van der Perre, F.Catthoor, M.Engels, "Memory power reduction for the high-speed implementation of turbo codes", *Proc. IEEE Wsh. on Signal Processing Systems (SIPS)*, Antwerp, Belgium, IEEE Press, pp.16-24, Sep. 2001.

[339] A.Malik, B.Moyer, D.Cermak, "A low power unified cache architecture providing power and performance flexibility", *Proc. IEEE Intnl. Symp. on Low Power Design*, Rapallo, Italy, pp.241-243, Aug. 2000.

[340] N.Manjiakian, T.Abdelrahman, "Array data layout for reduction of cache conflicts", *Intnl. Conf. on Parallel and Distributed Computing Systems*, 1995.

[341] N.Manjiakian, T.Abdelrahman, "Fusion of loops for parallelism and locality", Technical report CSRI-315, Comp. Systems Res. Inst. Univ. of Toronto, Canada, Feb. 1995.

[342] P.Marchal, C.Wong, A.Prayati, N.Cossement, F.Catthoor, R.Lauwereins, D.Verkest, H.De Man "Impact of task-level concurrency transformations on the MPEG4 IM1 player for weakly parallel processor platforms", *Wsh. on Compilers and Operating Systems for Low Power (COLP'00)* in conjunction with *Intnl. Conf. on Parallel Arch. and Compilation Techniques (PACT)*, Philadelphia PN, Oct. 2000.

[343] P.Marchal, C.Wong, A.Prayati, N.Cossement, F.Catthoor, R.Lauwereins, D.Verkest, H.De Man "Dynamic memory oriented transformations in the MPEG4 IM1-player on a low power platform", *Proc. Intnl. Wsh. on Power Aware Computing Systems (PACS)*, Cambridge MA, pp.31-40, Nov. 2000.

[344] P.Marwedel, G.Goossens (eds.), "Code Generation for Embedded Processors", Kluwer, Boston, 1995.

[345] M.Martonosi, K.Shaw, "Interactions between application write performance and compilation techniques: a preliminary view", *IEEE TC on Computer Architecture Newsletter*, special issue on "Interaction between Compilers and Computer Architectures", pp.16-18, June 1997.

[346] G.F.Marchioro, J.-M.Daveau, A.A.Jerraya, "Transformational partitioning for co-design of multiprocessor systems", *Proc. IEEE/ACM Intnl. Conf. on Computer-Aided Design*, pp.508-15, Nov. 1997.

[347] K.Masselos, K.Danckaert, F.Catthoor, N.Zervas, C.E.Goutis, H.De Man, "A specification refinement methodology for power efficient partitioning of data-dominated algorithms within performance constraints", *J. of VLSI Signal Processing*, Vol.26, No.3, Kluwer, Boston, pp.291-318, Nov. 2000.

[348] K.Masselos, F.Catthoor, C.E.Goutis, H.De Man, "A systematic methodology for the application of data transfer and storage optimizing code transformations for power consumption and execution time reduction in realisations of multimedia algorithms on programmable processors", accepted for *IEEE Trans. on VLSI Systems*, Vol.8, No., pp., 2001.

[349] K.Masselos, F.Catthoor, C.E.Goutis, "Effect of Data Transfer and Storage Optimization on Design Quality Factors of Multimedia Algorithms Realized on Instruction Set Processors", accepted for *Proc. IEEE Wsh. on Power and Timing Modeling, Optimization and Simulation (PATMOS)*, Yverdon-les-bains, Switzerland, pp., Sep. 2001.

[350] K.Masselos, F.Catthoor, C.E.Goutis, H.De Man, "Combined application of low-power code transformations and subword parallelism exploitation for VLIW multi-media processors", accepted for *IEEE Trans. on VLSI Systems*, Vol.8, No., pp., 2001.

[351] K.Masselos, K.Danckaert, F.Catthoor, C.E.Goutis, H.De Man, "A methodology for power efficient partitioning of data-dominated algorithm specifications within performance constraints", *Proc. IEEE Intnl. Symp. on Low Power Design*, San Diego CA, pp.270-272, Aug. 1999.

[352] K.Masselos, F.Catthoor, C.E.Goutis, H.De Man, "System-level power optimizing data-flow transformations for multimedia applications realized on programmable multimedia processors", *Proc. Intnl. Conf. on Electronic Circuits and Systems*, Paphos, Cyprus, Vol.III, pp.1733-1736, Sep. 1999.

[353] K.Masselos, F.Catthoor, C.E.Goutis, H.De Man, "A performance oriented use methodology of power optimizing code transformations for multimedia applications realized on programmable multimedia processors", *Proc. IEEE Wsh. on Signal Processing Systems (SIPS)*, Taipeh, Taiwan, IEEE Press, pp.261-270, Oct. 1999.

[354] K.Masselos, F.Catthoor, C.E.Goutis, H.De Man, "Code size effects of power optimizing code transformations for embedded multimedia applications", *Proc. IEEE Wsh. on Power and Timing Modeling, Optimization and Simulation (PATMOS)*, Kos, Greece, pp.61-70, Oct. 1999.

[355] K.Masselos, F.Catthoor, C.E.Goutis, H.De Man, "Low Power Mapping of Video Processing Applications on VLIW Multimedia Processors", *IEEE Alessandro Volta Memorial Intnl. Wsh. on Low Power Design (VOLTA)*, Como, Italy, pp.52-60, March 1999.

[356] K.Masselos, "Performance-efficient application of power optimizing code transformations on programmable multimedia processors", *Doctoral Dissertation*, Univ. Patras, March 2000.

[357] C.Mauras, P.Quinton, S.Rajopadhye, Y.Saouter, "Scheduling Affine Parameterized Recurrences by means of Variable Dependent Timing Functions", *Proc. Intnl. Conf. on Applic.-Spec. Array Processors*, Princeton NJ, Sep. 1990.

[358] D.McCrackin, "Eliminating interlocks in deeply pipelined processors by delay enforced multistreaming", *IEEE Trans. on Computers*, Vol.C-40, No.10, pp.1125-1132, Oct. 1991.

[359] M.C.McFarland, A.C.Parker, R.Camposano, "The high-level synthesis of digital systems", *Proc. of the IEEE*, special issue on "The future of computer-aided design", Vol.78, No.2, pp.301-318, Feb. 1990.

[360] K.McKinley, M.Hall, T.Harvey, K.Kennedy, N.McIntosh, J.Oldham, M.Paleczny, and G.Roth, "Experiences using the ParaScope editor: an interactive parallel programming tool", in *4th ACM SIGPLAN Symp. on Principles and Practice of Parallel Programming*, San Diego, USA, May 1993.

[361] K.McKinley, S.Carr, C-W.Tseng, "Improving data locality with loop transformations", *ACM Trans. on Programming Languages and Systems*, Vol.18, No.4, pp.424-453, July 1996.

[362] K.McKinley, "A compiler optimization algorithm for shared-memory multi-processors", *IEEE Trans. on Parallel and Distributed Systems*, Vol.9, No.8, pp.769-787, Aug. 1998.

[363] H.Mehta, R.Owens, M.J.Irwin, R.Chen, D.Ghosh, "Techniques for low energy software", *Proc. IEEE Intnl. Symp. on Low Power Design*, Monterey, CA, Aug. 1997.

[364] G.Mei, W.Liu, "Parallel algorithms for computer vision primitives", *Proc. IEEE Intnl. Conf. on Computer Design*, Port Chester NY, pp.506-509, Oct. 1986.

[365] T.H.Meng, B.Gordon, E.Tsern, A.Hung, "Portable video-on-demand in wireless communication", special issue on "Low power electronics" of the *Proc. of the IEEE*, Vol.83, No.4, pp.659-680, April 1995.

[366] B.Paul, "The Mesa 3D Graphics Library", Online document available at http://www.mesa3d.org/, 1999.

[367] S.Meyers, "More effective C++", Addison Wesley, 1996.

[368] P.Middelhoek, G.Mekenkamp, B.Molenkamp, T.Krol, "A transformational approach to VHDL and CDFG based high-level synthesis: a case study", *Proc. IEEE Custom Integrated Circuits Conf.*, Santa Clara CA, pp.37-40, May 1995.

[369] M.Miranda, C.Ghez, C.Kulkarni, F.Catthoor, "Systematic Speed-Power Memory Data-Layout Exploration for Cache Controlled Embedded Multimedia Applications", *Proc. 14th ACM/IEEE Intnl. Symp. on System-Level Synthesis* (ISSS), Montreal, Canada, pp.107-112, Oct. 2001.

[370] W.Miranker, A.Winkler, "Space-time representation of computational structures", *Computing*, pp.93-114, 1984.

[371] M.Miranda, F.Catthoor, M.Janssen, H.De Man, "ADOPT: Efficient Hardware Address Generation in Distributed Memory Architectures", *Proc. 9th ACM/IEEE Intnl. Symp. on System-Level Synthesis* (ISSS), La Jolla CA, pp.20-25, Nov. 1996.

[372] M.Miranda, F.Catthoor, M.Janssen, H.De Man, "High-level Address Optimisation and Synthesis Techniques for Data-Transfer Intensive Applications", *IEEE Trans. on VLSI Systems*, Vol.6, No.4, pp.677-686, Dec. 1998.

[373] N.Mitchell, L.Carter, J.Ferrante, "A compiler perspective on architectural evolutions", *IEEE TC on Computer Architecture Newsletter*, special issue on "Interaction between Compilers and Computer Architectures", pp.7-9, June 1997.

[374] T.Mitra, T.Chiueh, "Dynamic 3D graphics workload characterization and the architectural implications", *Proc. Intnl. Symp. on Microarchitecture*, pp.62-71, Haifa, Israel, Nov 1999.

[375] D.Moldovan, "On the design of algorithms for VLSI systolic arrays", *Proc. of the IEEE*, Vol.71, No.1, pp.113-120, Jan. 1983.

[376] M.Moonen, P.Van Dooren, J.Vandewalle, "An SVD updating algorithm for subspace tracking", *SIAM J. Matrix Anal. Appl.*, Vol.13, No.4, pp.1015-1038, 1992.

[377] S-M.Moon, K.Ebcioglu, "A study on the number of memory ports in multiple instruction issue machines", *Micro'26*, pp.49-58, Nov. 1993.

[378] D.Moolenaar, L.Nachtergaele, F.Catthoor, H.De Man, "System-level power exploration for MPEG-2 decoder on embedded cores : a systematic approach", *Proc. IEEE Wsh. on Signal Processing Systems (SIPS)*, Leicester, UK, Nov. 1997. Also in VLSI Signal Processing X, M.Ibrahim et al. (eds.), IEEE Press, New York, pp.395-404, 1997.

[379] T.C.Mowry, M.Lam, A.Gupta, "Design and evaluation of a compiler algorithm for prefetching", *Proc. of Fifth Intnl. Conf. on Architectural Support for Programming Languages and Operating Systems*, ACM Press, pp.62-73, New York, 1992.

[380] —, The ISO/IEC Moving Picture Experts Group Home Page, http://www.cselt.it/mpeg/

[381] S.Muchnick, "Advanced compiler design and implementation", *Morgan Kaufmann Publishers Inc.* , ISBN 1-55860-320-4, 1997.

[382] J.M.Mulder, N.T.Quach, M.J.Flynn, "An Area Model for On-Chip Memories and its Application", *IEEE J. of Solid-state Circ.*, Vol.SC-26, No.1, pp.98-105, Feb. 1991.

[383] P.Murthy, S.Bhattacharyya, "A buffer merging technique for reducing memory requirements of synchronous dataflow specifications", *Proc. 12th ACM/IEEE Intnl. Symp. on System-Level Synthesis* (ISSS), San Jose CA, pp.78-84, Dec. 1999.

[384] L.Nachtergaele, V.Tiwari, N.Dutt, "System and architecture-level power reduction of microprocessor-based communication and multi-media applications", *Proc. IEEE Intnl. Conf. on Comp. Aided Design*, Santa Clara CA, pp.569-573, Nov. 2000.

[385] L.Nachtergaele, F.Catthoor, C.Kulkarni, "Random access data storage components in customized architectures", *IEEE Design and Test of Computers*, Vol.18, No.2, pp.40-55, June 2001.

[386] L.Nachtergaele, F.Catthoor, F.Balasa, F.Franssen, E.De Greef, H.Samsom, H.De Man, "Optimisation of memory organisation and hierarchy for decreased size and power in video and image processing systems", *Proc. Intnl. Wsh. on Memory Technology, Design and Testing*, San Jose CA, pp.82-87, Aug. 1995.

[387] L.Nachtergaele, F.Catthoor, B.Kapoor, D.Moolenaar, S.Janssens, "Low power storage exploration for H.263 video decoder", *IEEE Wsh. on VLSI signal processing*, Monterey CA, Oct. 1996. Also in *VLSI Signal Processing IX*, W.Burleson, K.Konstantinides, T.Meng, (eds.), IEEE Press, New York, pp.116-125, 1996.

[388] L.Nachtergaele, D.Moolenaar, B.Vanhoof, F.Catthoor, H.De Man, "System-level power optimization of video codecs on embedded cores : a systematic approach", special issue on *Future directions in the design and implementation of DSP systems* (eds. Wayne Burleson, Konstantinos Konstantinides) of *J. of VLSI Signal Processing*, Vol.18, No.2, Kluwer, Boston, pp.89-110, Feb. 1998.

[389] L.Nachtergaele, T.Gijbels, J.Bormans, F.Catthoor, M.Engels, "Power and speed-efficient code transformation of multi-media algorithms for RISC processors", *IEEE Intnl. Wsh. on Multi-media Signal Proc.*, Los Angeles CA, pp.317-322, Dec. 1998.

[390] J.Navarro, T.Juan, T.Lang, "MOB forms : A class of multilevel block algorithms for dense linear algebra operations", *Intnl. Conf. on Supercomputing*, pp.354-363, July 1994.

[391] M.Neeracher, R.Rühl, "Automatic parallelization of LINPACK routines on distributed memory parallel processors", *Proc. IEEE Intnl. Parallel Proc. Symp.*, Newport Beach CA, April 1993.

[392] G.L.Nemhauser, L.A.Wolsey, "Integer and Combinatorial Optimization", J.Wiley&Sons, New York, N.Y., 1988.

[393] S.Y.Ohm, F.J.Kurdahi, N.Dutt, "Comprehensive lower bound estimation from behavioral descriptions", *IEEE/ACM Intnl. Conf. on Computer-Aided Design*, pp.182-7, 1994.

[394] T.Omnes, T.Franzetti, F.Catthoor, "Interactive algorithms for low-cost scheduling in Acropolis", *Proc. Wsh. on Compilation and Automatic Parallelisation*, St.Jacques de St.Narbor, France, Nov. 1999.

[395] T.Omnes, T.Franzetti, F.Catthoor, "Interactive algorithms for minimizing bandwidth in high throughput telecom and multimedia", *Proc. 37th ACM/IEEE Design Automation Conf.*, Los Angeles CA, pp.328-331, June 2000.

[396] T.Omnes, T.Franzetti, F.Catthoor, "Multi-dimensional selection techniques for minimizing memory bandwidth in high-throughput embedded systems", *High Performance Comp. (HiPC)*, Bangalore, India, Dec. 2000. Also in *Lecture Notes in Computer Science*, Springer Verlag, Vol.1970, pp.323-334, 2000.

[397] T.Omnes, F.Catthoor, "Space-time memory acces patterns: a step towards the object-oriented design of low-cost distributed platforms", *5th Intnl. Wsh. on Software and COmPilers for Embedded Systems (SCOPES)*, St Goar, Germany, March 2001.

[398] T.Omnes, T.Franzetti, F.Catthoor, "Dynamic and adaptive algorithms for minimizing memory bandwdth in high-throughput telecom networks, speech, image and video embedded systems", *Technique et Science Informatique (TSI)*, Ed. Hermès, Paris, France, Special Issue on "Parallel Compilation", Vol.20, No.8, pp.1-25, 2001.

[399] T.Omnes, E.Brockmeyer, C.Kulkarni, K.Danckaert, "Low-power software for System-on-a-Chip (SOC)", Special Session on "Low-power System-on-Chip design", *Proc. 4th ACM/IEEE Design and Test in Europe Conf.*, Munich, Germany, pp.488-494, March 2001.

[400] T.Omnes, "Acropolis: un precompilateur de specification pour l'exploration du transfert et du stockage des donnees en conception de systemes embarques a haut debit", *Doctoral Dissertation*, Ecole des Mines de Paris, May 2001.

[401] J.O'Rourke, "Computational geometry in C", Cambridge University Press, Cambridge NY, 1994.

[402] D.A.Padua, M.J.Wolfe. "Advanced compiler optimizations for supercomputers", *Communications of the ACM*, Vol.29, No.12, pp.1184-1201, 1986.

[403] M.Palis, J.Liou, D.Wei, "Task clustering and scheduling for distributed memory parallel architectures", *IEEE Trans. on Parallel and Distributed Systems*, Vol.7, No.1, pp.46-55, Jan. 1996.

[404] P.Panda, F.Catthoor, N.Dutt, K.Danckaert, E.Brockmeyer, C.Kulkarni, A.Vandecappelle, P.G.Kjeldsberg, "Data and Memory Optimizations for Embedded Systems", *ACM Trans. on Design Automation for Embedded Systems (TODAES)*, Vol.6, No.2, pp.142-206, April 2001.

[405] P.R.Panda, N.D.Dutt, A.Nicolau, " Memory data organization for improved cache performance in embedded processor applications", *Proc. 9th ACM/IEEE Intnl. Symp. on System-Level Synthesis (ISSS)*, La Jolla CA, pp.90-95, Nov. 1996.

[406] P.Panda, N.Dutt, "Low power mapping of behavioral arrays to multiple memories", *Proc. IEEE Intnl. Symp. on Low Power Design*, Monterey CA, pp.289-292, Aug. 1996.

[407] P.R.Panda, H.Nakamura, N.D.Dutt and A.Nicolau, "A data alignment technique for improving cache performance", *Proc. IEEE Intnl. Conf. on Computer Design*, Santa Clara CA, pp.587-592, Oct. 1997.

[408] P.R.Panda, N.D.Dutt, A.Nicolau, "Efficient utilization of scratch-pad memory in embedded processor applications", *Proc. 5th ACM/IEEE Europ. Design and Test Conf.*, Paris, France, pp., March 1997.

[409] P.R.Panda, "Memory optimizations and exploration for embedded systems", *Doctoral Dissertation*, U.C.Irvine, April 1998.

[410] P.R.Panda, N.D.Dutt, A.Nicolau, "Incorporating DRAM access modes into high-level synthesis", *IEEE Trans. on Comp.-aided Design*, Vol.CAD-17, No.2, pp.96-109, Feb. 1998.

[411] P.R.Panda, N.D.Dutt, A.Nicolau, "Data cache sizing for embedded processor applications", *Proc. 1st ACM/IEEE Design and Test in Europe Conf.*, Paris, France, pp.925-926, Feb. 1998.

[412] P.R.Panda, N.D.Dutt, A.Nicolau, "Memory issues in embedded in systems-on-chip: optimization and exploration", Kluwer Acad. Publ., Boston, 1999.

[413] P.R.Panda, N.D.Dutt, A.Nicolau, "Local memory exploration and optimization in embedded systems", *IEEE Trans. on Comp.-aided Design*, Vol.CAD-18, No.1, pp.3-13, Jan. 1999.

[414] P.R.Panda, H.Nakamura, N.D.Dutt, A.Nicolau, "Augmenting loop tiling with data alignment for improved cache performance", *IEEE Trans. on Computers*, Vol.48, No.2, pp.142-149, Feb. 1999.

[415] P.R.Panda, "Memory bank customization and assignment in behavioral synthesis", *Proc. IEEE Intnl. Conf. Comp. Aided Design*, Santa Clara CA, pp.477-481, Nov. 1999.

[416] K.Parhi, "Algorithmic transformation techniques for concurrent processors", *Proc. of the IEEE*, Vol.77, No.12, pp.1879-1895, Dec. 1989.

[417] K.Parhi, "Rate-optimal fully-static multiprocessor scheduling of data-flow signal processing programs", *Proc. IEEE Intnl. Symp. on Circuits and Systems*, Portland OR, pp.1923-1928, May 1989.

[418] N.Passos, E.Sha, "Full parallelism of uniform nested loops by multi-dimensional retiming", *Proc. Intnl. Conf. on Parallel Processing*, Vol.2, pp.130-133, Aug. 1994.

[419] N.Passos, E.Sha, "Push-up scheduling: optimal polynomial-time resource constrained scheduling for multi-dimensional applications", *Proc. IEEE Intnl. Conf. Comp. Aided Design*, San Jose CA, pp.588-591, Nov. 1995.

[420] N.Passos, E.Sha, L-F.Chao, "Multi-dimensional interleaving for time-and-memory design optimization", *Proc. IEEE Intnl. Conf. on Computer Design*, Austin TX, pp.440-445, Oct. 1995.

[421] N.Passos, E.Sha, "Achieving full parallelism using multidimensional retiming", *IEEE Trans. on Parallel and Distributed Systems*, Vol.7, No.11, pp.1150-1163, Nov. 1996.

[422] N.Passos, E.Sha, "Synchronous circuit optimization via multi-dimensional retiming", *IEEE Trans. on Circuits and Systems II: Analog and Digital Signal Processing*, Vol.CAS-43, No.7, pp.507-519, July 1996.

[423] D.A.Patterson, and J.L.Hennessy, "Computer Organisation and Design: the Hardware/Software Interface", *Morgan Kaufmann Publishers*, NY, 1994.

[424] D.Patterson, J.Hennessey, "Computer architecture : A quantitative approach", *Morgan Kaufmann Publ.*, San Francisco, 1996.

[425] P.G.Paulin, J.P.Knight, "Force-Directed Scheduling in Automatic Data Path Synthesis", *Proc. 24th ACM/IEEE Design Automation Conf.*, Miami, Florida, pp.195-202, June 1987.

[426] P.G.Paulin, J.P.Knight, "Force-directed scheduling for the behavioral synthesis of ASICs", *IEEE Trans. on Computer-Aided Design of Integrated Circuits and Systems*, Vol.8, No.6, pp.661-679, June 1989.

[427] "Pentium P-III data book", Intel Corporation, Santa Clara CA, 1999.

[428] K.Pettis and R.C.Hansen, "Profile guided code positioning", *In ACM SIGPLAN'90 Conf. on Programming Language and Design Implementation*, pp.16-27, June 1990.

[429] S.Pinter, "Register Allocation with Instruction Scheduling: a New Approach", *ACM SIGPLAN Notices*, Vol.28, pp.248-257, June 1993.

[430] C.Polychronopoulos, "Compiler optimizations for enhancing parallelism and their impact on the architecture design", *IEEE Trans. on Computers*, Vol.37, No.8, pp.991-1004, Aug. 1988.

[431] A.Porterfield, "Software methods for improvement of cache performance on supercomputer applications", *Ph.D. dissertation*, Rice Univ., May 1989.

[432] R.Potter, G.Steven, "Investigating the limits of fine-grained parallelism in a statically scheduled superscalar architecture", *Proc. EuroPar Conf.*, Lyon, France, Aug. 1996. "Lecture notes in computer science" series, Springer Verlag, pp.779-788, 1996.

[433] S.A.Przybylski, M.Horowitz, and J.Hennessy. "Performance tradeoffs in cache design", *Proc. 15th Annual Intnl. Symp. on Computer Architecture*, pp.290-298, Honolulu, Hawaii, 1988.

[434] S.Przybylski, "New DRAM architectures", tutorial at *IEEE Intnl. Solid-State Circ. Conf.*, San Francisco CA, Feb. 1997.

[435] W.Pugh, "The Omega Test: a fast and practical integer programming algorithm for dependence analysis", *Communications of the. ACM*, Vol.35, No.8, Aug. 1992.

[436] W.Pugh, D.Wonnacott, "An exact method for analysis of value-based array data dependences", *Proc. 6th Intnl. Wsh. on Languages and Compilers for Parallel Computing*, pp.546-566, Portland OR, Aug. 1993.

[437] F.Quillere, S.Rajopadhye, "Optimizing memory usage in the polyhedral model", *ACM Trans. on Programming Languages and Systems*, Vol.22, No.12, pp., Dec. 2000.

[438] P.Quinton. "Automatic synthesis of systolic arrays from recurrent uniform equations", *11th Intnl. Symp. Computer Architecture*, Ann Arbor, pp.208-214, June 1984.

[439] F.Quillere, S.Rajopadhye, "Optimizing memory usage in the polyhedral model", presented at *Massively Parallel Computer Systems Conf.*, April 1998. Also internal report IRISA, Rennes.

[440] J.Rabaey, H.De Man, J.Vanhoof, G.Goossens, F.Catthoor, "CATHEDRAL II: A Synthesis System for Multiprocessor DSP Systems", in *Silicon Compilation*, D.Gajski (ed.), Addison-Wesley, pp.311-360, 1988.

[441] J.Rabaey, M.Pedram, "Low Power Design Methodologies", Kluwer Acad. Publ., 1996.

[442] L.Rabiner, R.Schafer, "Digital processing of speech signals", Prentice Hall, Englewood Cliffs NJ, 1978.

[443] J.Ramanujam, J.Hong, M.Kandemir, A.Narayan, "Reducing memory requirements of nested loops for embedded systems", *38th ACM/IEEE Design Automation Conf.*, Las Vegas NV, pp.359-364, June 2001.

[444] L.Ramachandran, D.Gajski, V.Chaiyakul, "An algorithm for array variable clustering", *Proc. 5th ACM/IEEE Europ. Design and Test Conf.*, Paris, France, pp.262-266, Feb. 1994.

[445] S.Ravi, G.Lakshminarayana, N.Jha, "Removal of memory access bottlenecks for scheduling control-flow intensive behavioral descriptions", *Proc. IEEE Intnl. Conf. Comp. Aided Design*, Santa Clara CA, pp.577-584, Nov. 1998.

[446] C.Reffay, G-R.Perrin, "From dependence analysis to communication code generation: the "look-forwards" model", *Intnl. Wsh. on Algorithms and Parallel VLSI Architectures*, Leuven, Belgium, Aug. 1994. Also in "Algorithms and Parallel VLSI Architectures III" (eds. M.Moonen, F.Catthoor), Elsevier, pp.341-352, 1995.

[447] J.Rivers, E.Davidson, "Reducing conflicts in direct-mapped caches with a temporality-based design", *Proc. Intnl. Conf. on Parallel Processing (ICPP'96)*, pp.154-163, Aug. 1996.

[448] G.Rivera, C.Tseng, "Compiler optimizations for eliminating cache conflict misses", *Technical Report CS-TR-3819*, Dept of Computer Science, University of Maryland, July 1997.

[449] J.Robinson, "Efficient General-Purpose Image Compression with Binary Tree Predictive Coding", *IEEE Trans. on Image Processing*, Vol.6, No.4, pp.601-608, Apr. 1997.

[450] K.Roenner, J.Kneip, "Architecture and applications of the HiPar video signal processor", to appear in *IEEE Trans. on Circuits and Systems for Video Technology*, special issue on "VLSI for video signal processors" (eds. B.Ackland, T.Nishitani, P.Pirsch), 1998.

[451] D.Roose, R.V.Driessche, "Distributed memory parallel computers and computational fluid dynamics", Internal Report TW186, Dept. Computer Science, K.U.Leuven, March 1993.

[452] J.Rosseel, F.Catthoor, H.De Man, "The exploitation of global operations in affine space-time mapping", *Proc. IEEE Wsh. on VLSI signal processing*, Napa Valley CA, Oct. 1992. Also in *VLSI Signal Processing V*, K.Yao, R.Jain, W.Przytula (eds.), IEEE Press, New York, pp.309-319, 1992.

[453] K.H.Rosen, "Discrete Mathematics and its Applications", McGraw-Hill, Inc., New York, USA, 1995 (Third edition).

[454] B.Rouzeyre, G.Sagnes, "A new method for the minimization of memory area in high level synthesis", *Proc. Euro-ASIC Conf.*, Paris, France, pp.184-189, May 1991.

[455] J.Saltz, H.Berrymann, J.Wu, "Multiprocessors and runtime compilation", *Proc. Intnl. Wsh. on Compilers for Parallel Computers*, Paris, France, 1990.

[456] H.Samsom, L.Claesen, H.De Man, "SynGuide: an environment for doing interactive correctness preserving transformations", *IEEE Wsh. on VLSI signal processing*, Veldhoven, The Netherlands, Oct. 1993. Also in *VLSI Signal Processing VI*, L.Eggermont, P.Dewilde, E.Deprettere, J.van Meerbergen (eds.), IEEE Press, New York, pp.269-277, 1993.

[457] H.Samsom, F.Franssen, F.Catthoor, H.De Man, "Verification of loop transformations for real time signal processing applications", *IEEE Wsh. on VLSI signal processing*, La Jolla CA, Oct. 1994. Also in *VLSI Signal Processing VII*, J.Rabaey, P.Chau, J.Eldon (eds.), IEEE Press, New York, pp.208-217, 1994.

[458] H.Samsom, F.Franssen, F.Catthoor, H.De Man, "System-level Verification of Video and Image Processing Specifications", *Proc. 8th ACM/IEEE Intnl. Symp. on System-Level Synthesis* (ISSS), Cannes, France, pp.144-149, Sep. 1995.

[459] H.Samsom, F.Catthoor, "SynGuide, Reference Manual, Version 2.0", IMEC, Internal report, Aug. 1996.

[460] R.Schaffer, F.Catthoor, R.Merker, "Combining background memory management and regular array co-partitioning, illustrated on a full motion estimation kernel", special issue on *Advanced Regular Array Design* (T.Plaks, ed.), in *J. of Parallel Algorithms and Applications*, Gordan and Beach Sc. Publ., Vol.15, No.3-4, pp.201-228, Dec. 2000.

[461] R.Schaffer, F.Catthoor, R.Merker, "Combining background memory management and regular array co-partitioning, illustrated on a full motion estimation kernel", *Proc. Intnl. Conf. on VLSI Design*, Calcutta, India, pp.104-109, Jan. 2000.

[462] P.Schelkens, F.Decroos, G.Lafruit, J.Cornelis, F.Catthoor, "Implementation of an integer wavelet transform on a parallel TI TMS320C40 platform", *Proc. IEEE Wsh. on Signal Processing Systems (SIPS)*, Taipeh, Taiwan, IEEE Press, pp.81-89, Oct. 1999.

[463] H.Schmit and D.Thomas. "Synthesis of application-specific memory designs", *IEEE Trans. on VLSI Systems*, Vol.5, No.1, pp.101-111, March 1997.

[464] M.Schönfeld, M.Schwiegershausen, P.Pirsch, "Synthesis of intermediate memories for the data supply to processor arrays," in *Algorithms and Parallel Architectures II*, P.Quinton, Y.Robert (eds.), Elsevier, Amsterdam, pp.365-370, 1992.

[465] C.Schurgers, F.Catthoor, M.Engels, "Optimized MAP Turbo Decoder", *Proc. IEEE Wsh. on Signal Processing Systems (SIPS)*, Lafayette LA, IEEE Press, pp.245-254, Oct. 2000.

[466] C.Schurgers, F.Catthoor, M.Engels, "Memory optimization of MAP turbo decoder algorithms", *IEEE Trans. on VLSI Systems*, Vol.9, No.2, pp.305-312, June 2001.

[467] C.Schurgers, F.Catthoor, M.Engels, "Energy efficient data transfer and storage organisation for an optimized MAP turbo decoder", *Proc. IEEE Intnl. Symp. on Low Power Design*, San Diego CA, pp.76-81, Aug. 1999.

[468] D.A.Schwartz, T.P.Barnwell, "Cyclo-static multiprocessor scheduling for the optimal realization of shift-invariant flow graphs", *Proc. IEEE Intnl. Conf. on Acoustics, Speech and Signal Processing*, Tampa, Florida, pp.1384-1387, March 1985.

[469] M.Segal, K.Akeley, "The Design of the OpenGL Graphics Interface", *In Proc. of SIGGRAPH 94*, 1994.

[470] M.Segal, K.Akeley, "The OpenGL Graphics System : A Specification", *Version 1.1*, Silicon Graphics Inc. , March 1997.

[471] T.Seki, E.Itoh, C.Furukawa, I.Maeno, T.Ozawa, H.Sano, N.Suzuki, "A 6-ns 1-Mb CMOS SRAM with Latched Sense Amplifier", *IEEE J. of Solid-state Circuits*, Vol.SC-28, No.4, pp.478-483, Apr. 1993.

[472] O.Sentieys, D.Chillet, J.P.Diguet, J.Philippe, "Memory module selection for high-level synthesis", *Proc. IEEE Wsh. on VLSI signal processing*, Monterey CA, Oct. 1996.

[473] K.C.Shashidar, A.Vandecappelle, F.Catthoor, "Low Power Design of Turbo Decoder Module with Exploration of Energy-Performance Trade-offs", *Wsh. on Compilers and Operating Systems for Low Power (COLP'01)* in conjunction with *Intnl. Conf. on Parallel Arch. and Compilation Techniques (PACT)*, Barcelona, Spain, pp. 10.1-10.6, Sep. 2001.

[474] W.Shang, M.O'Keefe, J.Fortes, "Generalized cycle shrinking", presented at Wsh. on "Algorithms and Parallel VLSI Architectures II", Bonas, France, June 1991. Also in *Algorithms and parallel VLSI architectures II*, P.Quinton and Y.Robert (eds.), Elsevier, Amsterdam, pp.131-144, 1992.

[475] W.Shang, J.Fortes, "Independent partitioning of algorithms with uniform dependencies", *IEEE Trans. on Computers*, Vol.41, No.2, pp.190-206, Feb. 1992.

[476] W.Shang, Z.Shu, "Data alignment of loop nests without nonlocal communications", *Proc. Intnl. Conf. on Applic.-Spec. Array Processors*, San Francisco, CA, pp.439-451, Aug. 1994.

[477] W.Shang, E.Hodzic, Z.Chen, "On uniformization of affine dependence algorithms", *IEEE Trans. on Computers*, Vol.45, No.7, pp.827-839, July 1996.

[478] B.Shackleford, M.Yasuda, E.Okushi, H.Koizumi, H.Tomiyama, and H.Yasuura, "Memory-CPU size optimization for embedded system designs", *Proc. 34th ACM/IEEE Design Automation Conf.*, Anaheim CA, June 1997.

[479] W-T.Shiue, S.Tadas, C.Chakrabarti, "Low power multi-module, multi-port memory design for embedded systems", *Proc. IEEE Wsh. on Signal Processing Systems (SIPS)*, Lafayette LA, IEEE Press, pp.529-538, Oct. 2000.

[480] W-T.Shiue, C.Chakrabarti, "Memory design and exploration for low power embedded systems", *Proc. IEEE Wsh. on Signal Processing Systems (SIPS)*, Taipeh, Taiwan, IEEE Press, pp.281-290, Oct. 1999.

[481] W-T.Shiue, C.Chakrabarti, "Memory exploration for low power embedded systems", *Proc. 36th ACM/IEEE Design Automation Conf.*, New Orleans LA, pp.140-145, June 1999.

[482] T.Sikora, "The MPEG-4 video standard verification model", *IEEE Trans. on Circuits and Systems for Video Technology*, Vol.7, No.1, pp.19–31, Feb. 1997.

[483] P.Slock, S.Wuytack, F.Catthoor, G.de Jong, "Fast and extensive system-level memory exploration for ATM applications", *Proc. 10th ACM/IEEE Intnl. Symp. on System-Level Synthesis (ISSS)*, Antwerp, Belgium, pp.74-81, Sep. 1997.

[484] S.Smith and J.Brady, "Susan - a new approach to low level image processing", *Intnl. J. of Computer Vision*, Vol.23, No.1, pp.45-78, May 1997.

[485] D.Soudris, N.Zervas, A.Argyriou, M.Dasygenis, K.Tatas, C.Goutis, A.Thanailakis, "Data reuse and parellel embedded architectures for low power, real-time multimedia applications", *Proc. IEEE Wsh. on Power and Timing Modeling, Optimization and Simulation (PATMOS)*, Goettingen, Germany, pp.343-354, Oct. 2000.

[486] A.Stammermann, L.Kruse, W.Nebel, A.Pratsch, E.Schmidt, M.Schulte, A.Schulz, "System-level optimization and design-space exploration for low power", *Proc. 14th ACM/IEEE Intnl. Symp. on System-Level Synthesis (ISSS)*, Montreal, Canada, pp.142-146, Oct. 2001.

[487] Q.Stout, "Mapping vision algorithms to parallel architectures", *Proc. of the IEEE*, Vol.76, No.8, pp.982-995, Aug. 1988.

[488] L.Stok, J.Jess, "Foreground memory management in data path synthesis" *Intnl. J. on Circuit Theory and Appl.*, Vol.20, pp.235-255, 1992.

[489] P. Strobach, "QSDPCM – A New Technique in Scene Adaptive Coding," *Proc. 4th Eur. Signal Processing Conf.*, EUSIPCO-88, Grenoble, France, Elsevier Publ., Amsterdam, pp.1141–1144, Sep. 1988.

[490] J.Subhlok, D.O'Hallaron, T.Grosss, P.Dinda, J.Webb, "Communication and memory requirements as the basis for mapping task and data parallel programs", *Proc. Supercomputing*, Washington DC, Nov. 1994.

[491] A.Sudarsanam, S.Malik, "Memory bank and register allocation in software synthesis for ASIPs',, *Proc. IEEE Intnl. Conf. Comp. Aided Design*, San Jose CA, pp.388–392, Nov. 1995.

[492] A.Sudarsanam, S.Liao, S.Devadas, "Analysis and evaluation of address arithmetic capabilities in custom DSP architectures", *Proc. 34th ACM/IEEE Design Automation Conf.*, Anaheim CA, June 1997.

[493] S.Sudharsanan, "MAJC-5200: A High Performance Microprocessor for Multimedia Computing", *Lecture Notes in Computer Science (PDIVM/IPDPS 2000)*, Vol.1800, pp. 161-170, May 2000.

[494] ..., "Synopsys Digital Signal Processing – COSSAP Home Page", http://www.synopsys.com/products/dsp/dsp.html

[495] O.Temam, C.Fricker, W.Jalby, "Cache interference phenomena", *Proc. of ACM SIGMETRICS'94 Conf. on Measurement and Modeling of Computer Systems*, 1994.

[496] O.Temam, "An algorithm for optimally exploiting spatial and temporal locality in upper memory levels", *IEEE Trans. on Computers*, Vol.48, No.2, pp.150-158, Feb. 1999.

[497] Y.Therasse, G.H.Petit, M.Delvaux, "VLSI architecture of a SMDS/ATM router", *Annales des Télécommunications*, Vol.48, No.3-4, pp.166-180, 1993.

[498] L.Thiele, "On the design of piecewise regular processor arrays", *Proc. IEEE Intnl. Symp. on Circuits and Systems*, Portland OR, pp.2239-2242, May 1989.

[499] D.E.Thomas, E.Dirkes, R.Walker, J.Rajan, J.Nestor, R.Blackburn, "The system architect's workbench", *Proc. 25th ACM/IEEE Design Automation Conf.*, San Francisco CA, pp.337-343, June 1988.

[500] F.Thoen, F.Catthoor, "Modeling, Verification and Exploration of Task-level Concurrency in Real-Time Embedded Systems", ISBN 0-7923-7737-0, Kluwer Acad. Publ., Boston, 1999.

[501] Texas Instruments TMX320C55x01 DSP Data Book, Texas Instruments, Dallas, 2000.

[502] Texas Instruments TMX320C6201 DSP Data Book, Texas Instruments, Dallas, 1998.

[503] V.Tiwari, S.Malik, A.Wolfe, "Power analysis of embedded software: a first step towards software power minimization", *Proc. IEEE Intnl. Conf. Comp. Aided Design*, Santa Clara CA, pp.384-390, Nov. 1994.

[504] N.Topham, A.Gonzalez, "Randomized cache placement for eliminating conflicts", *IEEE Trans. on Computers*, Vol.48, No.2, pp.185-191, Feb. 1999.

[505] E.Torrie, M.Martonosi, C-W.Tseng, M.Hall, "Characterizing the memory behavior of compiler-parallelized applications", *IEEE Trans. on Parallel and Distributed Systems*, Vol.7, No.12, pp.1224-1236, Dec. 1996.

[506] R.Touzeau, "A Fortran compiler for the FPS-164 scientific computer", in *ACM SIGPLAN Symp. on Compiler Construction*, pp.48-57, June 1984.

[507] TriMedia TM1000 data book, Philips Semiconductors, Sunnyvale, CA, 1997.

[508] A.Nene, S.Talla, B.Goldberg, H.Kim, R.M.Rabbah, "Trimaran - An infrastructure for compiler research in instruction level parallelism", Online document available via http://www.trimaran.org/, 1998.

[509] D.N.Truong, F.Bodin, A.Seznec, "Accurate data distribution into blocks may boost cache performance", *IEEE TC on Computer Architecture Newsletter*, special issue on "Interaction between Compilers and Computer Architectures", pp.55-57, June 1997.

[510] C-J.Tseng, D.Siewiorek, "Automated synthesis of data paths in digital systems", *IEEE Trans. on Comp.-aided Design*, Vol.CAD-5, No.3, pp.379-395, July 1986.

[511] T.Van Achteren, M.Ade, R.Lauwereins, M.Proesmans, L.Van Gool, J.Bormans, F.Catthoor, "Transformations of a 3D Image Reconstruction Algorithm for Data Transfer and Storage Optimisation", *Design Autom. for Embedded Systems*, Kluwer Acad. Publ., Boston, Vol.5, No.3, pp.313-327, Aug. 2000.

[512] T.Van Achteren, R.Lauwereins, F.Catthoor, "Systematic data reuse exploration methodology for irregular access patterns", *Proc. 13th ACM/IEEE Intnl. Symp. on System-Level Synthesis* (ISSS), Madrid, Spain, pp.115-121, Sep. 2000.

[513] T.Van Achteren, M.Ade, R.Lauwereins, M.Proesmans, L.Van Gool, J.Bormans, F.Catthoor, "Global Memory Organisation Optimisations for a 3D Image Reconstruction Algorithm", *IEEE Intnl. Conf. on Signal Processing Appl. and Technology (ICSPAT)*, Orlando FL, (only CDROM), Nov. 1999.

[514] T.Van Achteren, M.Ade, R.Lauwereins, M.Proesmans, L.Van Gool, J.Bormans, F.Catthoor, "Transformations of a 3D image reconstruction algorithm for data transfer and storage optimisation", *IEEE Proc. 10th Intnl. Wsh. on Rapid System Prototyping*, Clearwater FA, pp.81-86, June 1999.

[515] G.Tyson, M.Farrens, J.Matthews, A.Pleszkun, "Managing data caches using selective cache line replacement", *Intnl. J. of Parallel Programming*, Vol.25, No.3, pp.213-242, June 1997.

[516] T.Tzen, L.Ni, "Trapezoid self-scheduling: a practical scheduling scheme for parallel compilers", *IEEE Trans. on Parallel and Distributed Systems*, Vol.4, No.1, pp.87-98, Jan. 1993.

[517] T.Tzen, L.Ni, "Dependence uniformization: a loop parallelization technique", *IEEE Trans. on Parallel and Distributed Systems*, Vol.4, No.5, pp.547-557, May 1993.

[518] S.Udayanarayanan, C.Chakrabarti, "Address Code Generation for Digital Signal Processors", *38th ACM/IEEE Design Automation Conf.*, Las Vegas NV, pp.353-358, June 2001.

[519] I.Verbauwhede, F.Catthoor, J.Vandewalle, H.De Man, "Background memory management for the synthesis of algebraic algorithms on multi-processor DSP chips", *Proc. VLSI'89, Intnl. Conf. on VLSI*, Munich, Germany, pp.209-218, Aug. 1989.

[520] I.Verbauwhede, F.Catthoor, J.Vandewalle, H.De Man, "High-level memory management for real-time signal processing of algebraic algorithms on application-specific micro-coded processors", *Proc. Intnl. Wsh. on Algorithms and Parallel VLSI Architectures*, Pont-a-Mousson, France, June 1990.

[521] I.Verbauwhede, F.Catthoor, J.Vandewalle, H.De Man, "In-place memory management of algebraic algorithms on application-specific IC's", in *Algorithms and Parallel VLSI Architectures*, Vol.B, E.Deprettere, A.Van der Veen (eds.), Elsevier, Amsterdam, pp.353-362, 1991.

[522] I.Verbauwhede, F.Catthoor, J.Vandewalle, H.De Man, "In-place memory management of algebraic algorithms on application-specific IC's", *J. of VLSI signal processing*, Vol.3, Kluwer, Boston, pp.193-200, 1991.

[523] I.Verbauwhede, C.Scheers, J.Rabaey, "Memory estimation for high-level synthesis", *Proc. 31st ACM/IEEE Design Automation Conf.*, San Diego, CA, pp.143-148, June 1994.

[524] A.Vandecappelle, M.Miranda, E.Brockmeyer F.Catthoor, D.Verkest, "Global Multimedia System Design Exploration using Accurate Memory Organization Feedback" *Proc. 36th ACM/IEEE Design Automation Conf.*, New Orleans LA, pp.327-332, June 1999.

[525] S.Van der Wiel, D.Lilja, "When caches aren't enough: data prefetching techniques", *IEEE Computer*, Vol.30, No.7, pp.23-30, July 1997.

[526] T.Van Meeuwen, A.Vandecappelle, A.van Zelst, F.Catthoor, D.Verkest, "System-level interconnect architecture exploration for custom memory organisations", *Proc. 14th ACM/IEEE Intnl. Symp. on System-Level Synthesis (ISSS)*, Montreal, Canada, pp.13-18, Oct. 2001.

[527] W.Verhaegh, P.Lippens, E.Aarts, J.Korst, J.van Meerbergen, A.van der Werf, "Modelling periodicity by PHIDEO streams", *Proc. 6th Intnl. Wsh. on High-Level Synthesis*, Laguna Beach CA, Nov. 1992.

[528] W.Verhaegh, P.Lippens, E.Aarts, J.Korst, J.van Meerbergen, A.van der Werf, "Improved Force-Directed Scheduling in High-Throughput Digital Signal Processing", *IEEE Trans. on Computer-aided design*, Vol.14, No.8, Aug. 1995.

[529] W.Verhaegh, "Multi-dimensional periodic scheduling", *Doctoral dissertation*, T.U.Eindhoven, Dec. 1995.

[530] W.Verhaegh, P.Lippens, E.Aarts, J.van Meerbergen, A.van der Werf, "Multi-dimensional periodic scheduling: model and complexity", *Proc. EuroPar Conf.*, Lyon, France, Aug. 1996. "Lecture notes in computer science" series, Springer Verlag, pp.226-235, 1996.

[531] W.Verhaegh, E.Aarts, P.Van Gorp, "Period assignment in multi-dimensional periodic scheduling", *Proc. IEEE Intnl. Conf. Comp. Aided Design*, Santa Clara CA, pp.585-592, Nov. 1998.

[532] J.Vanhoof, I.Bolsens, H.De Man, "Compiling multi-dimensional data streams into distributed DSP ASIC memory", *Proc. IEEE Intnl. Conf. Comp. Aided Design*, Santa Clara CA, pp.272-275, Nov. 1991.

[533] F.Vermeulen, F.Catthoor, D.Verkest, H.De Man, "Formalized Three-Layer System-Level Model and Reuse Methodology for Embedded Data-Dominated Applications", *IEEE Trans. on VLSI Systems*, Vol.8, No.2, pp.207-216, April 2000.

[534] F.Vermeulen, F.Catthoor, D.Verkest, H.De Man, "Extended Design Reuse Trade-Offs in Hardware-Software Architecture Mapping", *Proc. ACM/IEEE Wsh. on Hardware/Software Co-Design (Codes)*, San Diego CA, pp.103-107, May 2000.

[535] F.Vermeulen, L.Nachtergaele, F.Catthoor, D.Verkest, H.De Man, "Flexible hardware acceleration for multimedia oriented microprocessors", *Proc. IEEE/ACM Intnl. Symp. on Microarchitecture*, MICRO-33, Monterey CA, Dec. 2000.

[536] F.Vermeulen, F.Catthoor, D.Verkest, H.De Man, "Formalized Three-Layer System-Level Model and Reuse Methodology for Embedded Data-Dominated Applications", *Proc. 3rd ACM/IEEE Design and Test in Europe Conf.*, Paris, France, pp.92-98, April 2000.

[537] F.Vermeulen, F.Catthoor, D.Verkest, H.De Man, "A System-Level Reuse Methodology for Embedded Data-Dominated Applications", *Proc. IEEE Wsh. on Signal Processing Systems (SIPS)*, Boston MA, IEEE Press, pp.551-560, Oct. 1998.

[538] M.van Swaaij, F.Catthoor, H.De Man, "Architectural alternatives for the Hough transform", *Proc. IFIP Wsh. on Parallel Arch. on Silicon from Systolic Arrays to Neural Networks*, Grenoble, France, Nov. 1989.

[539] M.van Swaaij, F.Franssen, F.Catthoor, H.De Man, "Modelling data and control flow for high-level memory management", *Proc. 3rd ACM/IEEE Europ. Design Automation Conf.*, Brussels, Belgium, pp.8-13, March 1992.

[540] M.van Swaaij, F.Catthoor, H.De Man, "Signal analysis and signal transformations for ASIC regular array synthesis", presented at Wsh. on "Algorithms and Parallel VLSI Architectures II", Bonas, France, June 1991. Also in *Algorithms and parallel VLSI architectures II*, P.Quinton and Y.Robert (eds.), Elsevier, Amsterdam, pp.223-232, 1992.

[541] M.van Swaaij, F.Franssen, F.Catthoor, H.De Man, "Automating high-level control flow transformations for DSP memory management", *Proc. IEEE Wsh. on VLSI signal processing*, Napa Valley CA, Oct. 1992. Also in *VLSI Signal Processing V*, K.Yao, R.Jain, W.Przytula (eds.), IEEE Press, New York, pp.397-406, 1992.

[542] M.van Swaaij, "Data-flow geometry: exploiting regularity in system-level synthesis", *Doctoral dissertation*, ESAT/EE Dept., K.U.Leuven, Belgium, Dec. 1992.

[543] Z.Wang, M.Kirkpatrick, E.Sha, "Optimal two level partitioning and loop scheduling for hiding memory latency for DSP applications", *Proc. 37th ACM/IEEE Design Automation Conf.*, Los Angeles CA, pp.556-559, June 2000.

[544] D.Wang, Y.Hu, "Multiprocessor implementation of real-time DSP algorithms", *IEEE Trans. on VLSI Systems*, Vol.3, No.3, pp.393-403, Sep. 1995.

[545] C.Y.Wang, K.Parhi, "High-level DSP synthesis using concurrent transformations, scheduling and allocation", *IEEE Trans. on Comp.-aided Design*, Vol.14, No.3, pp.274-295, March 1995.

[546] S.VanderWiel, D.J.Lilja, "When caches aren't enough : Data prefetching techniques", *IEEE Computer Magazine*, pp.23-30, July 1997.

[547] M.Wilkes, "The memory gap", *27th Annual Intnl. Symp. on Computer Architecture*, Keynote speech at Wsh. on "Solving the Memory Wall Problem", Vancouver BC, Canada, June 2000.

[548] D.Wilde, S.Rajopadhye, "Allocating memory arrays for polyhedra", Technical Report, IRISA/INRIA 749, Rennes, France, July 1993.

[549] D.Wilde, S.Rajopadhye, "Memory reuse analysis in the polyhedral model" *Proc. EuroPar Conf.*, Lyon, France, Aug. 1996. "Lecture notes in computer science" series, Springer Verlag, Vol.1128, pp.389-397, 1996.

[550] M.Wolfe, "Iteration space tiling for memory hierarchies", *Proc. 3rd SIAM Conf. on Parallel Processing for Scientific Computing*, Dec. 1987.

[551] M.Wolfe, U.Banerjee, "Data Dependence and its Application to Parallel Processing", *Intnl. J. of Parallel Programming*, Vol.16, No.2, pp.137-178, 1987.

[552] M.Wolf, M.Lam, "A loop transformation theory and an algorithm to maximize parallelism", *IEEE Trans. on Parallel and Distributed Systems*, Vol.2, No.4, pp.452-471, Oct. 1991.

[553] M.Wolf, M.Lam, "A data locality optimizing algorithm", *Proc. of the SIGPLAN'91 Conf. on Programming Language Design and Implementation*, Toronto ON, Canada, pp.30-43, June 1991.

[554] M.Wolf, "Improving locality and parallelism in nested loops", *Ph.D. dissertation*, Stanford University, Stanford CA, USA, Aug. 1992.

[555] M.R.Wolf, "Optimizing supercompilers for supercomputing", *MIT Press*, Cambridge, MA, 1989.

[556] M.Wolfe, "Data dependence and program restructuring", *J. of Supercomputing*, No.4, Kluwer, Boston, pp.321-344, 1990.

[557] M.Wolfe, "The Tiny loop restructuring tool", *Proc. of Intnl. Conf. on Parallel Processing*, pp.II.46-II.53, 1991.

[558] D.Wong, E.Davis, J.Young, "A software approach to avoiding spatial cache collisions in parallel processor systems", *IEEE Trans. on Parallel and Distributed Systems*, Vol.9, No.6, pp.601-608, June 1998.

[559] S.Wuytack, F.Catthoor, F.Franssen, L.Nachtergaele, H.De Man, "Global communication and memory optimizing transformations for low power systems", *IEEE Intnl. Wsh. on Low Power Design*, Napa CA, pp.203-208, April 1994.

[560] S.Wuytack, F.Catthoor, H.De Man, "Transforming Set Data Types to Power Optimal Data Structures", *Proc. IEEE Intnl. Wsh. on Low Power Design*, Laguna Beach CA, pp.51-56, April 1995.

[561] S.Wuytack, F.Catthoor, L.Nachtergaele, H.De Man, "Power Exploration for Data Dominated Video Applications", *Proc. IEEE Intnl. Symp. on Low Power Design*, Monterey CA, pp.359-364, Aug. 1996.

[562] S.Wuytack, F.Catthoor, G.De Jong, B.Lin, H.De Man, "Flow Graph Balancing for Minimizing the Required Memory Bandwidth", *Proc. 9th ACM/IEEE Intnl. Symp. on System-Level Synthesis* (ISSS), La Jolla CA, pp.127-132, Nov. 1996.

[563] S.Wuytack, J.P.Diguet, F.Catthoor, H.De Man, "Formalized methodology for data reuse exploration for low-power hierarchical memory mappings", *IEEE Trans. on VLSI Systems*, Vol.6, No.4, pp.529-537, Dec. 1998.

[564] S.Wuytack, "System-level power optimisation of data storage and transfer", *Doctoral dissertation*, ESAT/EE Dept., K.U.Leuven, Belgium, Oct. 1998.

[565] S.Wuytack, J.L.da Silva, F.Catthoor, G.De Jong, C.Ykman-Couvreur, "Memory management for embedded network applications", *IEEE Trans. on Comp.-aided Design*, Vol.CAD-18, No.5, pp.533-544, May 1999.

[566] S.Wuytack, F.Catthoor, G.De Jong, H.De Man, "Minimizing the Required Memory Bandwidth in VLSI System Realizations", *IEEE Trans. on VLSI Systems*, Vol.7, No.4, pp.433-441, Dec. 1999.

[567] Y.Yaacoby, P.Cappello, "Scheduling a system of nonsingular affine recurrence equations onto a processor array", *J. of VLSI Signal Processing*, No.1, Kluwer, Boston, pp.115-125, 1989.

[568] T.Yamada (Sony), "Digital storage media in the digital highway era", Plenary paper in *Proc. IEEE Intnl. Solid-State Circ. Conf.*, San Francisco CA, pp.16-20, Feb. 1995.

[569] C.Ykman-Couvreur, D.Verkest, F.Catthoor, B.Svantesson, Sh.Kumar, A.Hemani, F.Wolf, R.Ernst, "Stepwise Exploration and Specification Refinement of Telecom Network Systems", accepted for *IEEE Trans. on VLSI Systems*, 2002.

[570] C.Ykman-Couvreur, J.Lambrecht, D.Verkest, F.Catthoor, H.De Man, "Exploration and Synthesis of Dynamic Data Sets in Telecom Network Applications", *Proc. 12th ACM/IEEE Intnl. Symp. on System-Level Synthesis* (ISSS), San Jose CA, pp.125-130, Dec. 1999.

[571] J.Zeman, G.Moschytz, "Systematic design and programming of signal processors, using project management techniques", *IEEE Trans. on Acoustics, Speech and Signal Processing*, Vol.31, No.12, pp., Dec. 1983.

[572] Y.Zhao and S.Malik, "Exact Memory Size Estimation for Array Computation without Loop Unrolling", *36th ACM/IEEE Design Automation Conf.*, New Orleans LA, pp.811-816, June 1999.

Index